Advanced Nanomaterials for Energy Storage Devices

Advanced Nanomaterials for Energy Storage Devices

Guest Editor

Sudeshna Chandra

MDPI

Basel • Beijing • Wuhan • Barcelona • Belgrade • Novi Sad • Cluj • Manchester

Guest Editor
Sudeshna Chandra
Technology and Science
Hanse-Wissenschaftskolleg –
Institute for Advanced Study
(HWK)
Delmenhorst
Germany

Editorial Office
MDPI AG
Grosspeteranlage 5
4052 Basel, Switzerland

This is a reprint of the Special Issue, published open access by the journal *Molecules* (ISSN 1420-3049), freely accessible at: www.mdpi.com/journal/molecules/special_issues/O4BEB6G232.

For citation purposes, cite each article independently as indicated on the article page online and as indicated below:

Lastname, A.A.; Lastname, B.B. Article Title. *Journal Name* **Year**, *Volume Number*, Page Range.

ISBN 978-3-7258-4076-2 (Hbk)
ISBN 978-3-7258-4075-5 (PDF)
https://doi.org/10.3390/books978-3-7258-4075-5

Contents

About the Editor

Sudeshna Chandra

Dr. Sudeshna Chandra is a seasoned materials research scientist with over 20 years of experience in research and more than 7 years in academia. She is a Fellow at the Hanse-Wissenschaftskolleg (HWK) at the University of Oldenburg, Germany. Previously, she served as a Professor and Head of the Chemistry Department at the Sunandan Divatia School of Science, SVKM's NMIMS University in Mumbai, India. She also holds visiting researcher positions at the University of Regensburg and the Technical University of Chemnitz in Germany.

Dr. Chandra earned her Ph.D. in Chemistry from the Indian Institute of Technology (IIT) Roorkee, India. Her research primarily focuses on the development of smart multifunctional hybrid nanomaterials, including metal, metal oxide, and metal sulfide nanoparticles, dendrimers, and upconversion nanomaterials. These materials are designed with controlled morphology and targeted functionalities for applications in drug delivery, biosensing, and energy devices.

Throughout her career, Dr. Chandra has been honored with several prestigious awards and fellowships, such as the Alexander von Humboldt Fellowship, Peter Salamon Young Scientist Award, DST Women Scientist Award, and DST Young Scientist Award. She is an active member of the Indian Science Congress Association, the Alexander von Humboldt Stiftung/Foundation, and the Society of Materials Chemistry.

Preface

The Special Issue "Advanced Nanomaterials for Energy Storage Devices" in *Molecules* throws light on the recent development and application of nanomaterials to enhance energy storage technologies. This Special Issue addresses the global demand for efficient and sustainable energy solutions by investigating novel materials and innovative approaches.

The papers in this Special Issue delve into diverse aspects of nanomaterials in energy storage and key topics that encompass advancements in lithium ion batteries, supercapacitors, electrolytes, phase change energy storage microcapsules, and thermal energy storage materials, to name a few. Understanding the fundamental mechanism of the behaviour of nanomaterials in energy storage systems is also a part of the compilation. Studies on ion-transport dynamics, charge–discharge processes, the synergism of unique materials, novel binders and materials for electrolytes, and the evaluation of the structural stability of nanostructured materials are the key focuses of this Special Issue.

My heartfelt gratitude goes to the authors who have contributed their valuable research that provided deep insights into advanced nanomaterials. Each submission was rigorously peer-reviewed by expert researchers in their respective areas. Their constructive feedback and assessments have ensured the highest scientific standards in this collection. I express my sincere thanks and appreciation to the reviewers for their time, expertise, and valuable suggestions.

Finally, I thank the editorial team of *Molecules* for their unwavering support throughout the process. We hope that this Special Issue will be a valuable resource for researchers in nanomaterials for energy and inspire further innovations in this area.

Sudeshna Chandra
Guest Editor

molecules

MDPI

Article

Hybridizing Fabrications of Gd-CeO_2 Thin Films Prepared by EPD and SILAR-A+ for Solid Electrolytes

Taeyoon Kim [1], Yun Bin Kim [1], Sungjun Yang [2] and Sangmoon Park [1,3,4,*]

1 Department of Engineering in Energy Materials, Graduate School, Silla University, Busan 46958, Republic of Korea
2 UNIST Central Research Facilities, Ulsan National Institute of Science and Technology, Ulsan 44919, Republic of Korea
3 Department of Environmental Energy & Chemistry, College of Engineering, Silla University, Busan 46958, Republic of Korea
4 Department of Fire Protection and Safety Management, College of Health and Welfare, Silla University, Busan 46958, Republic of Korea
* Correspondence: spark@silla.ac.kr; Tel.: +82-51-999-5891

Abstract: Thin films of gadolinium-doped ceria (GDC) nanoparticles were fabricated as electrolytes for low-temperature solid oxide fuel cells (SOFCs) by combining electrophoretic deposition (EPD) and the successive ionic layer adsorption and reaction-air spray plus (SILAR-A+) method. The $Ce_{1-x}Gd_xO_{2-x/2}$ solid solution was synthesized using hydrothermal (HY) and solid-state (SS) procedures to produce high-quality GDC nanoparticles suitable for EPD fabrication. The crystalline structure, cell parameters, and phases of the GDC products were analyzed using X-ray diffraction. Variations in oxygen vacancy concentrations in the GDC samples were achieved through the two synthetic methods. The ionic conductivities of pressed pellets from the HY, SS, and commercial $G_{0.2}DC$ samples, measured at 150 °C, were 0.6×10^{-6}, 2.6×10^{-6}, and 2.9×10^{-6} S/cm, respectively. These values were determined using electrochemical impedance spectroscopy (EIS) with a simplified equivalent circuit method. The morphologies of $G_{0.2}DC$ thin films prepared via EPD and SILAR-A+ processes were characterized, with particular attention to surface cracking. Crack-free GDC thin films, approximately 730–1200 nm thick, were successfully fabricated on conductive substrates through the hybridization of EPD and SILAR-A+, followed by hydrothermal annealing. EIS and ionic conductivity (1.39×10^{-9} S/cm) measurements of the $G_{0.2}DC$ thin films with thicknesses of 733 nm were performed at 300 °C.

Keywords: gadolinium doped ceria; thin film; SILAR-A+; EPD; hydrothermal annealing

1. Introduction

Since the first investigation of a solid electrolyte for solid oxide fuel cells (SOFCs) in 1937, oxide-ion (O^{2-})-conducting materials fueled by hydrogen have been extensively studied as solid electrolytes in SOFCs due to their appropriate electrochemical reactions at working temperatures [1–8]. One of the primary challenges in developing suitable O^{2-}-conducting solid electrolytes is overcoming the high operating temperature limitation to reduce maintenance costs and improve system stability. SOFCs are categorized into high (~1000 °C), intermediate (600–800 °C), and low (<600 °C) temperature ranges based on their operating conditions [1–8]. While yttrium-stabilized zirconia (YSZ) has demonstrated excellent O^{2-} conductivity along with thermal, chemical, and mechanical stability, its effectiveness is constrained to high operating temperatures (~1000 °C). Efforts to address this limitation have included fabricating ultrathin YSZ films via atomic layer deposition, which reduces operating temperatures by minimizing ohmic losses [9]. In parallel, intermediate-temperature electrolytes with superior O^{2-} conductivity have been developed, such as samarium-doped ceria (SDC) and gadolinium-doped ceria (GDC) [5–8]. The ceria (CeO_2) structure, which enables oxygen conductivity through the formation of oxygen vacancies, adopts a fluorite-type (F-structure, Fm-3m) crystal lattice, as shown in Figure 1 [10,11]. When Sm^{3+} or Gd^{3+} ions substitute Ce^{4+} in the CeO_2 lattice, the introduction of oxygen vacancies significantly enhances ionic conductivity [5–8]. However, as the concentration of Sm^{3+} or Gd^{3+} increases, the F-structure gradually transitions to a cubic C-structure (Ia-3) typical of Sm_2O_3 or Gd_2O_3, as depicted in Figure 1 [5–8,10,11]. Additionally, the Ce-O_8 bonds in the F-structure transform into Ce(Gd,Sm)-O_6 bonds with reduced coordination as Sm^{3+} or Gd^{3+} ions accumulate.

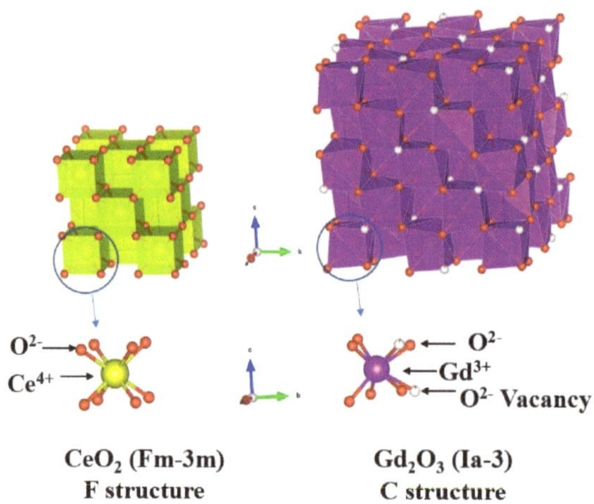

Figure 1. The structures of CeO_2 (F-structure, Fm-3m) and Gd_2O_3 (C-structure, Ia-3).

Thin films of these electrolytes have been prepared using techniques such as sputtering, spin-coating, pulsed laser deposition, and chemical vapor deposition [9,12–17]. In a previous study, crack-free YSZ thin films were achieved using the successive ionic layer adsorption and reaction-air spray plus (SILAR-A+) method, which integrates SILAR with air spray steps [18,19]. This method enables the formation of thin films composed of nanoparticles via dehydration and crystallization during post-hydrothermal annealing at temperatures below 200 °C. Repeated cycles of adsorption, rinsing, precipitation, and air spray deposition on oxidized substrates allow for the fabrication of ultrathin films. In this study, thin films of crystallized GDC nanoparticles were successfully fabricated using a hybrid deposition process combining electrophoretic deposition (EPD) with the SILAR-A+ method, followed by post-hydrothermal annealing. The structure, surface morphology, and ionic conductivity of the GDC films were characterized using X-ray diffraction, scanning electron microscopy, and electrochemical impedance spectroscopy, respectively.

2. Results and Discussion

Gd^{3+} ions were doped into the CeO_2 host lattice ($Ce_{1-x}Gd_xO_{2-x/2}$ (G_xDC), x = 0.1–0.5) to prepare solid electrolytes for low-temperature solid oxide fuel cells (SOFCs) using hydrothermal (HY) and solid-state (SS) methods. As shown in Figure 2a,b, the gradual increase in Gd^{3+} ion content in CeO_2 caused shifts in the X-ray diffraction (XRD) patterns to lower and higher angles, particularly in the Bragg reflection positions around 2θ = 28–29° for both HY and SS methods. The calculated lattice parameter (a) of GDC showed a nonlinear trend, initially increasing and then decreasing with the increase in Gd^{3+} ion content, as depicted in Figure 2c and Table S1. R_p values, indicating good agreement between the observed and calculated reflections, ranging from 6.9 to 11.1%, are observed in all samples prepared using both HY and SS methods. The *a* values reached their maximum at $x = 3$ for the SS method and $x = 4$ for the HY method. Interestingly, the unit cell of GDC synthesized via the SS method was significantly smaller than that prepared by the HY method. This difference is likely due to the formation of more oxygen vacancies in the SS GDC lattice, which are induced by the high-temperature conditions of the SS synthesis process.

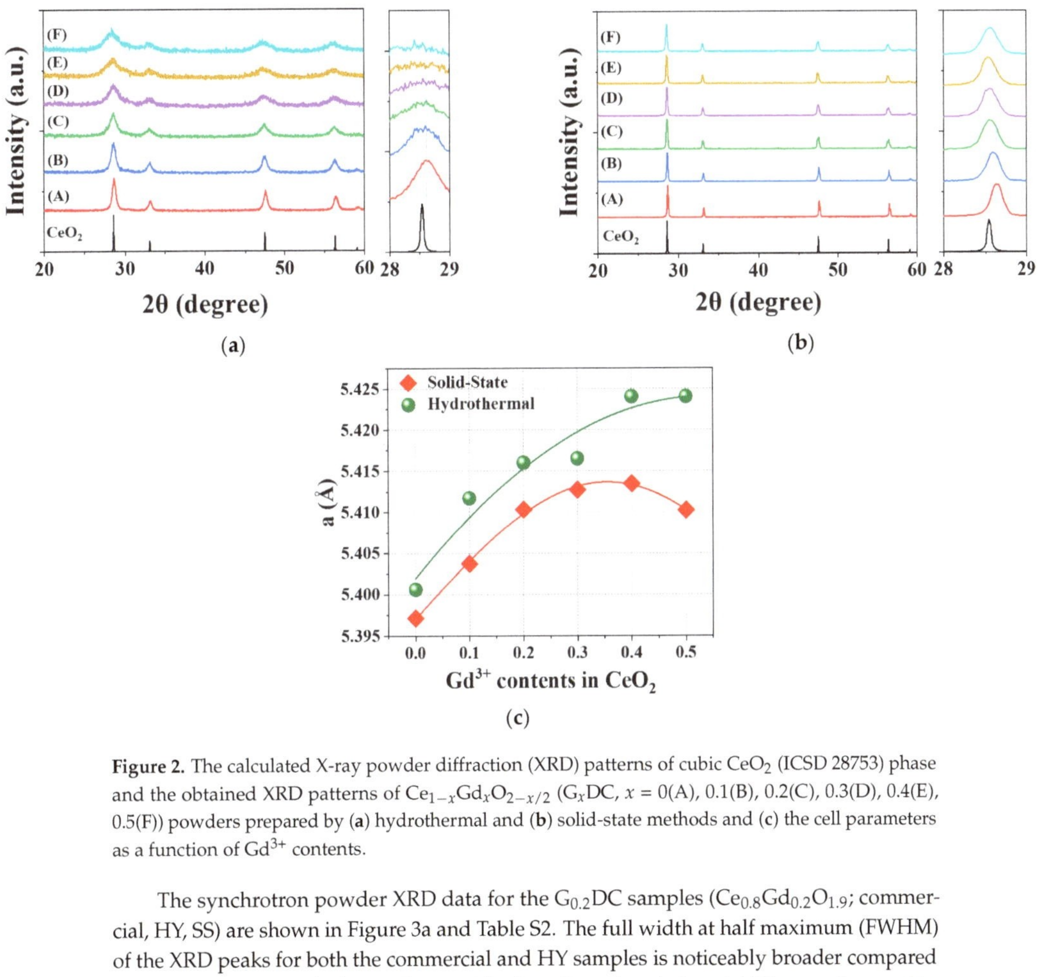

Figure 2. The calculated X-ray powder diffraction (XRD) patterns of cubic CeO_2 (ICSD 28753) phase and the obtained XRD patterns of $Ce_{1-x}Gd_xO_{2-x/2}$ (G_xDC, x = 0(A), 0.1(B), 0.2(C), 0.3(D), 0.4(E), 0.5(F)) powders prepared by (**a**) hydrothermal and (**b**) solid-state methods and (**c**) the cell parameters as a function of Gd^{3+} contents.

The synchrotron powder XRD data for the $G_{0.2}DC$ samples ($Ce_{0.8}Gd_{0.2}O_{1.9}$; commercial, HY, SS) are shown in Figure 3a and Table S2. The full width at half maximum (FWHM) of the XRD peaks for both the commercial and HY samples is noticeably broader compared to that of the SS sample due to the particle sizes. R_p values below 4.44% were observed in all the $G_{0.2}DC$ samples. The chi-square (χ^2) values of the refined $G_{0.2}DC$ samples represent goodness of fit, as shown in Table S2. The χ^2 values for HY and commercial $G_{0.2}DC$ samples (7.16 and 10.5, respectively) are larger than that of the SS $G_{0.2}DC$ sample (0.979). Figure 3b presents the cell parameter (*a*) and oxygen occupancy, along with SEM images, for the HY, SS, and commercial $G_{0.2}DCs$, respectively.

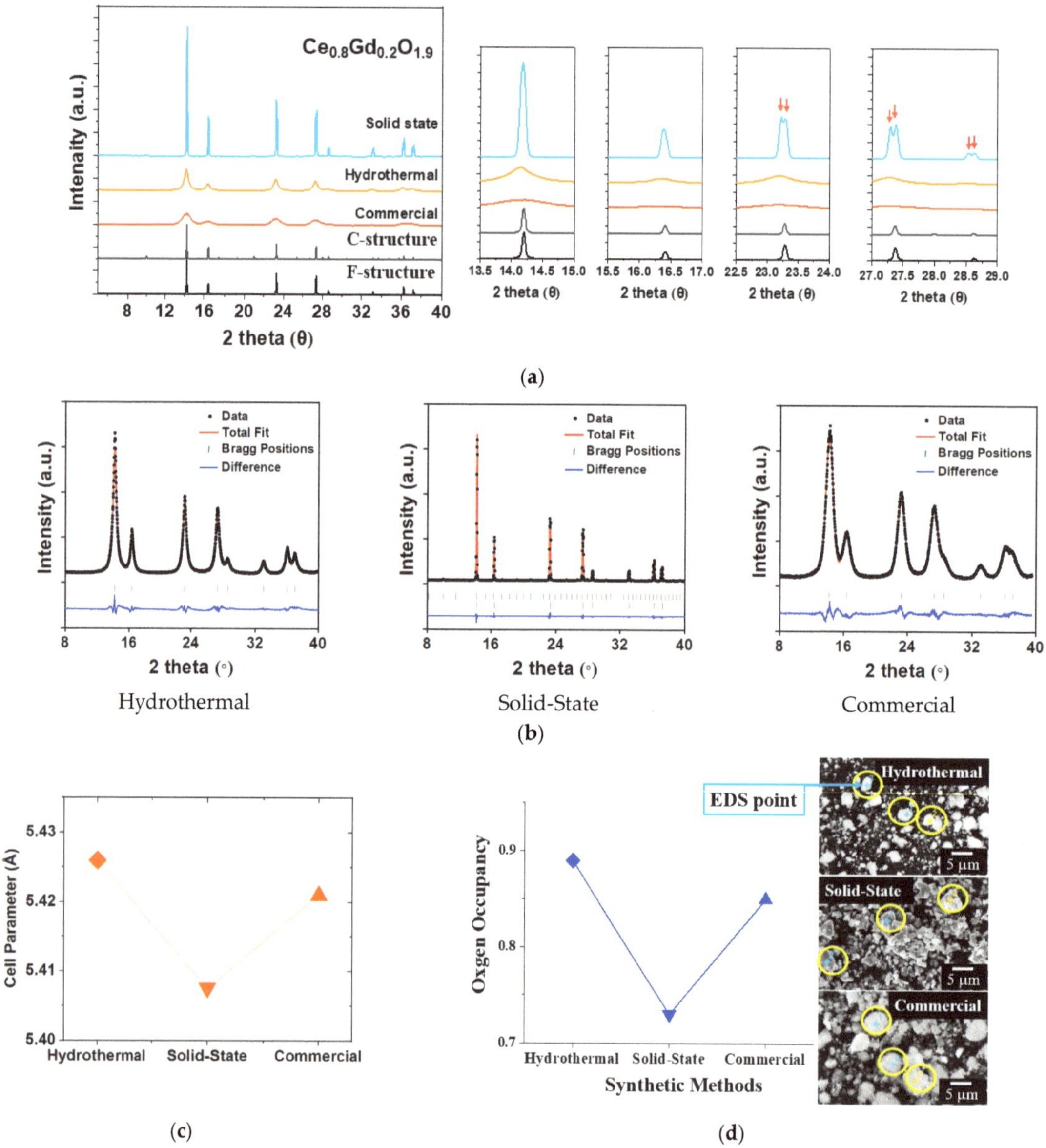

Figure 3. (**a**) The calculated XRD patterns of F- and C-structures and the obtained synchrotron XRD patterns and (**b**) structure refinements of $Ce_{0.8}Gd_{0.2}O_{1.9}$ (**c**) cell parameters and (**d**) oxygen occupancy with SEM images vs. HY, SS, and commercial sources.

The *a* values and oxygen occupancies were derived from synchrotron powder XRD analysis and energy dispersive X-ray spectroscopy (EDS), respectively. The cell parameter and oxygen occupancy values for HY and commercial $G_{0.2}DCs$ were similar, measured as 5.426(2) Å and 5.421(3) Å, and 0.89 and 0.85, respectively. However, the SS $G_{0.2}DC$ sample exhibited significantly lower values, with a = 5.40761(4) Å and oxygen occupancy of 0.73. In the SS $G_{0.2}DC$ sample, peak splitting was observed in the XRD patterns at higher angular resolution (2θ), as shown in Figure 3a. This peak splitting suggests the coexistence of mixed phases: the F-structure of CeO_2 and the C-structure of Gd_2O_3, resulting in distinct cell parameter and oxygen vacancy characteristics. For the GDC prepared using the HY method, the lattice parameter (*a*) as a function of Gd^{3+} ion content exhibited nonlinear behavior, as shown in Figure 2c. This trend may be attributed to a reduction in the average connectivity of Ce-O_8; in the F-structure, driven by a decrease in the coordination number around Ce^{4+} cations, which generates oxygen anion vacancies. Electrochemical impedance spectroscopies (EISs) for pressed pellets of HY and SS G_xDC (x = 0.2, 0.3, 0.4), and commercial $G_{0.2}DC$ powders are shown in Figure 4a,b, respectively. The total (σ_{Total}), which is combination of bulk (σ_{bulk}) and grain boundary (σ_{gb}) contributions, conductivity was calculated from the EIS fits at 150 °C in Figures 4c and S1 and Table S3.

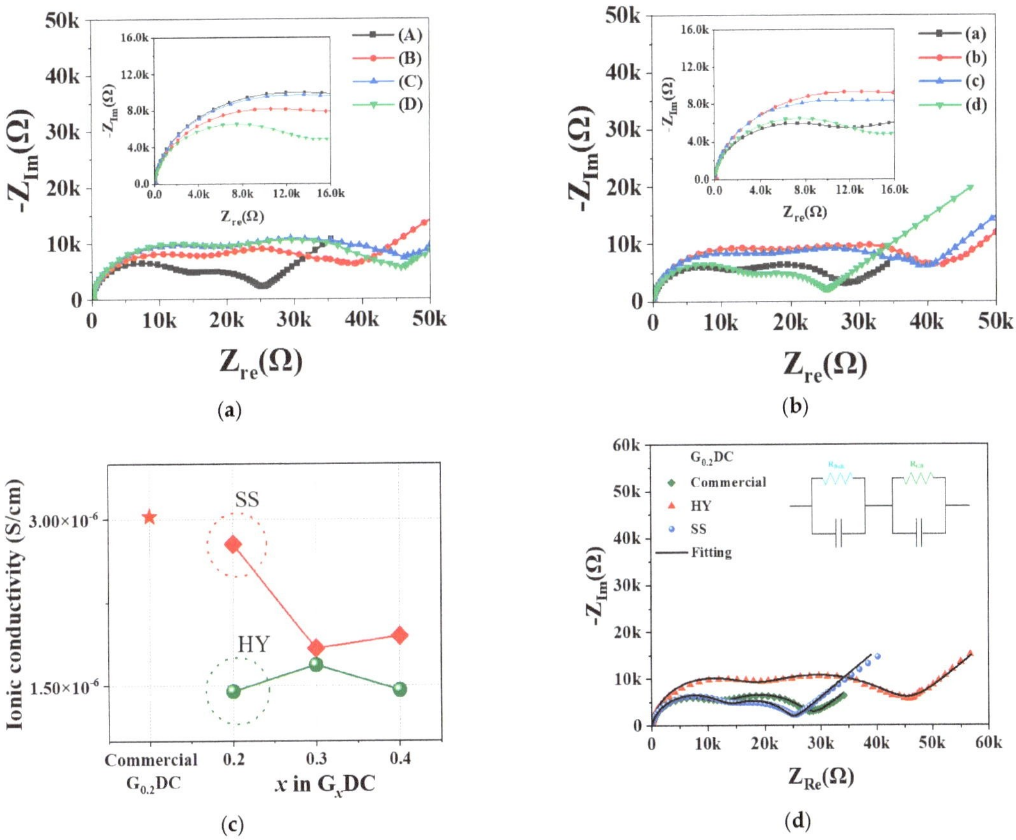

Figure 4. Electrochemical impedance spectroscopies (EISs) of (**a**) HY, (**b**) SS G_xDC (commercial (A), x = 0.2(B), 0.3(C), 0.4(D)), (**c**) their ionic conductivities with commercial sources at 150 °C, and (**d**) EIS data of $G_{0.2}DC$ samples fitted via an equivalent circuit.

The σ_{Total} values for HY and SS G_xDC (x = 0.2, 0.3, 0.4) were 1.4×10^{-6}, 1.7×10^{-6}, 1.4×10^{-6} and 2.8×10^{-6}, 1.8×10^{-6}, 1.9×10^{-6} S/cm, respectively (Figure S1 and Table S3). In addition, the σ_{Total} for the commercial $G_{0.2}DC$ sample was 3.1×10^{-6} S/cm (Figure S1 and Table S3). Among the HY, SS, and commercial samples, the HY $G_{0.2}DC$ exhibited a relatively low σ_{Total} value, which may be attributed to its smaller grain size (Figure S2), resulting in enhanced interfacial resistance [20]. The EIS raw data for $G_{0.2}DC$ samples were analyzed using a simplified equivalent circuit model, comprising a capacitor, bulk resistance (R_{Bulk}), and grain boundary resistance (R_{GB}). The analysis was performed using the EIS Spectrum Analyzer, as shown in Figures 4d and S1 and Table S3 [21–23]. The σ_{Total} values obtained from the fitting process for HY, SS, and commercial $G_{0.2}DC$ samples were 1.6×10^{-6}, 2.6×10^{-6}, and 2.9×10^{-6} S/cm, respectively (Figure S1 and Table S3). The ionic conductivity values derived from the raw and the fitted data exhibited differences in the range of 6–12%.

The HY, SS, and commercial $G_{0.2}DC$ particles were deposited on ITO glass using the electrophoretic deposition (EPD) method. The deposition utilized a suspension containing 0.1 g of $G_{0.2}DC$ and 0.02 g of iodine in 50 mL of acetone, with a DC power supply applied at 20 V for 15 s. Figure 5a(A–C) shows SEM images of the surface morphologies for HY, SS, and commercial $G_{0.2}DCs$, respectively. The particles were fully coated on the ITO substrate. HY $G_{0.2}DC$ particles were uniformly deposited, while the SEM images of SS and commercial $G_{0.2}DCs$ exhibited large grains and aggregation, respectively. Figure 5b illustrates $G_{0.2}DC$ thin films prepared using two methods: successive ionic layer adsorption and reaction (SILAR, A and B) and SILAR-air spray plus (SILAR-A+, C and D), with 150 cycles.

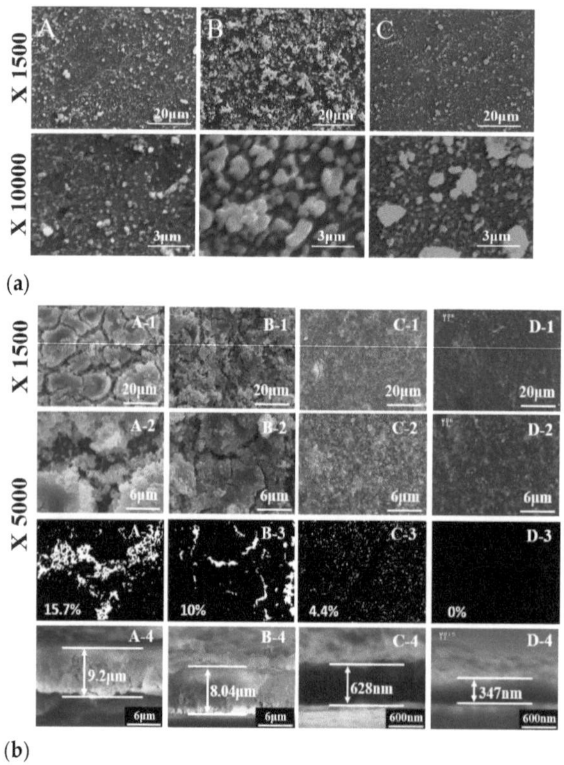

Figure 5. (**a**) Surface SEM images of the electrophoretic depositions (EPDs) of $G_{0.2}DC$ prepared by A. HY, B. SS, C. Commercial sources and (**b**) surface and cross-sectional SEM images of $G_{0.2}DC$ thin films by SILAR (A,B-1 to 4) and SILAR-A+ (C, D-1 to 4) methods (as-made (A, C) and subsequent hydrothermal annealing (B, D) at 175 °C).

During each cycle of the SILAR process, the glass substrate was sequentially immersed in a Gd^{3+}/Ce^{4+} cation solution for adsorption, rinsed in water to remove spectator anions, transferred to an OH^- anion solution for precipitation, and then rinsed again to eliminate residual layers. SEM images in Figure 5(A–D) represent $G_{0.2}DC$ films without post-annealing (A, C) and with additional post-hydrothermal annealing at 175 °C (B, D).

Each set of images (A, B, C, D-1 and D-2) was magnified by 1500× and 5000× to observe surface cracks. Figure 5 (A,B,C,D-3) shows crack and crack-free SEM images in white and black contrasts. In as-made films (A-3 and C-3), surface cracks were clearly reduced from 15.7% to 4.4%, and from 10% to 0% after annealing (B-3 and D-3), demonstrating the effectiveness of the SILAR-A+ process combined with post-hydrothermal annealing in producing crack-free $G_{0.2}DC$ films. Figure 5b (A,B,C,D-4) presents cross-sectional SEM images showing the thickness of $G_{0.2}DC$ thin films, accompanied by bar graphs. The thicknesses of films produced via the as-made SILAR and SILAR-A+ processes (A-4 and C-4) were 9.2 μm and 0.628 μm (628 nm), respectively, while those of post-hydrothermally annealed films (B-4 and D-4) were 8.04 μm and 0.347 μm (347 nm), respectively. Hydrothermal annealing removed crystallized hydroxyl and water layers, resulting in thinner cross-sections, as observed in SEM images. The reduced thickness and crack formation in the films were attributed to the efficient removal of diffusion layers during the air-spray process and the subsequent exclusion of hydroxyl and water layers through hydrothermal annealing at 175 °C. In a previous report, freestanding YSZ (Y-doped ZrO_2) membranes with thicknesses ranging from 500 nm to 2 μm were fabricated to test power density at 600 °C [24]. Further advancements reduced the thickness of YSZ membranes to 50 nm, enabling SOFCs to operate at lower temperatures, around 350 °C [24]. Deposition techniques such as pulsed laser deposition, chemical vapor deposition, spin coating, and physical vapor deposition were utilized to produce these solid membranes. For low-temperature SOFC applications, $G_{0.2}DC$ thin films were fabricated on conductive Pt/ITO substrates. These were prepared using the EPD method with commercial and HY $G_{0.2}DC$ powders, followed by the SILAR-A+ process and post-hydrothermal annealing. The SILAR-A+ technique proved effective in producing thin, smooth solid membranes, although it is primarily efficient in oxidizing surface environments due to its successive ionic adsorption mechanism. To address this limitation, EPD on the Pt metal surface was conducted prior to SILAR-A+ deposition to form GDC thin films. Figure 6a shows SEM images of hybridized films produced through EPD (using commercial and HY $G_{0.2}DC$ powders) followed by the SILAR-A+ process and post-hydrothermal annealing. The surface morphology of films fabricated using EPD with HY $G_{0.2}DC$ and the SILAR-A+/hydrothermal method exhibited a much smoother and more homogeneous surface compared to other methods. The cross-sectional SEM images reveal the film thicknesses from the hybridization processes: 2.6 μm for EPD (commercial)/SILAR-A+-as-made, 1.5 μm for EPD (HY)/SILAR-A+-as-made, 1.2 μm after post-hydrothermal annealing of the EPD (commercial)/SILAR-A+ film, and 0.733 μm (733 nm) for the annealed EPD (HY)/SILAR-A+ film. Figure 6b presents the EIS and ionic conductivity measurements for Pt/$G_{0.2}DC$/Pt-ITO films (1 cm × 1 cm).

The total conductivity (σ_{Total}) values of the films with thicknesses of 1.2 μm and 733 nm were 2.68×10^{-10} S/cm and 1.39×10^{-9} S/cm at 300 °C, respectively (Figure S1 and Table S3). The successful fabrication of $G_{0.2}DC$ thin films using a hybrid approach combining EPD (HY GDC) and SILAR-A+ with post-hydrothermal annealing highlights the potential of these membranes as promising solid electrolyte candidates for low-temperature SOFC applications.

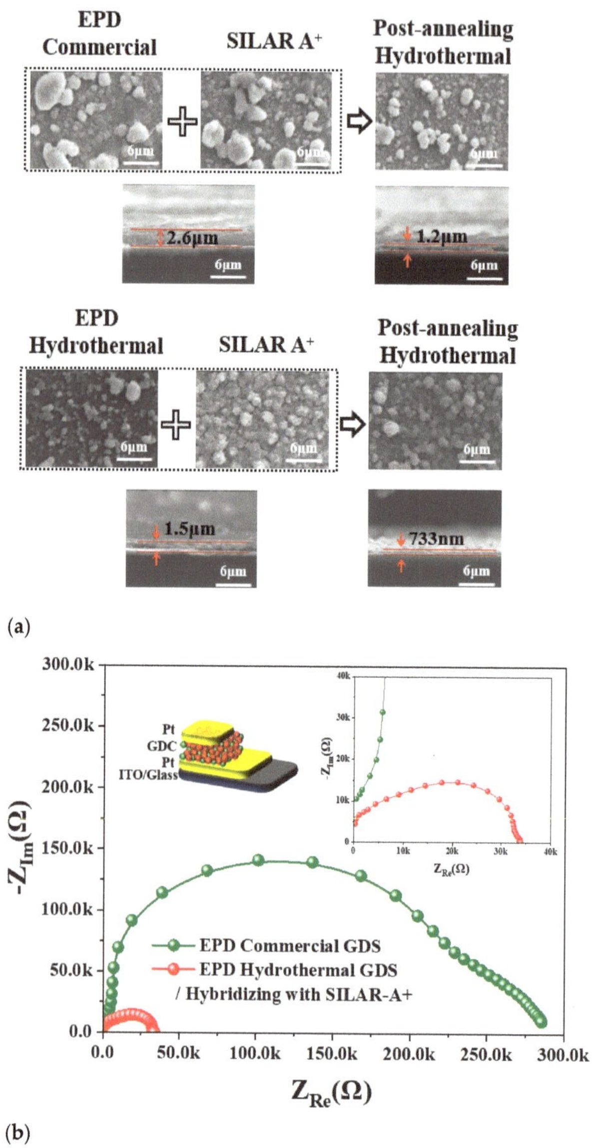

Figure 6. (**a**) SEM images and (**b**) EIS of $G_{0.2}DC$ thin films by hybridizing fabrications of EPD (commercial, HY) and SILAR-A+ process.

3. Experimental Procedure

The GDC powders were prepared using the solid-state (SS) method by mixing CeO_2 (Alfa, 99.9%; Thermo Fisher Scientific Inc. Seoul, Republic of Korea) and Gd_2O_3 (Alfa, 99.9%), followed by annealing at 1400 °C for 3 h in air. Additionally, GDC powders were synthesized via the hydrothermal (HY) technique using precursors of 0.1 M $Ce(CH_3COO)_3$ (Alfa, 99.9%), 0.1 M $Gd(NO_3)_3{\cdot}6H_2O$ (Alfa, 99.9%), and 0.1 M NaOH (Alfa, 98%) at 180 °C for 18 h. Phase identification of the synthesized powders was conducted using a Shimadzu XRD-6000 powder diffraction system (SHIMADZU CORPORATION, Kyoto, Japan) with Cu-Kα radiation. High-resolution synchrotron X-ray powder diffraction data were collected at the PLS-II 8D-XRS POSCO beamline of the Pohang Accelerator Laboratory (PAL). For the crystallization process, the hydrothermal synthesis was performed at 175 °C for 18 h. GDC thin films were fabricated on substrates using three methods: EPD, SILAR, and SILAR-A+. For the EPD process, 0.1 g of GDC powders (prepared by SS, HY, and commercial sources) was dispersed in 50 mL of acetone containing 0.02 g of iodine as an electrolyte. The dispersion underwent ultrasonication for 8 min, followed by electrophoretic deposition (EPD) using a DC power supply of 20 V for 15 s. For the SILAR method, 0.03 M solutions of $Ce(CH_3COO)_3$ $Gd(NO_3)_3 \cdot 6H_2O$, and NaOH were used as precursors. The substrate was attached to the arm of a Gilson 223 XYZ robotic sample changer for sequential adsorption (0.1 min), rinsing (0.1 min), and precipitation (0.1 min) reactions. Substrate surface pre-treatment involved 10 min of UVC exposure using an Omniscience UVC-150 system. The SILAR-A+ process incorporated an air-spraying step (0.1 min) at an air speed of 150 m/s as the final stage of the SILAR cycle. Thin film formation involved repeating the cycle 150 times in a sequence of cation adsorption, rinsing, anion reaction, rinsing, and air spraying. Hybridization of the EPD and SILAR-A+ methods was conducted on Pt/ITO substrates. The morphology of the GDC powders and films was characterized using scanning electron microscopy (SEM, Hitachi S-4200, Tokyo, Japan). The proportion of cracked versus crack-free areas in SEM images was quantified using the ImageJ (ver. 1.54k) program. Electrochemical impedance spectroscopy (EIS) was performed using a potentiostat (Ametek VERSASTAT3-500, AMETEK PAR(Princeton Applied Research), Oak, Ridge, TN, USA) over a frequency range of 1 MHz to 100 mHz. The ionic conductivity (σ) of GDC samples was calculated using the following equation:

$$\sigma = L/R \cdot A$$

where σ, L, R, and A represent the ionic conductivity, the thickness of the GDC samples, the total resistance, and the contact area of the electrode with the samples, respectively [3,8].

4. Conclusions

Gd-doped CeO_2 (GDC) nanoparticles were successfully prepared using three methods: hydrothermal synthesis, solid-state synthesis, and commercial sourcing. X-ray diffraction analysis and energy-dispersive X-ray spectroscopy were employed to characterize the F- and C-phase structures of CeO_2/Gd_2O_3 system, along with the presence of oxygen vacancies. Thin, crack-free GDC films were fabricated using a combined approach of electrophoretic deposition (EPD) and the successive ionic layer adsorption and reaction-air spray plus (SILAR-A+) methods. Hydrothermal treatment further enhanced the film quality, as confirmed by SEM surface and cross-sectional imaging. Oxide-ion conductivities of the GDC pellets and thin films were measured using electrochemical impedance spectroscopy. The GDC pellets exhibited conductivity at 150 °C, while the thin films demonstrated conductivity at 300 °C. Additionally, $G_{0.2}DC$ thin films, approximately 730 nm thick, were successfully fabricated on Pt-ITO substrates, demonstrating their potential for use as solid electrolytes in low-temperature SOFC applications.

Supplementary Materials: The following supporting information can be downloaded at: https://www.mdpi.com/article/10.3390/molecules30030456/s1, Table S1. Cell parameters of $Ce_{1-x}Gd_xO_{2-x/2}$ synthesized by hydrothermal and solid-stat methods and commercial $G_{0.2}DC$; Table S2. Rietveld refinement and crystal data and refined atomic coordinates for $Ce_{0.8}Gd_{0.2}O_{1.9}$ prepared by solid-state, hydrothermal, and commercial source; Figure S1. Ionic conductivities of $Ce_{1-x}Gd_xO_{2-x/2}$ pellets prepared by HY, SS, and commercial sources at 150 °C and fitted via an equivalent circuit, and hybridizing $G_{0.2}DC$ thin-films at 300 °C; Table S3. Ionic conductivities of $Ce_{1-x}Gd_xO_{2-x/2}$ pellets prepared by HY, SS, and commercial sources at 150 °C and fitted via an equivalent circuit, and hybridizing $G_{0.2}DC$ thin-films at 300 °C; Figure S2. SEM images of $G_{0.2}DC$ powders prepared by HY(A), SS(B), Commercial(C) sources.

Author Contributions: Conceptualization, S.P.; methodology, S.P. and T.K.; software, S.P. and T.K.; validation, S.P., T.K. and Y.B.K.; formal analysis, S.P., T.K. and S.Y.; investigation, S.P.; resources, S.P.; data curation, S.P., T.K., Y.B.K. and S.Y.; writing—original draft preparation, S.P. and T.K.; writing—review and editing, S.P.; visualization, S.P. and T.K.; supervision, S.P.; project administration, S.P.; funding acquisition, S.P. All authors have read and agreed to the published version of the manuscript.

Funding: This research was supported by the "Regional Innovation Strategy (RIS)" through the National Research Foundation of Korea (NRF) funded by the Ministry of Education (MOE) (2023RIS-007).

Data Availability Statement: The original contributions presented in this study are included in the article/Supplementary Materials. Further inquiries can be directed to the corresponding author(s).

Acknowledgments: The authors appreciate the Pohang Accelerator Laboratory (Pohang, Republic of Korea) for providing the synchrotron radiation sources at the 8D beamlines used in this study.

Conflicts of Interest: The author declares no conflicts of interest.

References

1. Baur, E.; Preis, H. Über Brennstoff-Ketten mit Festleitern. *Z. Elektrochem. Angew. Phys. Chem.* **1937**, *43*, 727–732. [CrossRef]
2. Yauchi, H.; Morikawa, D.; Saito, T.; Tsuda, K.; Yashima, M. High oxide-ion conductivity through the interstitial oxygen site i sillen oxychorides. *Adv. Funct. Mater.* **2023**, *33*, 2214082. [CrossRef]
3. Yashima, M.; Tsujiguchi, T.; Sakuda, Y.; Yasui, Y.; Zhou, Y.; Fujii, K.; Toril, S.; Kamiyama, T.; Skinner, S. High oxide-ion conductivit through the interstitial oxygen site in $Ba_7Nb_4MoO_{20}$-based hexagonal perovskite related oxides. *Nat. Commun.* **2021**, *12*, 55 [CrossRef] [PubMed]
4. Fop, S.; McCombie, K.; Smith, R.I.; Mclaughlin, A.C. Enhanced oxygen ion conductivity and mechanistic understanding i $Ba_3Nb_{1-x}V_xMoO_{8.5}$. *Chem. Mater.* **2020**, *32*, 4724. [CrossRef]
5. Yang, G.; Loubani, M.E.; Chalaki, H.R.; Kim, J.; Keum, J.K.; Rouleau, C.M.; Lee, D. Tuning ionic conductivity in fluorite Gd-dope CeO_2-bixbyite RE_2O_3 (RE = Y and Sm) multilayer thin films by controlling interfacial strain. *ACS Appl. Electron. Mater.* **2023**, 4556–4563. [CrossRef]
6. Rehman, S.U.; Song, H.-S.; Kim, H.-S.; Hassan, M.H.; Joh, D.-W.; Song, R.-H.; Lim, T.-H.; Hong, J.-E.; Park, S.-J.; Lee, S.- A dynamic infiltration technique to synthesize nanolayered cathodes for high performance and robust solid oxide fuel cell *J. Energy Chem.* **2022**, *70*, 201–210. [CrossRef]
7. Acuña, L.M.; Cabezas, M.D.; Fuentes, R.O. Improvement on the electrochemical performance by morphology control o nanostructured $La_{0.6}Sr_{0.4}CoO_{3-\delta}$-$Gd_{0.1}Ce_{0.9}O_{1.95}$ cathodes for IT-SOFC. *Energy Adv.* **2022**, *1*, 344–356. [CrossRef]
8. Ghaemi, N.; Sharifianjazi, F.; Irandoost, M.; Esmaeikhanian, A. Electrical and microstructural characteristics of GDC electrolyt synthesized by benzoate copercipitation for solid oxide fuel cells. *Ceram. Int.* **2024**, *50*, 41780–41791. [CrossRef]
9. Shim, J.H.; Chao, C.-C.; Huang, H.; Prinz, F.B. Atomic layer deposition of yttria-stabilized zirconia for solid oxide fuel cells. *Chem. Mater.* **2007**, *19*, 3850–3854. [CrossRef]
10. Artinai, C.; Pani, M.; Carnasciali, M.M.; Buscaglia, M.T.; Plaisier, J.R.; Costa, G.A. Structural features of Sm- and Gd-doped ceri studied by synchrotron X-ray diffraction and μ-Raman spectroscopy. *Inorg. Chem.* **2015**, *54*, 4126–4137. [CrossRef]
11. Artini, C.; Costa, G.A.; Pani, M.; Lausi, A.; Plaisier, J. Structural characterization of the CeO_2/Gd_2O_3 mixed system by synchrotror X-ray diffraction. *J. Solid State Chem.* **2012**, *190*, 24–28. [CrossRef]
12. Mickan, M.; Coddet, P.; Vulliet, J.; Caillard, A.; Sauvage, T.; Thomann, A.-L. Optimized magnetron sputtering process for the deposition of gadolinia doped ceria layers with controlled structural properties. *Surf. Coat. Technol.* **2020**, *398*, 126095. [CrossRef]
13. Solovyev, A.A.; Rabotkin, S.V.; Shipilova, A.V.; Ionov, I.V. Magnetron sputtering of gadolinium-doped ceria electrolyte fo intermediate temperature solid oxide fuel cells. *Int. J. Electrochem. Sci.* **2019**, *14*, 575–584. [CrossRef]
14. Bernadet, L.; Segura-Ruiz, J.; Yedra, L.; Estrade, S.; Peiró, F.; Montinaro, D.; Torrell, M.; Morata, A.; Tarancón, A. Enhanced diffusion barrier layers for avoiding degradation in SOFCs aged for 14,000 h during 2 years. *J. Power Sources* **2023**, *555*, 232400 [CrossRef]
15. Bernadet, L.; Buzi, F.; Baiutti, F.; Segura-Ruiz, J.; Dolado, J.; Montinaro, D.; Torrell, M.; Morata, A.; Tarancón, A. Thickness effec of thin-film barrier layers for enhanced long-term operation of solid oxide fuel cells. *APL Energy* **2023**, *1*, 036101. [CrossRef]
16. Develos-Bagarinao, K.; Ishiyama, T.; Kishimoto, H.; Shimada, H.; Yamaji, K. Nanoengineering of cathode layers for solid oxide fuel cells to achieve superior power densities. *Nat. Commun.* **2021**, *12*, 3979. [CrossRef]

17. Meng, G.; Song, H.; Dong, Q.; Peng, D. Application of novel aerosol-assisted chemical vapor deposition techniques for SOFC thin films. *Solid State Ion.* **2004**, *175*, 29–34. [CrossRef]
18. Park, S.; Clark, B.L.; Keszler, D.A.; Bender, J.P.; Wager, J.F.; Reynolds, T.A.; Herman, G.S. Low-temperature thin-film deposition and crystallization. *Science* **2002**, *297*, 65. [CrossRef]
19. Kim, T.; Park, S. Crack-free fabrications of yttria-stabilized zirconia films using successive-ionic-layer-adsorption-and-reaction and air-spray plus method. *Korean J. Mater. Res.* **2024**, *34*, 79–84. [CrossRef]
20. Hong, T.; Zhang, Y.; Brinkman, K. Enhance oxygen electrocatalysis in heterostructured ceria electrolytes for intermediate-temperature solid oxide fuel cells. *ACS Omega* **2018**, *3*, 13559–13566. [CrossRef]
21. Bondarenko, A.S.; Ragoisha, G.A. *Progress in Chemometrics Research*; Pomerantsev, A.L., Ed.; Nova Science Publishers: New York, NY, USA, 2005; p. 89.
22. Yoon, J.; Choi, Y.; Shin, C. Grain-size adjustment in $Hf_{0.5}Zr_{0.5}O_2$ ferroelectric film to improve the switching time in $Hf_{0.5}Zr_{0.5}O_2$-based ferroelectric capacitor. *Nanotechnology* **2024**, *35*, 135203. [CrossRef] [PubMed]
23. Ohta, S.; Kihira, Y.; Asaoka, T. Grain Boundary Analysis of the Garnet-Like Oxides $Li_{7+X-Y}La_{3-X}A_XZr_{2-Y}Nb_YO_{12}$ (A = Sr or Ca). *Front. Energy Res.* **2016**, *4*, 30. [CrossRef]
24. Su, P.-C.; Chao, C.-C.; Shim, J.H.; Fasching, R.; Prinz, F.B. Solid oxide fuel cell with corrugated thin film electrolyte. *Nano Lett.* **2008**, *8*, 2289–2292. [CrossRef] [PubMed]

molecules

MDPI

Article

Enhanced Electrochemical Performance of Dual-Ion Batteries with T-Nb_2O_5/Nitrogen-Doped Three-Dimensional Porous Carbon Composites

Chen Qi [1], Duo Ying [1], Cheng Ma [1], Wenming Qiao [1], Jitong Wang [1,2,*] and Licheng Ling [1]

[1] State Key Laboratory of Chemical Engineering, East China University of Science and Technology, Shanghai 200237, China; 18919095863@163.com (C.Q.); shapeofvoice@163.com (D.Y.); mac@ecust.edu.cn (C.M.); qiaowm@ecust.edu.cn (W.Q.); lchling@ecust.edu.cn (L.L.)

[2] Guangxi Key Laboratory of Petrochemical Resource Processing and Process Intensification Technology, School of Chemistry and Chemical Engineering, Guangxi University, Nanning 530004, China

* Correspondence: wangjt@ecust.edu.cn

Abstract: Niobium pentoxide (T-Nb_2O_5) is a promising anode material for dual-ion batteries due to its high lithium capacity and fast ion storage and release mechanism. However, T-Nb_2O_5 suffers from the disadvantages of poor electrical conductivity and fast cycling capacity decay. Herein, a nitrogen-doped three-dimensional porous carbon (RMF) was prepared for loading niobium pentoxide to construct a composite system with excellent electrochemical performance. The obtained T-Nb_2O_5/RMF composites have a well-developed pore structure and a high specific surface area of 1568.5 $m^2\ g^{-1}$, which could effectively increase the contact area between the material and electrolyte, improving the electrode reaction and lithium-ion transfer diffusion. Nitrogen doping increased surface polarity, creating more active sites and accelerating the electrode reaction rate. The introduction of T-Nb_2O_5 imparted high power density and excellent cycling stability to the battery. The composites exhibited good electrochemical performance when used as dual-ion battery anode, with a stable cycle life of 207.2 mA h g^{-1} at 1 A g^{-1} current density after 650 cycles and great rate performance of 181.5 mA h g^{-1} at 5A g^{-1} was also obtained. This work provides the possibility for applying T-Nb_2O_5/RMF as an anode for a high-performance dual-ion battery.

Keywords: dual-ion batteries; Nb_2O_5; anode; structure design; Nitrogen doping

1. Introduction

The current depletion of fossil fuel reserves and the growing problem of environmental pollution have prompted extensive research into the development and storage of new green energy sources [1,2]. The intermittent nature of renewable solar and wind energy necessitates the development of energy storage solutions [3,4]. Lithium-ion batteries have gained widespread commercial applications due to their good energy storage performance and stability [5]. However, limited and unsustainable lithium and cobalt resources, high costs and safety concerns remain [6–10]. Consequently, it is essential to investigate new electrochemical energy storage devices that offer high safety, low cost, and environmental sustainability, as well as high energy density and long cycle life.

Dual-ion batteries (DIBs) represent a promising class of energy storage technologies, distinct from traditional alkali-metal ion batteries that operate on the 'rocking chair' mechanism [11]. In this case, cations shuttle between the positive and negative electrodes during charging and discharging. In contrast, DIBs involve both cations and anions during charging and discharging. Cations and anions enter the anode and cathode simultaneously during charging, and then released from the electrodes simultaneously during discharging [12–14]. This unique mechanism contributes to the high energy density of DIBs, similar to that of lithium-ion batteries, while also offering a higher operating voltage and improved environmental friendliness [15,16]. Consequently, DIBs have become an active area of research in recent years. The electrode material is a crucial factor limiting the development of DIBs, with the anode material typically responsible for the intercalation and deintercalation of cations. Graphite is commonly utilized as the anode material in DIBs due to its layered structure, which facilitates the reversible intercalation of various cations. While graphite helps reduce the overall cost of DIBs, it has limitations in terms of lithium intercalation capacity and cycling stability. Specifically, graphite has a relatively low lithium intercalation capacity of 372 mAh g^{-1} and a low intercalation potential (~0.1 V vs. Li/Li^+), which is close to the lithium deposition potential. This proximity increases the risk of dendrite formation during cycling, which can compromise battery safety by causing short circuits [17,18]. Furthermore, the cycling stability of graphite is also a concern, as the structural integrity of its layered framework is prone to degradation during the insertion and extraction of lithium ions [19,20]. Therefore, in this work, it is desired to prepare a DIBs anode material to improve the lithium intercalation capacity, cycling stability and structural recovery.

The anode ion storage mechanism of dual-ion batteries is similar to that of lithium-ion batteries [21], and is also divided into intercalation [22,23], alloying [23,24], conversion [25,26], and organic [27,28] anodes materials [29]. For dual-ion batteries, an ideal anode material should demonstrate a favorable cation reduction reaction and rapid ion transport kinetics, aligning with the swift insertion kinetics of anions on the cathode side to maintain robust electrochemical performance, even under high current densities [30–32]. Additionally, the anode material must exhibit a strong binding capacity to prevent the spontaneous release of active ions, thereby reducing the self-discharge rate [33]. Good compatibility with the electrolyte is also necessary, allowing for the rapid formation of a solid electrolyte interface (SEI) that enhances cation transport efficiency and extends the electrode's operational lifespan [34,35]. The electrolyte can be stabilized and concentrated under high pressure. The orthorhombic crystal system of Nb_2O_5 (T-Nb_2O_5) has been identified as a conversion anode with a unique pseudo-capacitance effect [36,37]. The insertion and detachment of lithium ions in this material are characterized by a fast charge storage process, which can produce high energy density without phase transition [38]. Nevertheless, it is important to acknowledge that pure Nb_2O_5 also exhibits certain limitations as an anode material. These include large potential lag during cycling, low initial cycling coulombic efficiency, rapid capacity decay, and poor rate performance [36,39,40]. To compensate for these problems, Nb_2O_5-based nanocomposites require more conductive networks to achieve higher performance.

Based on these, in this work, T-Nb_2O_5/nitrogen-doped three-dimensional porous carbon composites were synthesized by in-situ growth of T-Nb_2O_5 nanoparticles on the surface and inside the pores of a nitrogen-doped mesoporous carbon (RMF). The loading could be easily controlled by adjusting the ratio of ammonium niobium oxalate and RMF. The nitrogen-doped mesoporous carbon has a highly developed pore structure, which possesses a large amount of space for the embedding of T-Nb_2O_5. The constructed three-dimensional mesh skeleton provides excellent structural stability. Meanwhile, the growth of T-Nb_2O_5 in the pore channels makes the larger pores into micropores with smaller pore diameters and larger specific surface area, which can effectively increase the contact area between the material and electrolyte. This is conducive to the electrode reaction and lithium-ion transfer diffusion at the electrode interface. Meanwhile, the doping of nitrogen breaks the uniform arrangement of electrons on the surface of the carbon structure, which enhances the surface polarity of the material, helps the conduction of charge in the composite structure, and accelerates the reaction kinetics. In addition, T-Nb_2O_5 has a unique pseudocapacitance property that provides a high-speed ion embedding mechanism and good electrochemical stability. Therefore, the dual-ion battery with T-Nb_2O_5/RMF as the anode was able to achieve an initial specific capacity of 332.8 mA h g^{-1} at a high current density of 1 C, and still retained a specific capacity of 207.2 mA h g^{-1} after 650 cycles, with a decay rate of only 0.058% per cycle. The T-Nb_2O_5/RMF composites prepared in this work are able to achieve high electrochemical performance as a simple and effective strategy for the practical application of dual-ion battery anodes.

2. Results and Discussion

2.1. *Synthesis and Characterization of Materials*

The approximate synthetic route of T-Nb_2O_5/RMF is shown in Figure 1. The RMF was synthesized using SiO_2 as a template, resorcinol-formaldehyde resin as a carbon source, and melamine as a nitrogen source. Subsequently, T-Nb_2O_5 was loaded on the RMF by a simple water bath using niobium ammonium oxalate as a niobium source. A series of materials were prepared by varying the ratios of ammonium niobium oxalate and RMF. As a comparison, in addition to T-Nb_2O_5/RMF loaded with T-Nb_2O_5 on 3D porous carbon RMF, T-Nb_2O_5/CNTs loaded with T-Nb_2O_5 on 1D carbon nanotubes and T-Nb_2O_5/GO loaded with T-Nb_2O_5 on 2D graphene were also prepared in this work. The synthesized three-dimensional porous carbon RMF has a high specific surface area and abundant pore structure, which can effectively load T-Nb_2O_5. Nitrogen doping destroys the surface polarity of the carbon material, which makes it exhibit more active sites and accelerates the electrode reaction rate. The T-Nb_2O_5/RMF loaded with T-Nb_2O_5 has a higher specific surface area and a developed porechannel structure, which provides a place for lithium-ion storage and transport and improves the battery performance.

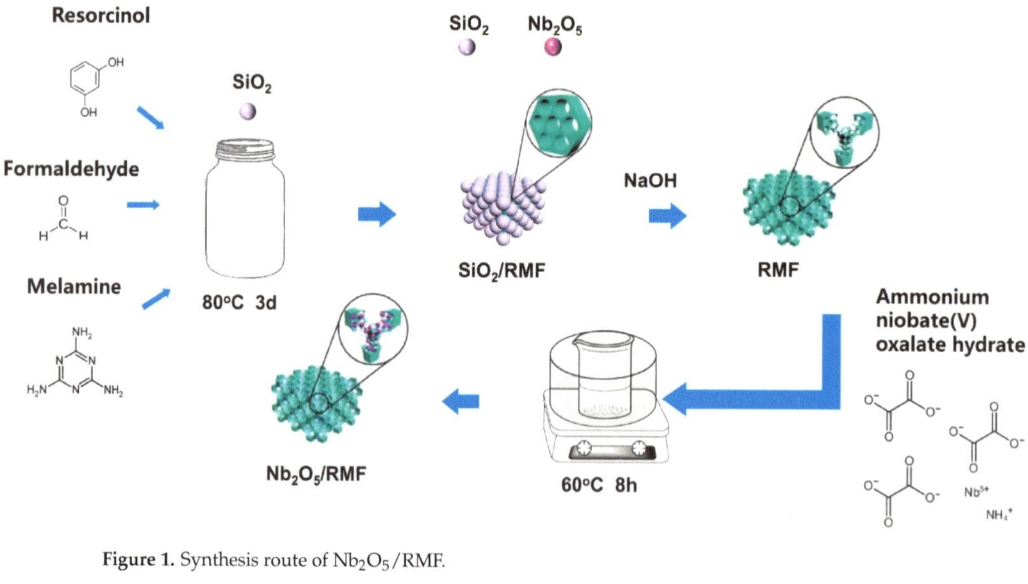

Figure 1. Synthesis route of Nb_2O_5/RMF.

In order to clearly observe the microscopic morphology of the samples in different scales and the dispersion state of T-Nb_2O_5 in the mesoporous carbon materials. The scanning electron micrographs (SEM) and transmission electron micrographs (TEM) were taken for the samples. As can be seen from Figure 2a, the RMFs prepared after selective etching of the silicon templates with NaOH solution show a three-dimensional reticulated structure with fairly well-developed pores. Through Figure 2b–e, it can be found that the pore size of the material tends to decrease gradually with the increasing proportion of niobium source added, which is due to the fact that the addition of more T-Nb_2O_5 fills up the pores with larger internal diameters of the carbon material. This also shows that T-Nb_2O_5 can effectively grow on the surface and inside the pores of the mesoporous carbon material. Further observation of the micromorphology by transmission electron microscopy in Figure 2f–i shows that T-Nb_2O_5 nanoparticles can be found to be uniformly dispersed in the composite with a diameter of about 10 nm, and their lattice can be clearly observed. T-Nb_2O_5 grows uniformly on the surface and inside the pores of mesoporous carbon, which provides a more stable spatial structure for T-Nb_2O_5 nanoparticles. Meanwhile, the three-dimensional reticulated carbon skeleton enhances the charge transfer rate, and the developed pore structure offers a place for the storage and transport of lithium ions, which is conducive to improving the performance and dynamics of the battery. T-Nb_2O_5/GO and T-Nb_2O_5/CNTs were also characterized by SEM and TEM, as shown in Figures S1 and S2. From Figure S1, the layered structure of graphene and T-Nb_2O_5 nanoparticles dispersed on the surface of graphene can be clearly seen. The layered structure of graphene has been curled to a certain extent and even formed a more complex reticular structure. Figure S2 shows that the carbon nanotubes are irregularly interwoven, and the T-Nb_2O_5 nanoparticles are bound on the surface and inside of the carbon nanotubes. Thus, the as-prepared T-Nb_2O_5/CNTs composites construct a complex interwoven mesh carbon skeleton with carbon nanotubes randomly interwoven.

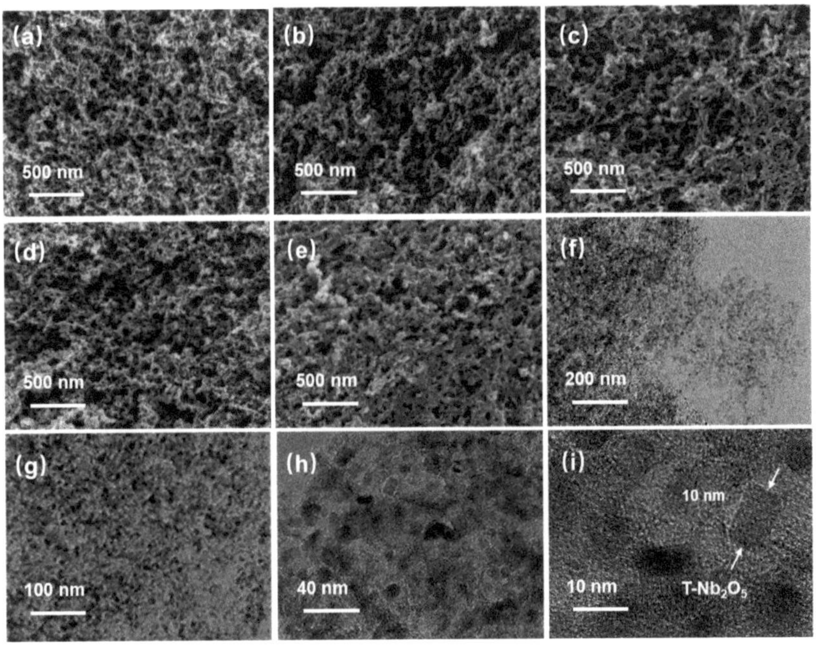

Figure 2. SEM images of (**a**) RMF, (**b**) T-Nb_2O_5/RMF 1:1, (**c**) T-Nb_2O_5/RMF 1:2, (**d**) T-Nb_2O_5/RMF 1:6, (**e**) T-Nb_2O_5/RMF 1:10, and (**f–i**) TEM images of T-Nb_2O_5/RMF 1:6.

In order to further analyze the composition and lattice conditions of the composites, the samples with different amounts of T-Nb_2O_5 were subjected to XRD and Raman characterization, and the results are shown in Figure 3. Figure 3a presents a comparative XRD analysis of T-Nb_2O_5/RMF. The XRD curves of RMF exhibit a wide range of disordered carbon characteristic peaks in the range of 20°–30°, while the diffraction peaks of T-Nb_2O_5 gradually become more pronounced with the increase in the introduction of T-Nb_2O_5. At a T-Nb_2O_5/RMF ratio of 1:1, the characteristic peaks exhibit a clear correspondence with the standard card of T-Nb_2O_5 (PDF:30-0873), indicating that T-Nb_2O_5 had been effectively synthesized and loaded into the mesoporous carbon material. As illustrated in Figure 3b, the Raman spectra of the composites exhibit two broad carbon peaks, which are the disordered carbon peak (D peak) and the ordered carbon peak (G peak). Generally, the D peak is attributable to the defects resulting from heteroatoms and functional groups, while the G peak is related to the in-plane tensile vibration of sp2-hybridized carbon atoms [41,42]. The intensity ratio of the D peak to the G peak can be employed as an indicator of the degree of defectification of the material surface. The calculation of the peak intensities of the Raman spectra of different materials reveals that the I_D/I_G values of the composites gradually increase with the ratio of T-Nb_2O_5/RMF. This indicates that the introduction of T-Nb_2O_5 effectively increases the degree of defectivation on the surface of the material. The presence of metal oxides breaks the neatly distributed electronic arrays on the surface of the carbon, enhances the surface polarity of the carbon material, and thus improves the electrical conductivity of the composite material. Figure S3a,b illustrate the XRD and Raman analysis profiles of T-Nb_2O_5 when loaded on 1D, 2D, and 3D carbon materials. As illustrated in Figure S3a, the 2D T-Nb_2O_5/GO exhibits the most pronounced T-Nb_2O_5 characteristic peaks, followed by the 1D T-Nb_2O_5/CNTs. However, the T-Nb_2O_5/CNTs display the most prominent 26° carbon peaks. Figure S3b illustrates that the 2D T-Nb_2O_5/GO has the largest I_D/I_G, the 3D T-Nb_2O_5/RMF the second largest, and the 1D T-Nb_2O_5/CNTs the smallest. It has been demonstrated that an appropriate degree of carbon surface defects enhances the surface polarity of the carbon material and improves the overall electrical conductivity of the composite.

Figure 3c–f demonstrate the nitrogen adsorption/desorption curves with pore size distribution data for the samples. Based on the nitrogen isothermal adsorption/desorption curves of the materials, the specific surface area, total pore volume and average pore diameter of the composites were calculated, as shown in Table 1. It can be seen that RMF exhibits an obvious type IV adsorption/desorption curve and H3 hysteresis loop, indicating that it has a rich mesoporous structure. The specific surface area of RMF is 1008.6 $m^2\ g^{-1}$, and its pore size is mainly distributed around 15 nm, with a pore volume of 4.51 $cm^3\ g^{-1}$. After the introduction of T-Nb_2O_5 nanoparticles, the specific surface area of T-Nb_2O_5/RMF increased to 1568.5 $m^2\ g^{-1}$, the pore size was mainly distributed around 4 nm, and the pore volume was reduced to 2.79 $cm^3\ g^{-1}$, which may be due to the fact that T-Nb_2O_5 nanoparticles split the larger pores into smaller pores with a more complex structure and larger internal surface area. The smaller pores can improve the contact effect between the composite and the electrolyte, as well as the lithium-ion transport rate at the electrode material–electrolyte interface. Comparing the BET data of the T-Nb_2O_5 composites with different carbon supports, it can be observed that the specific surface areas show an increasing trend from 1D, 2D to 3D composites, which are 135.3 $m^2\ g^{-1}$, 263.9 $m^2\ g^{-1}$ and 1568.5 $m^2\ g^{-1}$, respectively. The main pore structure size of all three composites is mesoporous, and the pore volume and main pore size distribution also tend to increase with dimension.

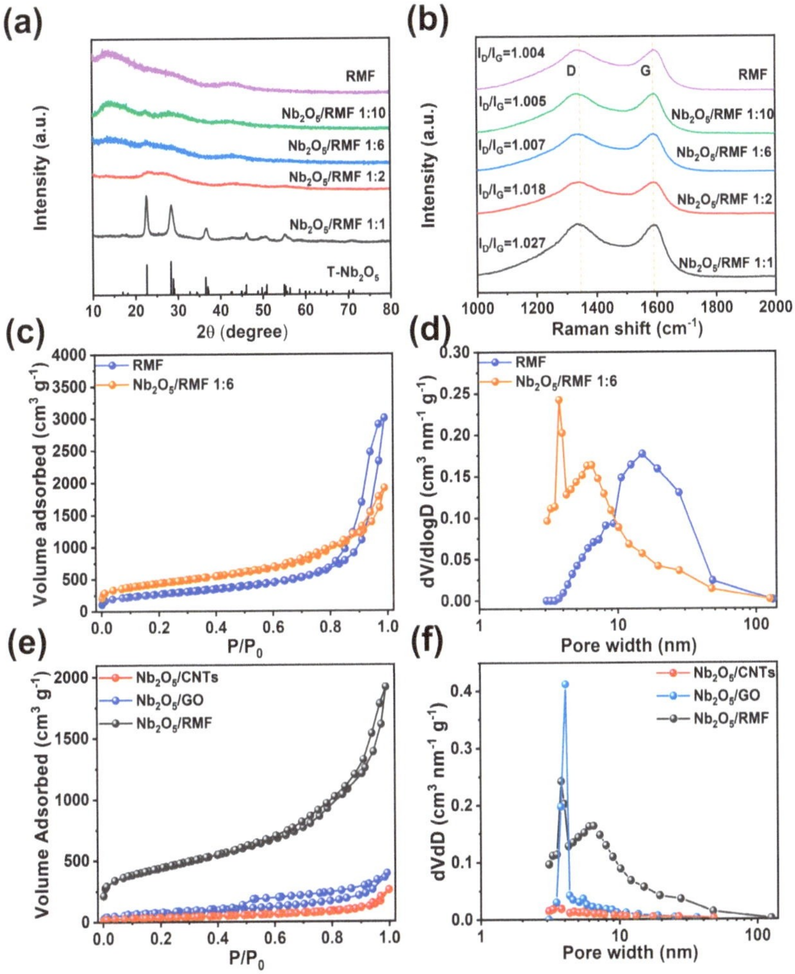

Figure 3. (**a**) XRD pattern (**b**) Raman pattern of composites with different T-Nb_2O_5/RMF ratios and RMF. (**c**) Nitrogen isothermal adsorption–desorption curves and (**d**) DFT pore size distribution of T-Nb_2O_5/RMF composite with RMF. (**e**) Nitrogen isothermal adsorption–desorption curves and (**f**) DFT pore size distribution of different T-Nb_2O_5/C composites.

Table 1. Porosity parameters of T-Nb_2O_5/C composite and RMF.

Samples	S_{BET} (m^2/g)	V_t (cm^3/g)	D_{meso} (nm)
RMF	1008.6	4.51	15.13
T-Nb_2O_5/RMF	1568.5	2.79	7.56
T-Nb_2O_5/GO	263.9	0.61	4.11
T-Nb_2O_5/CNTs	135.3	0.40	3.78

X-ray photoelectron spectroscopy (XPS) characterization determines the elemental composition and chemical bonding energy of the T-Nb_2O_5/RMF composites, as shown in Figure 4. XPS analysis reveals the presence of four elements in the composite sample: carbon (C), nitrogen (N), oxygen (O) and niobium (Nb). It can be seen that the C, N and O elements are present in the RMF, and the Nb element is introduced by the compositing process with T-Nb_2O_5. By analyzing the fine spectra of the different elements with peak fitting, it is possible to obtain the different valence patterns of the element. As illustrated in Figure 4c for Nb 3d, the Nb 3d spectrum is primarily comprised of two peaks, corresponding to the Nb $3d_{5/2}$ orbital at 207.3 eV and the Nb $3d_{3/2}$ orbital at 210.1 eV, respectively. By fitting the analysis in Figure 4d, it can be demonstrated that there are C–C/C=C, C–O, C=C, and C–N bonding modes for the C elements in the material. In Figure 4e, the N 1s are categorized into four forms: N–O, C–N, C=N, and graphitic N. Among these, C–N and C=N confirm that the N element was successfully doped on the surface of the carbon material, which effectively enhances the surface polarity. In the fine spectrum of O 1s, the split-peak fitting yields C–O/N–O, Ov, and Nb–O, corresponding to binding energies at 532.9 eV, 532.1 eV, and 530.8 eV, respectively; Ov represents the presence of oxygen vacancies. Figure S5 demonstrates the XPS spectra of T-Nb_2O_5/GO, and Nb_2O_5/CNTs; comparing the three composites reveals that T-Nb_2O_5/GO has a stronger O peak. This may be due to the presence of oxygen-containing functional groups hydroxyl and epoxy groups in graphene oxide.

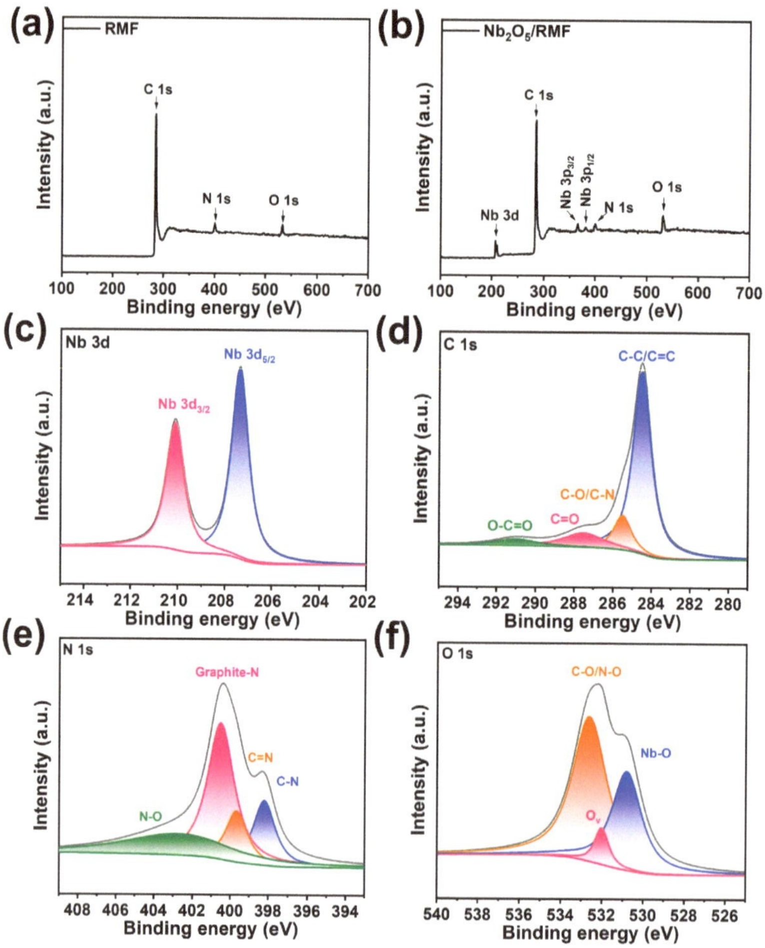

Figure 4. (**a**) XPS full spectral of RMF, (**b–f**) XPS spectrum of T-Nb_2O_5/RMF composite.

In order to accurately determine the mass ratio of carbon to metal elements in the T-Nb_2O_5/RMF composites, thermogravimetric analyses were performed in air, and the thermogravimetric data are presented in Figure S4. From the thermogravimetric curves, it is observed that the composites exhibit a significant weight loss at approximately 550 °C. It is also noted that the peak temperature of weight loss increased with the increase in metal content.

2.2. Battery Dynamics Analysis

In order to investigate the electrochemical performance of the T-Nb_2O_5 composite electrode, a half-cell was prepared using a 2025 button cell with lithium metal as the counter electrode and reference electrode. As illustrated in Figure 5a, the electrochemical properties of T-Nb_2O_5/RMF show a trend of 1:6 > 1:10 > RMF > 1:2 > 1:1, which proves that the appropriate amount of T-Nb_2O_5 loading plays an important role in the electrochemical properties of the composites. The introduction of excessive metal oxides leads to the blockage of the pores of RMF itself, which seriously reduces the migration rate of lithium ions and deteriorates the wettability of the electrolyte, resulting in a significant reduction of the capacity, while the introduction of too little metal oxides results in less active material and a decrease in the lithium storage capacity. Therefore, the best electrochemical properties were obtained at the ratio 1:6, exhibiting an initial capacity of 1320.8 mA h g^{-1} at 0.1 C and subsequently stabilizing at 945.2 mA h g^{-1} after 100 cycles. Figure 5b shows that when charging and discharging at a high current density of 1 C, the initial specific capacity of 332.8 mA h g^{-1} is obtained. The specific capacity of 207.2 mA h g^{-1} is still maintained after 650 cycles, with a decay rate of 0.058% per cycle, which shows good cyclability. In addition, the material also exhibits good rate performance, as shown in Figure 5c. The discharge-specific capacities are found to be 663.3 mA h g^{-1}, 463.5 mA h g^{-1}, 379.6 mA h g^{-1}, 291.3 mA h g^{-1}, 228.1 mA h g^{-1}, and 181.5 mA h g^{-1} at 0.1 C, 0.2 C, 0.5 C, 1 C, 2 C and 5 C current densities, respectively. When the current density drops from 5 C to 0.1 C, the battery shows a great recovery capacity of 817.6 mA h g^{-1}, which is even better than the initial discharge specific capacity. This may be due to the fact that with the increase of the cycling time, the active material of the electrode can be fully contacted with the electrolyte to form a thicker and more stable SEI film. The SEI film effectively improves the capacity of the composite material of the deep-embedded lithium and the diffusion rate of lithium ions inside the electrode. Figure 5d shows the capacity–voltage curves of T-Nb_2O_5/RMF 1:6 composites at different current rates. The result demonstrates that the batteries assembled with these materials can effectively adjust their reaction rates to accommodate the accelerated charge transfer rates associated with variations in charge and discharge current densities. From Table 2, by comparing the electrochemical properties with other anode materials prepared in the literature for dual-ion batteries, it can be seen that the T-Nb_2O_5/RMF composites prepared in this work have a very high specific capacity at low current densities, which provides a great advantage for the preparation of high-performance dual-ion batteries.

Table 2. Comparison of electrochemical performances of different dual-ion battery anodes.

Electrolyte	Specific Capacity	Reference
T-Nb_2O_5/RMF	945 mA h g^{-1} after 100 cycles at 100 mA g^{-1}	This work
Co_3O_4/carbon fiber paper (CFP)	488 mA h g^{-1} after 40 cycles at 200 mA g^{-1}	[43]
$MoSe_2$/nitrogen-doped carbon (NC)	759 mA h g^{-1} after 100 cycles at 100 mA g^{-1}	[44]
MoO_2/carbon matrix	734 mA h g^{-1} after 350 cycles at 50 mA g^{-1}	[45]
SnO_2	613 mA h g^{-1} after 60 cycles at 60 mA g^{-1}	[46]
$Li_6C_{12}O_{12}$	730 mA h g^{-1} after 100 cycles at 210 mA g^{-1}	[47]
Sb_2O_3@MCNF	626.9 mA h g^{-1} after 80 cycles at 100 mA g^{-1}	[42]

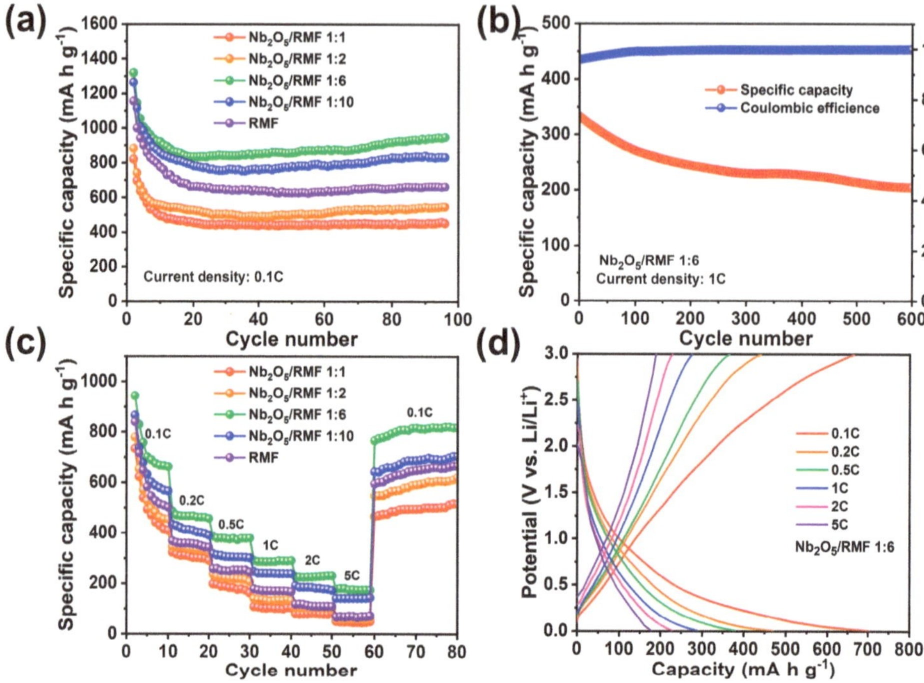

Figure 5. (**a**) Comparison of cycle performance of batteries assembled with different T-Nb_2O_5/RMF ratio composites at 0.1 C, (**b**) long cycle performance of batteries assembled with T-Nb_2O_5/RMF 1:6 composite at 1 C. (**c**) Comparison of rate performance of batteries assembled with different T-Nb_2O_5/RMF ratio composites, (**d**) Rate charge/discharge curves of T-Nb_2O_5/RMF 1:6.

The CV curves of T-Nb_2O_5/RMF 1:6 were tested in the range of 0.01–3 V at a scan rate of 0.1–5 mV s^{-1}, as illustrated in Figure 6a. It is observed that the cathodic peaks and anodic peaks at 1.48 V and 1.98 V, respectively, correlated with the lithium embedding and delithiation behaviors of T-Nb_2O_5 during electrochemical processes. This indicates that the composites exhibit a highly reversible lithium storage process. Figure 6b presents the AC impedance spectra and the fitted equivalent circuit diagrams for T-Nb_2O_5 loaded on carbon substrates of different dimensions. As shown in Figure 6b, all three composite materials exhibit a semicircular feature in the high-frequency region and a linear feature in the low-frequency region. The resistance between the electrode and the electrolyte, known as the solution resistance (R_b), is reflected by the intersection point between the EIS curve and the x-axis. The high-frequency semicircle corresponds to the charge transfer resistance (R_{ct}), while the low-frequency linear portion is associated with the Warburg impedance (Z_W), which reflects the diffusion of lithium ions in the active anode material. Among the three materials, Nb_2O_5/RMF shows the smallest impedance, while Nb_2O_5/CNTs shows the largest, further confirming the performance differences among the materials. Figure 6c shows a comparison of the cycling performance of the three composites at 0.5 C. The cells were run for 5 revolutions at 0.1 C and 0.2 C, respectively, and then elevated to the level of 0.5 C to promote the formation of the SEI film. Hence, the sudden drop is formed in the front part of the curve. According to the curves, it can be seen that T-Nb_2O_5/RMF has the best cycling performance of 411.4 mA h g^{-1} in 200 cycles. Figure 6d shows the comparison of the rate performance of the three composites. At 0.1 C, the specific capacity of T-Nb_2O_5/RMF was as high as 702.4 mA h g^{-1}, which was the highest among the three, followed by Nb_2O_5/GO, and Nb_2O_5/CNTs was the lowest. With the increase of the current density, the rate of Nb_2O_5/GO exhibits the least capacity weakening trend, and even the specific capacity is comparable to that of Nb_2O_5/RMF at high current densities, which is attributed to the extremely high electrical conductivity and electron mobility of graphene. All three materials showed good capacity recovery when the current density recovered from 5 C to 0.1 C. The highly developed pore structure and large specific surface area of Nb_2O_5/RMF provided enough space for contact between the active ingredient and the electrolyte. The three-dimensional reticulated carbon skeleton gives the composite material excellent structural stability, and the introduced T-Nb_2O_5 nanoparticles are firmly anchored within the pores of the material. The composite material has a good degree of surface defects and a well-connected carbon skeleton, which offers a more convenient path for the conduction of lithium ions and charge and accelerates the overall reaction kinetics of the battery. Therefore, among the three kinds of spatial structures, the three-dimensional mesh carbon skeleton structure has a better nanoparticle composite effect and excellent electrochemical performance, which has a good prospect for the application of dual-ion battery electrode materials.

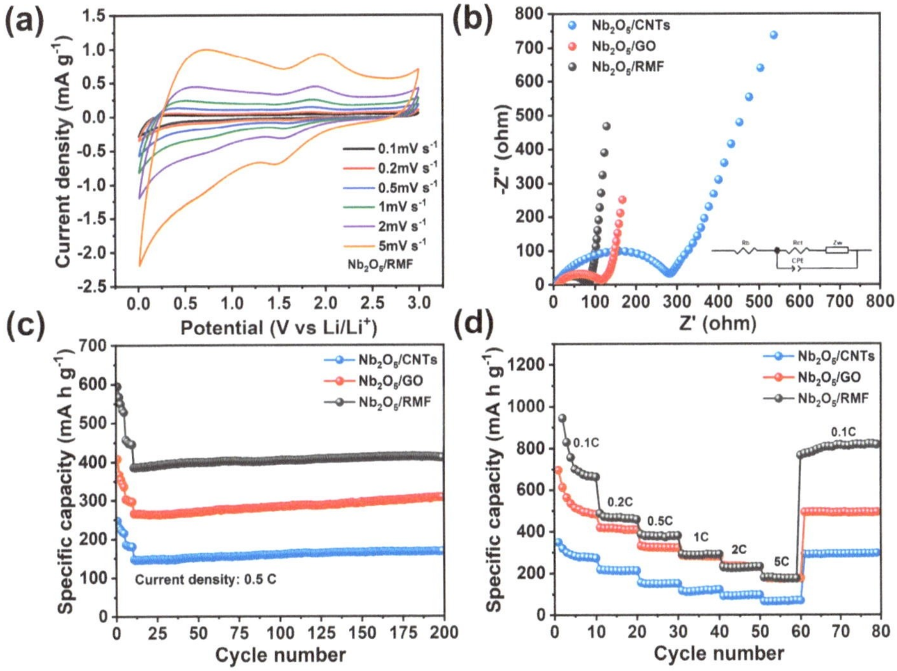

Figure 6. (**a**) CV curves of T-Nb_2O_5/RMF composite at different scanning rates (**b**) EIS curves of different T-Nb_2O_5/C composites (**c**) cycle performance of different T-Nb_2O_5/C composites (**d**) rate performance of different T-Nb_2O_5/C composites.

In order to cope with the various conditions of practical applications, this work also carried out low-temperature electrochemical performance tests at 0 °C. T-Nb_2O_5/RMF 1:6 composites were selected as the electrode materials to be assembled into a half cell, which had previously shown the best performance. As shown in Figure 7a,b, the composite material can achieve an initial discharge specific capacity of 435 mA h g^{-1} at 0.1 C, and after 100 cycles, the specific capacity can be stabilized at 283.6 mA h g^{-1}. At various current densities from 0.1 to 5 C, the composite material can achieve 383.2 mA h g^{-1}, 293.2 mA h g^{-1}, 242.5 mA h g^{-1}, 203.6 mA h g^{-1}, 167. 6 mA h g^{-1}, 123.3 mA h g^{-1}, respectively. The specific capacity gradually stabilizes at 311.2 mA h g^{-1} when the current density is returned to 0.1 C. The specific capacity did not increase greatly when the specific capacity was restored from high to low current density at low temperatures, suggesting that temperature may affect the electrochemical behavior of the materials and the formation of SEI films to some extent. This indicates that the composite electrodes can still achieve stable cycling performance and excellent rate performance even at low ambient temperatures.

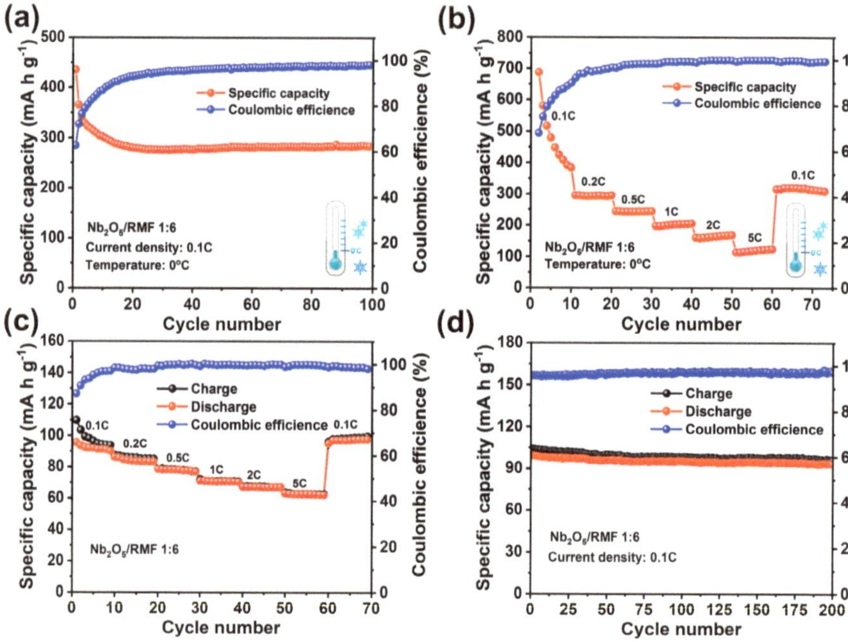

Figure 7. (**a**) Cycling performance curve at 0.1 C of battery assembled with T-Nb_2O_5/RMF 1:6 composite in 0 °C low-temperature environment (**b**) rate performance curve of battery assembled with T-Nb_2O_5/RMF 1:6 composite in 0 °C low-temperature environment (**c**) Cycling performance curve at 0.1 C of dual-ion battery with T-Nb_2O_5/RMF 1:6 as anode and graphite as cathode (**d**) rate performance curve of dual-ion battery with T-Nb_2O_5/RMF 1:6 as anode and graphite as cathode.

In this work, the T-Nb_2O_5/RMF 1:6 composite was also used as the anode material of the dual-ion battery full cell, matched with graphite cathode, and the full-cell performance was tested, in which the voltage range of the test was 3–5 V. The full-cell cycling performance curve at 0.1 C is shown in Figure 7c. It shows that the initial capacity of this dual-ion full-cell battery at 0.1 C was 99.6 mA h g^{-1}. The capacity stability was 93.3 mA h g^{-1} after 200 cycles, with a capacity decay rate of 0.031% per cycle, which shows the dual-ion battery has excellent cycling stability. Figure 7d shows the rate performance of the full cell at 0.1 C to 5 C. The specific capacities at different current density platforms are 93.8 mA h g^{-1}, 83.3 mA h g^{-1}, 77.3 mA h g^{-1}, 70.5 mA h g^{-1}, 66.9 mA h g^{-1}, and 62.4 mA h g^{-1}, respectively. The specific capacity is 97.2 mA h g^{-1} when getting back to 0.1 C; the overall capacity change trend is not obvious with the change of current density, indicating that the battery has excellent rate capability. Figure S6 shows the results of AC impedance testing of a dual-ion battery with T-Nb_2O_5/RMF 1:6 as the anode material and graphite as the cathode material before and after 50 charge/discharge cycles. The result indicates a noticeable increase in the radius of the semicircle in the high-frequency region and a change in the slope of the line in the low-frequency region, suggesting that both the charge transfer resistance and lithium-ion diffusion resistance within the full-cell battery increase after cycling. This effect is attributed to the irreversible phase transformation of the electrode material during charge/discharge cycles, leading to a gradual decline in battery capacity over time. This is related to the irreversible phase transition of the electrode material in the charge/discharge cycle so that the battery will have a certain capacity decay in the cycling process.

In order to clearly observe the structural stability of T-Nb_2O_5/RMF composites during charge/discharge cycling, the electrode sheets coated with T-Nb_2O_5/RMF composites before and after cycling were characterized by SEM images from Figure 8. It can be seen that the surface of the electrode is relatively flat and the pore structure on the surface can be observed after magnification in Figure 8a,b. Compared with the SEM image of the surface of the electrode pole piece after cycling in Figure 8c, the amorphous porous carbon structure of the composite was not obviously damaged, in the porous structure of the material can still be clearly observed from the magnified Figure 8d. The T-Nb_2O_5 nanoparticles are still uniformly dispersed in the pore channels, which effectively avoids the phenomenon of pulverization of the metal material. Meanwhile, the large specific surface area provided by the porous structure is conducive to the contact of the active substance with the electrolyte, which ensures the structural stability of the composite material and improves the transport rate of lithium ions at the electrode material–electrolyte interface.

Figure 8. SEM images of electrodes coated with T-Nb_2O_5/RMF composites, (**a**,**b**) before cycling, (**c**,**d**) after cycling.

3. Materials and Methods

Ammonium niobate oxalate hydrate was purchased from Shanghai Dibber Chemical Technology Co., Ltd. (Shanghai, China), resorcinol, melamine, formaldehyde and sodium hydroxide were purchased from Sinopharm Chemical Reagent Co., Ltd. (Shanghai, China), and silica sol was purchased from Shanghai McLean Biochemical Technology Co. (Shanghai, China). Graphene oxide solution and carbon nanotubes were purchased from Shanxi Institute of Coal Chemistry, Chinese Academy of Sciences (Taiyuan, China); hydrochloric acid solution and N-methylpyrrolidone were purchased from Shanghai Taitan Science and Technology Co Ltd. (Shanghai, China); polypropylene (PP) diaphragm was purchased from Celgard Corporation (Concord, NC, USA); and polyvinylidene fluoride was purchased from Arkema France (Paris, France).

The hard template method for the preparation of nitrogen-doped mesoporous carbon materials (RMF) was carried out as follows: in 20 mL of deionized water, 4.85 g of resorcinol and 7.15 g of formaldehyde solution were added and dissolved with stirring in a water bath at 40 °C to obtain the RF solution. In 20 mL of deionized water, 5.56 g of melamine and 10.73 g of formaldehyde were added again and dissolved with stirring in a water bath at 70 °C, then cooled to room temperature to obtain the MF solution. The RF solution was mixed with the MF solution, and 70 g of SM-30 (silica sol) was added, and then mixed, stirred again for 5 min at room temperature, poured into a plastic bottle, and then sealed and reacted at 80 °C for 3 days, after which it was freeze-dried to obtain the nitrogen-doped mesoporous carbon precursor containing silicon.

The precursor was placed in a high-temperature tube furnace with nitrogen venting at a rate of 2 °C min^{-1} to 400 °C for 1 h and then continued to 800 °C for 3 h. The precursor was then cooled down to room temperature naturally. The carbonized samples were placed in plastic bottles and etched with 2 mol L^{-1} NaOH solution for about 3 days, during which time the etched supernatant was removed and replaced with a freshly prepared 2 mol L^{-1} NaOH solution to ensure adequate removal of the SiO_2 template. After the etching, the sample was washed several times with deionized water and ethanol until neutral pH. Subsequently, it was freeze-dried in a lyophilizer to obtain nitrogen-doped mesoporous carbon materials (RMFs) with well-developed pore structures, which were abbreviated as RMFs because they were prepared from resorcinol melamine-formaldehyde.

In order to reach Nb/C mass ratios of 1:1, 1:2, 1:6, and 1:10, 0.6 g, 0.3 g, 0.1 g, and 0.06 g of niobium ammonium oxalate hydrate were added to 10 mL of deionized water and stirred thoroughly for 30 min. After that, 0.4 g of the previously prepared nitrogen-doped mesoporous carbon material (RMF) was dispersed in the above aqueous solution and ultrasonicated for 20 min. The resulting mixture was stirred magnetically in a water bath heated at 60 °C for 8 h and then washed and freeze-dried. After drying, the samples were carbonized in a tube furnace under nitrogen ventilation at a temperature of 700 °C and a residence time of 3 h. The T-Nb_2O_5/RMF composites were finally obtained.

4. Conclusions

In summary, nitrogen-containing mesoporous carbon materials were synthesized by hard template method using resorcinol, melamine and formaldehyde solutions as raw materials. The prepared composites present a three-dimensional reticulated skeleton structure with excellent stability and a highly developed pore structure, which facilitates the diffusion and transfer of lithium ions on the electrode surface. The nitrogen-doped porous carbon and T-Nb_2O_5 show good synergistic effects, and the Nb_2O_5 nanoparticles can not only effectively shorten the ion/electron transport distance in RMF but also alleviate the strain and stress caused by the volume change in the lithiation/delithiation process. The doping of nitrogen and the introduction of T-Nb_2O_5 enhance the surface polarity of the composites by destroying the uniform electronic arrangement on the surface of the carbon material, making it possess more active sites and accelerating the electrode reaction rate. The synergistic effect of nitrogen doping with carbon substrate can further improve the electronic conductivity and pseudocapacitive behavior of the active materials. The composite material exhibits a high specific capacity of 945.2 mA h g^{-1} at 0.1 C. At a high current density of 1 C, the initial specific capacity of the battery assembled with the T-Nb_2O_5/RMF composite material is 332.8 mA h g^{-1}. The specific capacity of 207.2 mA h g^{-1} is still retained after 650 cycles, with a decay rate of 0.058% per cycle, which is very excellent in cycle stability. At a low temperature of 0 °C, the battery still achieves an initial discharge specific capacity of 435 mA h g^{-1} at 0.1 C, which is stabilized at 283.6 mA h g^{-1} after 100 cycles, and a specific capacity of 123.3 mA h g^{-1} at a high current density of 5 C, which recovers to 311.2 mA h g^{-1} when the current density returns to 0.1 C. The initial specific capacity of the full-cell dual-ion battery assembled with graphite cathode was 99.6 mA h g^{-1} at 0.1 C, and the capacity stabilized at 93.3 mA h g^{-1} after 200 cycles with a capacity decay rate of 0.031% per cycle. Good low-temperature performance and full-cell performance also open up possibilities for practical battery applications. This composite material provides a feasible idea for the synthetics of the anode of dual-ion batteries.

Supplementary Materials: The following are the Supporting Information to this article: https://www.mdpi.com/article/10.3390/molecules30020227/s1, The experimental part is described in detail, including the preparation methods of Nb_2O_5/GO, and Nb_2O_5/CNTs; instrumental parameters used for SEM, TEM, XRD, RAMAN, BET, and XPS characterization; electrode sheet preparation for assembling the button cell, and slurry ratios; instrumental parameters used for CV, EIS, and electrochemical performance testing; Nb_2O_5/GO and Nb_2O_5/CNTs; SEM and TEM plots; XRD and RAMAN spectra of Nb_2O_5/RMF, Nb_2O_5/GO, and Nb_2O_5/CNTs; weightlessness analysis curves of different ratios of Nb_2O_5/RMF in air atmosphere; XPS spectra of Nb_2O_5/GO and Nb_2O_5/CNTs; EIS curves before and after 50 cycles of dual-ion battery full cell with T-Nb_2O_5/RMF 1:6 as anode and graphite as cathode. Figure S1 (a,b) SEM images and (c–f) TEM images of T-Nb_2O_5/GO composites. Figure S2 (a,b) SEM images and (c–f) TEM images of T-Nb_2O_5/CNTs composites. Figure S3 (a) XRD pattern (b) Raman pattern of different T-Nb_2O_5/C composites. Figure S4 (a) TG curves and (b) the thermogravimetric rate curves of RMF and different proportions of T-Nb_2O_5/RMF composites in air atmosphere. Figure S5 (a–d) XPS spectrum of T-Nb_2O_5/GO composite. (e–f) XPS spectrum of T-Nb_2O_5/CNTs composite. Figure S6 EIS curves before and after 50 cycles of dual-ion battery full cell with T-Nb_2O_5/RMF 1:6 as anode and graphite as cathode.

Author Contributions: The manuscript was written through contributions of all authors. C.Q.: Investigation, methodology, software, data curation, formal analysis, visualization, and writing–original draft. D.Y.: Conceptualization, methodology, validation, formal analysis, visualization, and supervision. C.M.: Methodology. W.Q.: Methodology and supervision. J.W.: Methodology, conceptualization, supervision, project administration, funding acquisition, resources, and writing–review and editing. L.L.: Supervision, funding acquisition, project administration, and resources. All authors have read and agreed to the published version of the manuscript.

Funding: This work is partly supported by the National Natural Science Foundation of China (Nos. U21A2060, 22178107, and 22178116) and Natural Science Foundation of Shanghai (No. 22ZR1417400).

Institutional Review Board Statement: Not applicable.

Informed Consent Statement: Not applicable.

Data Availability Statement: Data are contained within the article and Supplementary Materials.

Acknowledgments: The authors sincerely acknowledge the State Key Laboratory of Chemical Engineering, East China University of Science and Technology for providing the experimental apparatus.

Conflicts of Interest: The authors declare no conflict of interest.

References

1. Semeraro, C.; Olabi, A.G.; Aljaghoub, H.; Alami, A.H.; Al Radi, M.; Dassisti, M.; Abdelkareem, M.A. Digital twin application i energy storage: Trends and challenges. *J. Energy Storage* **2023**, *58*, 106347. [CrossRef]
2. Silva, L.F.O.; Santosh, M.; Schindler, M.; Gasparotto, J.; Dotto, G.L.; Oliveira, M.L.S.; Hochella, M.F., Jr. Nanoparticles in fossil an mineral fuel sectors and their impact on environment and human health: A review and perspective. *Gondwana Res.* **2021**, *9* 184–201. [CrossRef]
3. Hassan, Q.; Algburi, S.; Sameen, A.Z.; Salman, H.M.; Jaszczur, M. A review of hybrid renewable energy systems: Solar an wind-powered solutions: Challenges, opportunities, and policy implications. *Results Eng.* **2023**, *20*, 101621. [CrossRef]
4. Feng, X.; Guo, Z.; Luo, G.; Wu, C.; Qin, W. Structure-modulation of flower-like VS2 via ammonium ion intercalation for hig performance potassium-ion batteries. *Vacuum* **2024**, *227*, 113393. [CrossRef]
5. Subramanyan, K.; Divya, M.L.; Aravindan, V. Dual-carbon Na-ion capacitors: Progress and future prospects. *J. Mater. Chem.* **2021**, *9*, 9431–9450. [CrossRef]
6. Zhang, L.; Wang, H.; Zhang, X.; Tang, Y. A Review of Emerging Dual-Ion Batteries: Fundamentals and Recent Advances. *Ad Funct. Mater.* **2021**, *31*, 2010958. [CrossRef]
7. Deng, X.; Li, L.; Zhang, G.; Zhao, X.; Hao, J.; Han, C.; Li, B. Anode chemistry in calcium ion batteries: A review. *Energy Storag Mater.* **2022**, *53*, 467–481. [CrossRef]
8. Rajagopalan, R.; Tang, Y.; Ji, X.; Jia, C.; Wang, H. Advancements and Challenges in Potassium Ion Batteries: A Comprehensiv Review. *Adv. Funct. Mater.* **2020**, *30*, 1909486. [CrossRef]
9. Das, A.; Balakrishnan, N.T.M.; Sreeram, P.; Fatima, M.J.J.; Joyner, J.D.; Thakur, V.K.; Pullanchiyodan, A.; Ahn, J.-H.; Raghavan, Prospects for magnesium ion batteries: A compreshensive materials review. *Coord. Chem. Rev.* **2024**, *502*, 215593. [CrossRef]
10. Tay, I.R.; Xue, J.; Lee, W.S.V. Methods for Characterizing Intercalation in Aqueous Zinc Ion Battery Cathodes: A Review. *Adv. Sc* **2023**, *10*, 2303211. [CrossRef] [PubMed]
11. Placke, T.; Fromm, O.; Lux, S.F.; Bieker, P.; Rothermel, S.; Meyer, H.-W.; Passerini, S.; Winter, M. Reversible Intercalation o Bis(trifluoromethanesulfonyl)imide Anions from an Ionic Liquid Electrolyte into Graphite for High Performance Dual-Ion Cell *J. Electrochem. Soc.* **2012**, *159*, A1755. [CrossRef]
12. Zhang, F.; Ji, B.; Tong, X.; Sheng, M.; Zhang, X.; Lee, C.-S.; Tang, Y. A Dual-Ion Battery Constructed with Aluminum Foil Anod and Mesocarbon Microbead Cathode via an Alloying/Intercalation Process in an Ionic Liquid Electrolyte. *Adv. Mater. Interface* **2016**, *3*, 1600605. [CrossRef]
13. Li, N.; Xin, Y.; Chen, H.; Jiao, S.; Jiang, H.; Song, W.-L.; Fang, D. Thickness evolution of graphite-based cathodes in the dual ior batteries via in operando optical observation. *J. Energy Chem.* **2019**, *29*, 122–128. [CrossRef]
14. Placke, T.; Schmuelling, G.; Kloepsch, R.; Meister, P.; Fromm, O.; Hilbig, P.; Meyer, H.-W.; Winter, M. In situ X-ray Diffraction Studies of Cation and Anion Intercalation into Graphitic Carbons for Electrochemical Energy Storage Applications. *Z. Anorg Allg. Chem.* **2014**, *640*, 1996–2006. [CrossRef]
15. Hao, J.; Li, X.; Song, X.; Guo, Z. Recent progress and perspectives on dual-ion batteries. *EnergyChem* **2019**, *1*, 100004. [CrossRef]
16. Wang, M.; Tang, Y. A Review on the Features and Progress of Dual-Ion Batteries. *Adv. Energy Mater.* **2018**, *8*, 1703320. [CrossRef]
17. Zhao, Z.; Alshareef, H.N. Sustainable Dual-Ion Batteries beyond Li. *Adv. Mater.* **2024**, *36*, 2309223. [CrossRef] [PubMed]
18. Rodríguez-Pérez, I.A.; Ji, X. Anion Hosting Cathodes in Dual-Ion Batteries. *ACS Energy Lett.* **2017**, *2*, 1762–1770. [CrossRef]
19. Wang, G.; Yu, M.; Feng, X. Carbon materials for ion-intercalation involved rechargeable battery technologies. *Chem. Soc. Rev* **2021**, *50*, 2388–2443. [CrossRef]
20. Zhao, W.; Zhao, C.; Wu, H.; Li, L.; Zhang, C. Progress, challenge and perspective of graphite-based anode materials for lithium batteries: A review. *J. Energy Storage* **2024**, *81*, 110409. [CrossRef]
21. Lu, J.; Chen, Z.; Pan, F.; Cui, Y.; Amine, K. High-Performance Anode Materials for Rechargeable Lithium-Ion Batteries. *Electrochem Energy Rev.* **2018**, *1*, 35–53. [CrossRef]
22. Majid, A.; Fatima, S.A.; Ud-Din Khan, S.; Almutairi, Z.A. Assessment of 2H–SiC based intercalation compound for use as anode in lithium ion batteries. *Ceram. Int.* **2020**, *46*, 5297–5305. [CrossRef]
23. Yang, C.; Xie, H.; Ping, W.; Fu, K.; Liu, B.; Rao, J.; Dai, J.; Wang, C.; Pastel, G.; Hu, L. An Electron/Ion Dual-Conductive Alloy Framework for High-Rate and High-Capacity Solid-State Lithium-Metal Batteries. *Adv. Mater.* **2019**, *31*, 1804815. [CrossRef] [PubMed]
24. Song, K.; Liu, C.; Mi, L.; Chou, S.; Chen, W.; Shen, C. Recent Progress on the Alloy-Based Anode for Sodium-Ion Batteries and Potassium-Ion Batteries. *Small* **2021**, *17*, 1903194. [CrossRef]
25. Fan, S.; Liu, H.; Bi, S.; Gao, C.; Meng, X.; Wang, Y. Insight on the conversion reaction mechanism of $NiCo_2S_4$@CNTs as anode materials for lithium ion batteries and sodium ion batteries. *Electrochim. Acta* **2021**, *388*, 138618. [CrossRef]
26. Hao, S.-M.; Qu, J.; Chang, W.; Zhang, Y.-J.; Tang, Y.; Yu, Z.-Z. A High-Performance Dual-Ion Battery Enabled by Conversion-Type Manganese Silicate Anodes with Enhanced Ion Accessibility. *ChemElectroChem* **2019**, *6*, 1040–1046. [CrossRef]

27. Liang, Y.; Luo, C.; Wang, F.; Hou, S.; Liou, S.-C.; Qing, T.; Li, Q.; Zheng, J.; Cui, C.; Wang, C. An Organic Anode for High Temperature Potassium-Ion Batteries. *Adv. Energy Mater.* **2019**, *9*, 1802986. [CrossRef]
28. Zhou, R.; Hou, Z.; Fan, K.; Wun, C.K.; Liu, Q.; Benedict Lo, T.W.; Huang, H.; Zhang, B. An advanced organic cathode for non-aqueous and aqueous calcium-based dual ion batteries. *J. Power Sources* **2023**, *569*, 232995. [CrossRef]
29. Liu, J.-H.; Wang, P.; Gao, Z.; Li, X.; Cui, W.; Li, R.; Ramakrishna, S.; Zhang, J.; Long, Y.-Z. Review on electrospinning anode and separators for lithium ion batteries. *Renew. Sustain. Energy Rev.* **2024**, *189*, 113939. [CrossRef]
30. Tang, B.; Wei, Y.; Jia, R.; Zhang, F.; Tang, Y. Rational Design of High-Loading Electrodes with Superior Performances Toward Practical Application for Energy Storage Devices. *Small* **2024**, *20*, 2308126. [CrossRef] [PubMed]
31. Chen, C.; Yao, W.; Tang, Y. Emerging Solutions to Enable the Efficient Use of Sodium Metal Anodes: Progress and Perspectives. *Adv. Funct. Mater.* **2024**, *34*, 2310833. [CrossRef]
32. Chen, C.; Lee, C.-S.; Tang, Y. Fundamental Understanding and Optimization Strategies for Dual-Ion Batteries: A Review. *Nano-Micro Lett.* **2023**, *15*, 121. [CrossRef]
33. Wu, H.; Luo, S.; Wang, H.; Li, L.; Fang, Y.; Zhang, F.; Gao, X.; Zhang, Z.; Yuan, W. A Review of Anode Materials for Dual-Ion Batteries. *Nano-Micro Lett.* **2024**, *16*, 252. [CrossRef] [PubMed]
34. Wang, R.; Wang, L.; Liu, R.; Li, X.; Wu, Y.; Ran, F. "Fast-Charging" Anode Materials for Lithium-Ion Batteries from Perspective of Ion Diffusion in Crystal Structure. *ACS Nano* **2024**, *18*, 2611–2648. [CrossRef] [PubMed]
35. Liu, W.; Liu, P.; Mitlin, D. Review of Emerging Concepts in SEI Analysis and Artificial SEI Membranes for Lithium, Sodium, and Potassium Metal Battery Anodes. *Adv. Energy Mater.* **2020**, *10*, 2002297. [CrossRef]
36. Chen, Y.; Pu, Z.; Liu, Y.; Shen, Y.; Liu, S.; Liu, D.; Li, Y. Enhancing the low-temperature performance in lithium ion batteries of Nb_2O_5 by combination of W doping and MXene addition. *J. Power Sources* **2021**, *515*, 230601. [CrossRef]
37. Chen, G.; Chen, J.; Zhao, S.; He, G.; Miller, T.S. Pseudohexagonal Nb_2O_5 Anodes for Fast-Charging Potassium-Ion Batteries. *ACS Appl. Mater. Interfaces* **2023**, *15*, 16664–16672. [CrossRef]
38. Ji, Q.; Chen, W.; Chen, X.; Wang, X.; Dong, Q.; Yin, S.; Shen, Y.; Müller-Buschbaum, P.; Cheng, Y.-J.; Xia, Y. Synergistic Effect of Dual Phases to Improve Lithium Storage Properties of Nb_2O_5. *ACS Appl. Mater. Interfaces* **2024**, *16*, 7232–7242. [CrossRef] [PubMed]
39. Cheong, J.Y.; Kim, C.; Jung, J.-W.; Yoon, K.R.; Cho, S.-H.; Youn, D.-Y.; Jang, H.-Y.; Kim, I.-D. Formation of a Surficial Bifunctional Nanolayer on Nb_2O_5 for Ultrastable Electrodes for Lithium-Ion Battery. *Small* **2017**, *13*, 1603610. [CrossRef] [PubMed]
40. So, S.; Ahn, Y.N.; Ko, J.; Kim, I.T.; Hur, J. Uniform and oriented zinc deposition induced by artificial Nb_2O_5 Layer for highly reversible Zn anode in aqueous zinc ion batteries. *Energy Storage Mater.* **2022**, *52*, 40–51. [CrossRef]
41. Das, P.; Ibrahim, S.; Chakraborty, K.; Ghosh, S.; Pal, T. Stepwise reduction of graphene oxide and studies on defect-controlled physical properties. *Sci. Rep.* **2024**, *14*, 294. [CrossRef] [PubMed]
42. Li, Y.; Song, Z.; Sun, T.; Shen, Y.; Lv, X.; Xu, D.; Wang, H.-G. Well-dispersed Sb_2O_3 nanoparticles encapsulated in multi-channel-carbon nanofibers as high-performance anode materials for Li/dual-ion batteries. *Int. J. Hydrogen Energy* **2021**, *46*, 26308–26317. [CrossRef]
43. Sui, L.; Shi, X.; Deng, T.; Yang, H.; Liu, H.; Chen, H.; Zhang, W.; Zheng, W. Integrated Co_3O_4/carbon fiber paper for high-performance anode of dual-ion battery. *J. Energy Chem.* **2019**, *37*, 7–12. [CrossRef]
44. Zheng, C.; Wu, J.; Li, Y.; Liu, X.; Zeng, L.; Wei, M. High-Performance Lithium-Ion-Based Dual-Ion Batteries Enabled by Few-Layer MoSe2/Nitrogen-Doped Carbon. *ACS Sustain. Chem. Eng.* **2020**, *8*, 5514–5523. [CrossRef]
45. Liu, X.; Liu, Y.; Yan, X.; Lan, J.-L.; Yu, Y.; Yang, X. Ultrafine MoO_3 nanoparticles embedded in porous carbon nanofibers as anodes for high-performance lithium-ion batteries. *Mater. Chem. Front.* **2019**, *3*, 120–126. [CrossRef]
46. Anjali, A.; Jayan, P.; Akshay, M.; Lee, Y.-S.; Aravindan, V. SnO_2 as an alloy anode for Li-based dual-ion battery with enhanced performance. *Next Mater.* **2024**, *2*, 100117. [CrossRef]
47. Zhang, F.; Wu, M.; Wang, X.; Xiang, Q.; Wu, Y.; Ding, J.; Sun, Y. Reversible multi-electron redox chemistry of organic salt as anode for high-performance Li-ion/dual-ion batteries. *Chem. Eng. J.* **2023**, *457*, 141335. [CrossRef]

molecules

Article

MUF-n-Octadecane Phase-Change Microcapsules: Effects of Core pH and Core–Wall Ratio on Morphology and Thermal Properties of Microcapsules

Lin Lin [1,2,†], Ziqi Li [1,2,†], Jian Zhang [1,2,*], Tonghua Ma [3], Renzhong Wei [3], Qiang Zhang [2] and Junyou Shi [1,2,*]

[1] School of Agricultural Engineering and Food Science, Shandong University of Technology, Zibo 255000, China
[2] Key Laboratory of Wooden Materials Science and Engineering of Jilin Province, Beihua University, Jilin 132013, China
[3] Treezo Group New Material Technology Co., Ltd., Hangzhou 311112, China
* Correspondence: zhangjian_beihua@126.com (J.Z.); bhsjy64@163.com (J.S.)
† These authors contributed equally to this work and co-first author.

Abstract: Phase change energy storage microcapsules were synthesized in situ by using melamine-formaldehyde–urea co-condensation resin (MUF) as wall material, n-octadecane (C18) as core material and styryl-maleic anhydride copolymer (SMA) as emulsifier. Fourier transform infrared spectroscopy, scanning electron microscopy, differential scanning calorimetry and thermogravimetric analysis were used to study the effects of emulsifier type, emulsifier dosage, core–wall ratio and pH on the morphology and thermal properties of microcapsules. The results show that the pH of core material and the ratio of core to wall have a great influence on the performance of microcapsules. SMA emulsifiers and MUF are suitable for the encapsulation of C18. When the pH is 4.5 and the core–wall ratio is 2/1, the latent heat and encapsulation efficiency of phase transition reaches 207.3 J g^{-1} and 84.7%, respectively. The prepared phase-change microcapsules also have good shape stability and thermal stability.

Keywords: phase change; microcapsules; emulsifier; latent heat; encapsulate

1. Introduction

In the process of phase transformation, organic solid–liquid phase variable materials can store or release a large amount of latent heat at almost isothermal temperature. At present, a variety of organic solid–liquid phase variable materials have been used for thermal energy storage, such as organic alkanes, organic fatty acids, polyols and so on. Among these materials, organic alkanes are very popular due to their many advantages, including high energy storage density, good chemical stability, non-corrosion, non-toxicity, non-phase separation and low subcooling, making them ideal materials for many heat storage applications [1,2]. However, leakage in the solid–liquid phase transformation process is an inherent defect of the phase change energy storage microcapsules (PCM), which greatly limits its application range. The use of microencapsulation technology to encapsulate phase-change materials in a sealed tiny container to make core–shell structure phase-change microcapsules can effectively prevent the leakage of phase-change materials and improve their shape stability [3].

Microencapsulation not only enables the PCM to maintain its macroscopic solid state when the solid–liquid phase changes, but also provides sufficient protection for the PCM from harmful interactions and interference of the surrounding environment [4–6]. In addition, microencapsulation can provide more heat transfer area for the phase-change material, resulting in significantly enhanced heat transfer and thermal response [7]. Therefore, microencapsulation has been recognized as a reliable phase-change material encapsulation technology. The shell material used to encapsulate the phase-change material should meet the requirements of no chemical reactivity with the phase-change material, good sealing, good chemical stability and thermal stability so as to provide good shape stability for the phase-change material. At present, many studies have been carried out on the encapsulation of phase-variant materials with various shell materials, including melamine-formaldehyde–urea co-condensation resin (MUF), melamine-formaldehyde resin (MF), urea-formaldehyde resin (UF), polystyrene (PS), polymethyl methacrylate (PMMA), silicon dioxide (SiO_2), titanium dioxide (TiO_2), calcium carbonate ($CaCO_3$), etc. [8–13]. Compared with the case materials, poor sealing of inorganic materials may lead to leakage of phase-change materials. Therefore, organic materials are ideal shells for encapsulating phase-change materials [14]. In the case of materials, MUF is widely used to encapsulate phase-change materials because of its good sealing, thermal stability and excellent waterproof performance. MUF also has good mechanical properties and a certain toughness [15–17]. In addition, the MUF synthesis process is simple and the price is reasonable, so it has a broader prospect in large-scale applications. Wang et al. [18] prepared microcapsule phase-change materials by coating paraffin with nano Fe_3O_4-modified melamine urea formaldehyde resin (MUF). Similarly, Han et al. [19] prepared phase-change microcapsules with paraffin as the core material and MUF as the shell.

In the preparation of phase-change microcapsules, the control of the pH value is very important because it affects the emulsification process, polymerization reaction and the properties of the final product. During the emulsification process, the pH value will affect the performance of the emulsifier and the stability of the emulsion. For different emulsifiers, there is a specific optimal pH range. For example, some ionic emulsifiers have better surface activity in a specific pH range so that a more stable emulsion can be formed. In acidic environments, MF prepolymers are positively charged due to the reaction of hydroxyl methyl groups with hydrogen ions. The positively charged prepolymer is strongly attracted to the nuclear particles with a negative electron field by an anionic SMA emulsifier, and under the action of acid and heat, a microencapsulated reaction occurs on the surface of the nuclear particles to form a microencapsulated film. In the system containing the cationic SMA emulsifier, MF prepolymer is repelled by core particles with the same positive charge and does not enclose the core material. In the non-ionic SMA emulsifier system, the positively charged MF prepolymer is partially attracted to the core particle surface, and the encapsulation is not complete. In the system without the emulsifier, due to the weak hydrogen bond, a small amount of MF prepolymer interacts with the hydroxyl group and is deposited on the surface of core particles [20]. In the interfacial polymerization process, the pH value affects the ionization state of the monomer, thus affecting the rate and efficiency of the polymerization reaction [21]. For example, when using polyamide or polyurethane as a wall material, the pH value needs to be controlled in the acidic or alkaline range to facilitate the reaction. In in situ polymerization, pH values affect the solubility and reaction rate of monomers [22]. Chang et al. [23] synthesized phase-change microcapsules using N-tetradecane as the core material and SDS as the emulsifier and concluded that the microcapsules prepared when the mass fraction of SDS was 1.2 wt%, the pH was 3.5 and the core–shell mass ratio was 2:1 had the best performance. Huang et al. [24] prepared carboxymethyl cellulose-modified phase-change microcapsules with urea–formaldehyde resin as the shell, and explored the effects of curing the pH value, emulsifier dosage and emulsification rate on the properties of microcapsules.

The core–wall ratio refers to the mass ratio of the phase-change material to the wall material in microcapsules. Generally speaking, the higher the core material content, the greater the latent heat of the phase transition microcapsules, and the higher core–wall ratio means that the microcapsules contain more PCM, thus increasing the energy storage capacity. However, too high a core–wall ratio may lead to too thin a wall material, which makes it easy for the PCM to break or leak during use, affecting the structural stability and mechanical strength of microcapsules [25,26]. The thermal conductivity of the phase-change material is usually higher than that of the wall material, and a higher core–wall ratio usually improves the thermal conductivity of the microcapsule, but if the wall material is too thin, it may not be able to effectively isolate the temperature change of the external environment, resulting in the phase-change material releasing or absorbing heat too quickly and failing to achieve the desired temperature control effect [27].

Many scholars have improved the performance of the phase-change microcapsules by modifying the shell material, adding thermal conductivity enhancers and changing the packaging method, but they may have overlooked the most basic synthesis process of microcapsules. Based on the synthesis process of microcapsules, it is possible to better improve their defects and seek the changing patterns of microcapsule morphology, latent heat, shape stability, etc. In this work, the effects of emulsifier type, emulsifier dosage, core–shell ratio and pH on the morphology and thermal properties of microcapsules were studied. As shown in Figure 1, phase-change microcapsules of C18 with a MUF shell were successfully prepared. The morphology of the C18 phase-change microcapsules prepared by different technologies was mainly observed by scanning electron microscopy, and their latent heat was carefully analyzed by DSC. The phase-change microcapsules prepared in this work are compatible with building materials and can be applied in building thermal storage fields such as exterior wall insulation and building roofs.

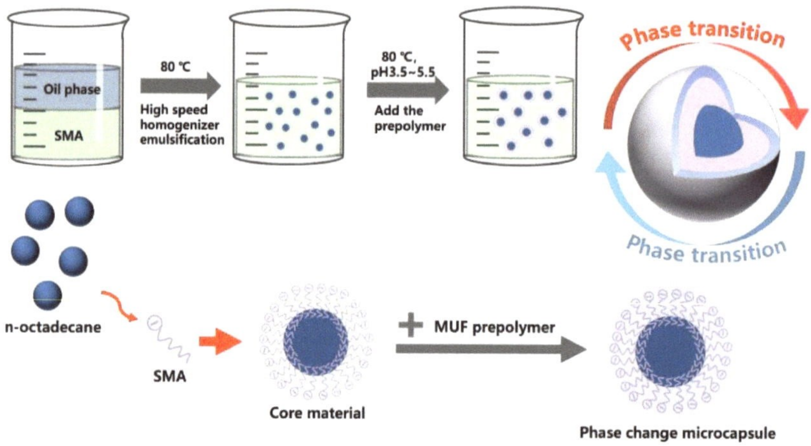

Figure 1. Schematic diagram of MUF-n-octadecane phase-change microcapsules.

2. Results

2.1. Microcapsule Morphology Analysis

Scanning electron microscopy (SEM) was used to observe the microcapsule morphology, as shown in Figure 2a,b,d,e,g,h. The microcapsule morphology is prepared when the pH is 3.5, 4.5 and 5.5. It can be seen from the figure that the microcapsule as a whole presents a regular spherical appearance. When the pH is 3.5, more irregular particles appear on the surface of microcapsules. As can be seen from the local magnification of Figure 2b, there is adhesion between the microcapsules. This is because when the initial pH is too low, the polycondensation rate of wall material prepolymers will accelerate, resulting in the self-polymerization of MUF resin. As a result, microcapsules adhere to each other and produce irregular particles on the surface. The wall material itself has no energy storage effect but occupies a part of the volume of the microcapsule. Therefore, the aggregation of excessively thick wall materials will reduce the overall latent heat value of the microcapsule and affect the energy storage effect. Figure 2d,e shows the microscopic morphology of the microcapsules prepared at pH 4.5. It can be seen from the figure that the surface of the microcapsules is smooth and well coated, and they are independent of each other without adhesion, indicating that the microcapsules have a good coating effect at pH 4.5. Figure 2g,h shows the microcapsule prepared at pH 5.5. At this pH value, a large number of microcapsules are broken, and it can be observed from the broken section of the wall material that the thickness of the wall material is very thin. This is because the polymerization speed of the prepolymer is very slow at this pH value, which leads to the instability in the synthesis process of the microcapsule. And the synthesis process has been accompanied by the shear force generated by the electric stirrer so that the microcapsule has a large number of broken phenomena and the core material cannot be effectively coated.

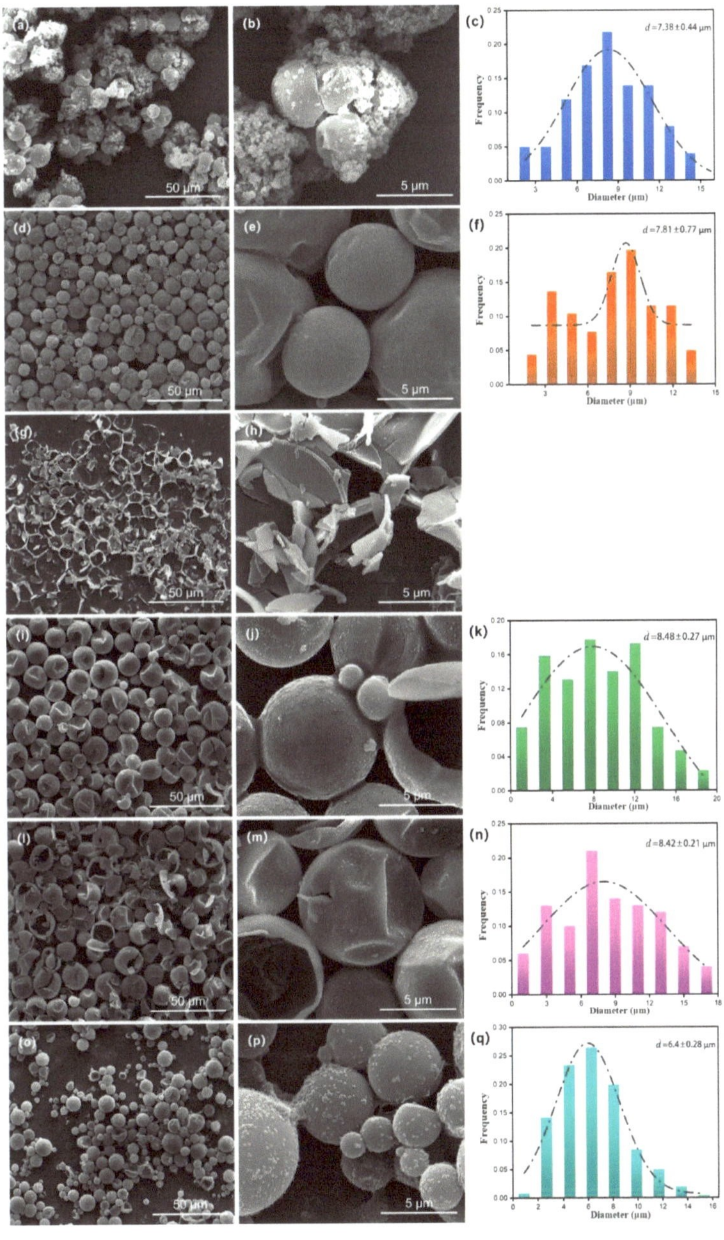

Figure 2. Microstructure and particle size distribution of microcapsules (**a–c**) pH3.5; (**d–f**) pH4.5; (**g,h**) pH5.5; (**i–k**) 1/1 (**l–n**) 4/1 (**o–q**) emulsifier dosage 15%.

Figure 2d,e,i,j,l,m shows the microscopic morphology of microcapsules prepared with a core–wall ratio of 2/1, 1/1 and 4/1, respectively. When the core–wall material ratio is 1/1, there is a slight adhesion between microcapsules, and the cross-section of the wall material shows that the wall material of the microcapsules is thicker, as can clearly be seen in Figure 2j. This is because the number of prepolymers is more than the number of core molecules. When the core–wall ratio reaches 2/1, the surface impurities of microcapsules are significantly reduced, the surface impurities are smoother, there is no adhesion between each other and the dispersion is enhanced. As the core–wall ratio continues to increase to 4/1, the wall material cannot completely cover the core material, and a large number of microcapsules appear damaged, which will cause the core material to flow out from the damaged area, resulting in the leakage of the core material. Therefore, when the core–wall ratio is 2/1, the microcapsule has better coating and dispersion. SMA was selected as the emulsifier and emulsified at different dosages (5%, 10% and 15%). When the emulsifier dosage was 5%, n-octadecane could not be emulsified to form a stable emulsion and could not be successfully synthesized into microcapsules, so its microscopic morphology could not be observed.

Figure 2d,e,o,p shows the micromorphology of the microcapsules when the dosage is 10% and 15%. It can be seen that both of them show a relatively complete spherical appearance without fracture phenomenon, indicating that 10% emulsifier dosage is enough to emulsify the core material into a stable emulsion, so 10% is the optimal dosage. SMA, SDBS and SDS were, respectively, used for emulsification, and SMA could form a uniform and stable emulsion system for the core material, while the emulsion formed by SDBS and SDS would produce demulsification, which was not conducive to the synthesis of microcapsules in the next step. Therefore, the microcapsules prepared by using SMA emulsification to form emulsions had the best effect. Figure 2c,f,k,n,q shows the particle size diagram of microcapsules under different conditions. The microcapsules synthesized at pH 5.5 were broken too much, so it was impossible to use software to analyze their particle size. It can be seen that the average particle size of microcapsules was 6–8 μm, and different experimental conditions did not regularly affect the particle size of microcapsules. In addition, almost all microcapsules show different degrees of depression, which is because when the core material inside the microcapsule is cooled below the phase transition point, the core material will solidify and shrink in volume, while the area of the wall material basically remains unchanged, resulting in different degrees of microcapsule shrinkage [28].

2.2. *Leak-Proof Analysis of Microcapsules*

The macro-leakage resistance of microcapsules at different pH values was tested. At 80 °C (higher than the melting temperature of n-octadecane), with the increase in time, the macro-leakage resistance of microcapsules is shown in Figure 3. From left to right is the leakage of the sample at 0 min, 10 min and 20 min. Among them, the microcapsules synthesized at pH 5.5 were blocky macroscopically. Combined with the SEM figure in the previous section, the microcapsules synthesized under this condition had poor coating properties and would solidify together when n-octadecane crystallized, resulting in the microcapsules being blocky macroscopically. As can be seen from Figure 3, n-octadecane began to melt gradually when heated at 80 °C for 10 min, resulting in leakage. At this time, the microcapsule materials could maintain good shape stability, and no leakage of n-octadecane was observed. When heated for 20 min, n-octadecane was completely melted. This instability will seriously affect the practical application of phase-change materials. While pH 3.5 and pH 4.5 microcapsules can still maintain good shape stability, no leakage of the internal core material was observed, but the bottom of pH 5.5 microcapsules also showed signs of leakage. Therefore, microcapsules with a pH of 3.5 and 4.5 have good leak-proof performance. Combined with the SEM diagram in the previous section, microcapsules have good microscopic morphology and excellent leak-proof performance when the pH is 4.5.

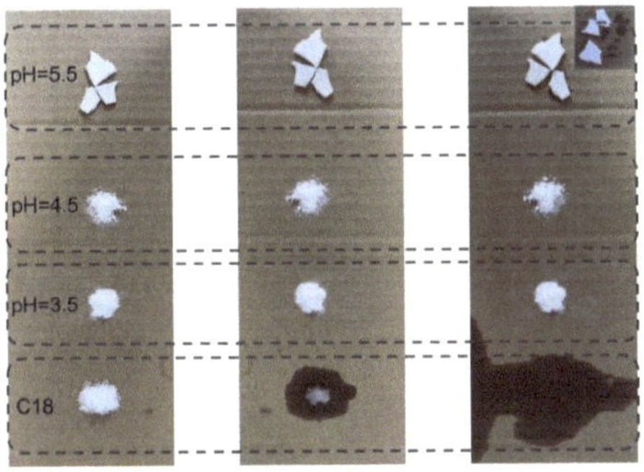

Figure 3. Leakage resistance of microcapsules.

2.3. Chemical Structure of Microcapsules

Fourier transform infrared spectroscopy was used to analyze the chemical structure of the microcapsules. The infrared spectra of MUF, n-octadecane and microcapsules are shown in Figure 4. Among them, the superposition of O-H and N-H tensile vibrations in MUF resin resulted in a strong and wide absorption peak near 3400 cm^{-1} [29]. The characteristic absorption peak at 1675 cm^{-1} is due to the C=O tensile vibration of urea in MUF resin. The tensile vibration of C-N is 1555 cm^{-1} [30]. The characteristic peak at 1365 cm^{-1} is caused by an O-H bending vibration. The characteristic peak at 1000 cm^{-1} is related to C-O tensile vibration. Due to the bending vibration of the triazine ring in MUF resin, there is a significant absorption peak at 1492 cm^{-1} and 812 cm^{-1}. The above peaks are basically consistent with the IR spectra of the polymerization products of urea, melamine and formaldehyde. In the spectrum of n-octadecane, there are three characteristic peaks at 2955, 2915 and 2847 cm^{-1}, which belong to the tensile vibrations of -CH, -CH_3 and -CH_2, respectively. The peaks at 1467 cm^{-1} and 1369 cm^{-1} are caused by flexural vibrations of -CH_2 and -CH_3, respectively. In addition, the peak at 721 cm^{-1} is considered to be a typical peak of the alkyl (CH_2) n (n $\geq$ 4) group [31]. In addition to the characteristic peaks of n-octadecane and MUF resin described above, no other characteristic peaks were observed in the spectrum of the microcapsule, indicating that the microcapsule has been successfully synthesized, the wall material is physically coated with the core material and there is no chemical reaction between the two.

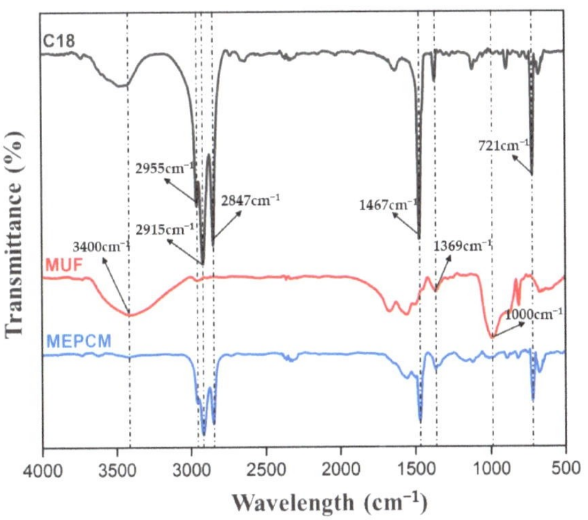

Figure 4. Infrared spectrum of n-octadecane, MUF wall material, microcapsule.

Figure 5 shows the X-ray diffraction pattern of n-octadecane, MUF resin and microcapsules. The diffraction curve of n-octadecane has four distinct characteristic diffraction peaks at 19.3°, 19.7°, 23.4° and 24.7°, corresponding to the four characteristic crystal faces (010), (011), (100) and (111) in the triclinic system [32]. In the XRD pattern of MUF resin, a wide band can be observed only at 2θ = 15–30°, indicating that MUF resin is amorphous. In the XRD pattern of the microcapsule in Figure 5, only the diffraction peak of n-octadecane is observed, which is caused by the amorphous structure of the MUF shell. Combined with SEM and IR spectra, it can be said that n-octadecane is successfully coated in the microcapsule.

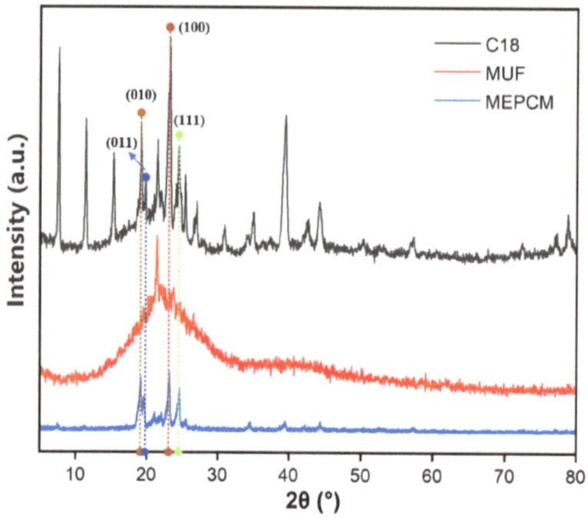

Figure 5. X-ray diffraction spectra of n-octadecane, MUF wall material and microcapsules.

2.4. *Thermal Properties Analysis of Microcapsules*

The thermal properties of n-octadecane, MUF and microcapsules were measured using a differential scanning calorimeter, as shown in Figure 6, and the results are shown in Table 1. In the range of 5–55 °C, MUF does not have an absorption/exothermic peak, which indicates that the thermal effect of the microcapsule is mainly caused by the n-octadecane inside. As can be seen from Figure 6b, bimodal crystallization behavior occurred in both bulk n-octadecane and microcapsule samples during the crystallization process. The presence of two exothermal peaks is due to the fact that melt surface crystallization reduces molecular interactions before the complete crystallization of n-octadecane in bulk and microcapsules, resulting in the generation of metastable rotating phases [33–35]. The melting process has only a significant endothermic peak because both solid-metastable solids and metastable solid–liquids occur at close temperatures [36].

Table 1. Thermal properties of microcapsules.

Sample	Melting				Cooling				E_{en} (%)
	T_{om}/(°C)	T_{pm}/(°C)	T_{em}/(°C)	ΔH_m/(J/g)	T_{oc}/(°C)	T_{pc}/(°C)	T_{ec}/(°C)	ΔH_c/(J/g)	
C18	26.4	28.8	35.7	244.8	25.4	23.5	17.1	244.2	
MUF	-	-	-	-	-	-	-	-	-
pH3.5	23.9	30.5	33.4	120.9	25.7	24.5	15.9	88.2	49.4
pH4.5	24.4	32.0	37.0	207.3	25.8	23.7	15.9	181.6	84.7
pH5.5	24.8	31.3	35.1	219.8	25.8	23.7	16.7	208.8	89.8
4/1	23.7	31.2	35.6	192.4	25.9	23.7	18.9	177.4	78.6
1/1	24.5	30.6	33.5	140.2	26.0	24.7	21.4	138.1	52.3

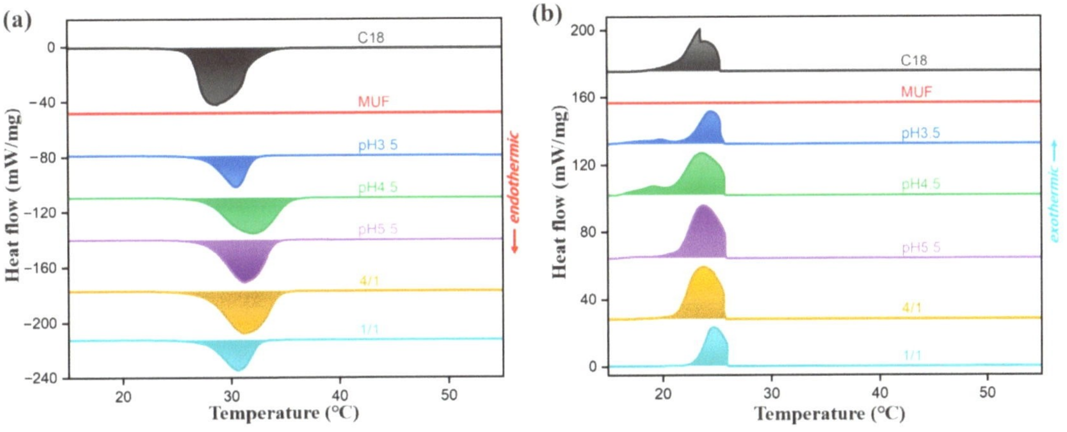

Figure 6. DSC curve of microcapsules: (**a**) melting, (**b**) cooling (−15–55 °C, heating/cooling rate = ±5 °C min^{-1}).

Table 1 shows that the latent heat of melting and crystallization (ΔH_m and ΔH_c) of n-octadecane are 244.8 J g^{-1} and 244.2 J g^{-1}, respectively. After encapsulation, the ΔH_m and ΔH_c of microcapsules decreased, because the wall material in microcapsules occupied a part of the volume but had no heat storage and release capacity, so the addition of wall material would reduce the volume of n-octadecane per unit volume of microcapsules, and thus reduce the latent heat. According to Formula (1), the encapsulation rate of microcapsules can be calculated. It can be found that with the decrease in pH, the encapsulation rate and latent heat value of microcapsules show a downward trend. This is because the polymerization of prepolymers is more rapid and sufficient under a low pH environment, and the self-polymerization of prepolymers may occur, which increases the volume ratio occupied by wall materials and thus reduces the latent heat value of microcapsules. Although the latent heat value of microcapsules is the highest at pH5.5, combined with the SEM image and anti-leakage performance test, the microcapsules at this time have a leakage risk and poor coating properties. When the pH was 4.5, the surface of the microcapsule was intact without leakage, the encapsulation rate reached 84.7%, and the latent heat value was still above 200 J g^{-1}, maintaining a high level, and the overall performance was good. With the increase in the core–wall ratio, the encapsulation rate and latent heat value of the microcapsule gradually increase. However, when the core–wall ratio reaches 4/1, there are defects on the surface of the microcapsule, and the core material cannot be completely coated, resulting in the waste of a part of n-octadecane. Therefore, the core–wall ratio of 2/1 is a relatively ideal condition.

Our work was compared with the phase transition temperature and latent heat of microencapsulated phase change materials with C18 as the core material reported in recent years (Table 2). The result shows that our work is excellent.

This is example 1 of an equation:

$$E_{en} = \frac{\Delta H_{m,MEPM}}{\Delta H_{m,\ PCM}} \times 100\% \tag{1}$$

where $\Delta H_{m,PCM}$, $\Delta H_{m,MEPCM}$ are the latent heat of the melting of massive n-octadecane and n-octadecane microcapsules, respectively.

2.5. Thermal Stability Analysis of Microcapsules

The range of applications and uses of microcapsules largely depend on their thermal stability [37,38]. The thermal degradation behavior of n-octadecane, MUF resin and microcapsules was studied using TGA, and the TGA and DTG curves are shown in Figure 7a,b.

The TGA curve shows that n-octadecane exhibits a typical one-step degradation curve under heating conditions, that is, from 99 °C to 221 °C, the weight decomposition of n-octadecane approaches 100%, which means that n-octadecane is almost completely decomposed. In contrast, the TGA curve of microcapsule samples showed a two-step thermal degradation process. The evaporation of n-octadecane and the degradation of MUF resin could be clearly seen in the TGA curve of the microcapsule. The evaporation of n-octadecane led to the first step of degradation, which occurred between 131 and 66 °C. MUF is broken, so the weightlessness of microcapsules in this step is significant. After this, a further weight loss occurs between 270 and 335 °C, which is due to the decomposition of the MUF shell. As can be seen from the DTG curve, due to the protection of the MUF shell on n-octadecane, the decomposition of n-octadecane in microcapsules has a temperature lag compared with that of n-octadecane in bulk. Therefore, the thermal stability of microcapsules is improved compared with that of bulk n-octadecane. When the pH is 4.5, the maximum degradation rate temperature reaches 224.3 °C, which is 28.1 °C higher than bulk n-octadecane.

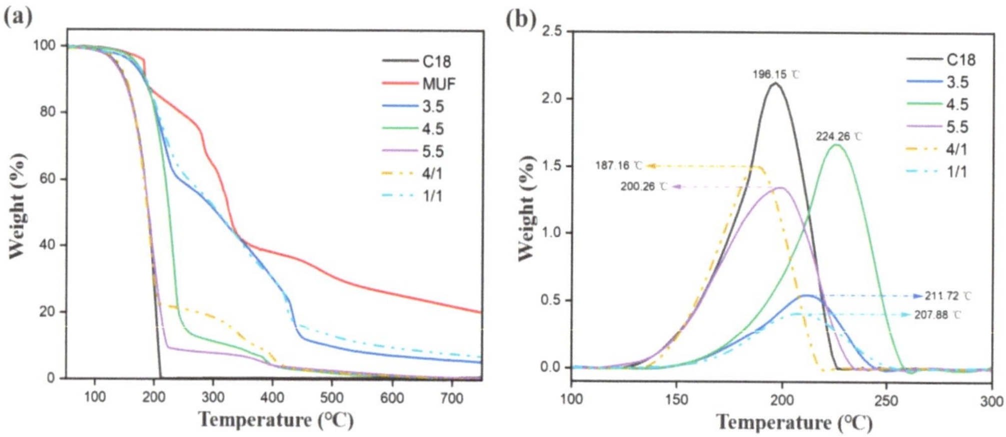

Figure 7. Thermogravimetric curve of microcapsules (**a**) TGA (**b**) DTG.

With the increase in the core–wall ratio and pH during the synthesis of microcapsules, the wall material cannot provide better barrier protection for the inner core material, and the thermal stability of microcapsules also decreases. In addition, the TGA curve of the microcapsule synthesized under the conditions of pH 3.5 and core–wall ratio of 1/1 shows that there is still about 5% weight residue when the temperature reaches 750 °C, which is because the MUF resin as the wall material will be left after high-temperature heating. The wall material synthesized under this condition accounts for a relatively large proportion of the microcapsule wall material, and the amount of residual carbon after high-temperature decomposition is correspondingly more. The operating temperature of the microcapsule is below 100 °C, at which time the microcapsule loses only 0.12% of its volume weight. From another perspective, the initial decomposition temperature of the microcapsule is significantly higher than its operating temperature. Therefore, it can be determined that the microcapsule has a high thermal stability in its operating temperature range.

Table 2. Comparison with the literature on phase-change microcapsules with n-octadecane as core material.

Core Component	Melting		Cooling		Ref
	T_{pm}/(°C)	ΔH_m/(J/g)	T_{pc}/(°C)	ΔH_c/(J/g)	
C18	20.5	184.6	30.7	185.6	[39]
C18/C28	51.9	187.9	48.8	187.7	[40]
C18	29.0	114.5	23.1	118.8	[41]
C18/C22	39.8	100.8	-	-	[42]
C18	29.9	116.5	27.0	124.7	[43]
C18/C-PODMA	24.1	111.6	16.7	112.6	[44]
C24-C18	26.0	156.4	26.0	152.8	[45]
C19-C18	26.4	114.1	26.5	113.4	[46]
C18	30.6	171.9	25.3	172.4	[47]
C18	-	125.9	-	-	[48]
C18	32.0	207.3	23.7	181.6	This work

3. Materials and Methods

3.1. Materials

N-octadecane (C18) (AR), melamine (AR), formaldehyde solution (37–40 wt%) and urea were purchased (AR) from Shanghai Aladdin Biochemical Technology (Shanghai, China) Co., Ltd. Styrene–Maleic anhydride (SMA) (95%) was purchased from Nanjing Yinxin (Nanjing, China) Chemical Co., Ltd. Sodium hydroxide (AR), triethanolamine (AR) and urea (AR) were also purchased from Aladdin Biochemical (Shanghai, China) Technology Co., Ltd.

3.2. Preparation of n-Octadecane-SMA Core Material

Add 0.05 g NaOH into the three-neck flask, stir with 40 mL deionized water until the NaOH is completely dissolved, then add 0.25 g styrene–maleic anhydride copolymer (SMA) into the flask and place the three-neck flask in a constant temperature water bath at 90 °C. Use an electric mixer at 600 rpm to stir until the SMA is completely dissolved, and then cool to room temperature to obtain the SMA solution. The pH of the SMA solution was adjusted to 3.5~5.5 with 10 wt% citric acid solution, and then 5 g n-octadecane was added to the SMA solution and transferred to a constant temperature water bath at 70 °C. When the n-octadecane was completely melted, a homogenized n-octadecane emulsion was obtained by emulsifying at 13,000 rpm for 30 min using a high-speed homogenizer.

3.3. Preparation of MUF Shell

Add 3.12 g formaldehyde solution to the three-mouth flask, add 19.9 g deionized water, adjust the pH to 8.5 with 10 wt% triethanolamine solution and then add 0.69 g melamine. The three-neck flask was transferred to a constant temperature water bath at 75 °C and the solution was stirred with an electric mixer at 600 rpm until the solution was clarified. Then, 0.58 g urea was added and the reaction was continued for 20 min under the above experimental conditions to obtain the MUF prepolymer solution.

3.4. Synthesis of Microcapsules

The n-octadecane emulsion was placed in a constant temperature water bath at 70 °C, and the MUF prepolymer solution was slowly added to the n-octadecane emulsion with an electric mixer at the stirring speed of 600 rpm. After the drip was completed for 15 min, the reaction was carried out at 70 °C and 600 rpm for 3 h, and then the reaction system was taken out and left at room temperature for 12 h. Due to the low density of n-octadecane, the upper layer is a n-octadecane microcapsule. The impurities in the microcapsule are removed by rinsing with deionized water and ethanol at 80 °C 3–5 times, and then the microcapsule is dried in a vacuum drying oven at 80 °C for 24 h to obtain n-octadecane microcapsule powder.

3.5. Characterization

3.5.1. Scanning Electron Microscopy (FE-SEM)

A scanning electron microscope (SEM, FEI Quanta 200, Fremont, CA, USA) was used to observe the micromorphology of microcapsules. Image J (v 1.8.0) software was used to analyze the size distribution and average particle size of microcapsules.

3.5.2. Fourier Transform Infrared Spectroscopy (FTIR)

The functional groups of the sample were determined using Fourier transform infrared spectroscopy (FTIR, IRAffinity-1S, SHIMADZU, Tokyo, Japan) with a wavelength range of 500–4000 cm^{-1}. The MUF, paraffin and paraffin/MUF microcapsules obtained under the best conditions were pressed into ultra-thin particles.

3.5.3. Thermogravimetric Analysis (TGA)

Thermogravimetric analysis of MUF, paraffin and paraffin/MUF microcapsules obtained under optimal conditions was performed using a thermogravimetric analyzer (TGA, NETZSCH, TG209 F3, Selb, Germany). Each sample (~8 mg) was measured with dry pure N_2 (50 mL min^{-1}) at elevated temperatures (25~750 °C, 10 °C min^{-1}).

3.5.4. X-ray Diffractometer (XRD)

X-ray diffractometry (XRD, ARL EQUINOX 1000, Waltham, MA, USA) was used to characterize the crystal structure of the sample with a scanning angle of 5–80 ° and a scanning speed of 4 °/min.

3.5.5. Differential Scanning Calorimetry (DSC)

A differential scanning calorimeter (DSC4000, PerkinElmer, Waltham, MA, USA) (−15–55 °C, heating/cooling rate = ±5 °C min^{-1}) was used to analyze the phase transition characteristics of paraffin and paraffin/MUF microcapsules. Nitrogen was used as a protection and purification gas. For DSC analysis, each sample (~6 mg) was measured in an aluminum crucible. The melting temperature (T_m) and latent heat of melting (ΔH_m) were determined by the heating scan, and the crystallization temperature (T_c) and latent heat of crystallization (ΔH_c) were determined by the cooling scan.

4. Conclusions

Using the pH of the core material and the core-to-wall ratio as entry points for studying the performance of phase-change microcapsules, octadecane core layer phase-change microcapsules with a MUF shell were prepared by the in situ polymerization method. The emulsifier content and the initial pH value of the core material have a decisive role in the shape and encapsulation efficiency of the microcapsule. The content of the core material and the core–wall ratio of the phase-change microcapsule also have a huge impact on the latent heat and structural stability of the microcapsule. By adjusting these parameters in the experimental system, we were surprised to find that when the pH = 4.5 and the core–wall ratio is 2/1, the prepared MUF shell-coated octadecane phase-change microcapsule has the most ideal performance. Specifically, the phase-change microcapsule at this time has a perfect spherical core–shell structure, the encapsulation efficiency is 84.7% higher than that of the unpacked octadecane phase-change material and it also has excellent thermal stability, shape stability and thermal conductivity (0.1288 W/m·K). We also found that with the decrease in the pH and core-to-wall ratio, although the leakproof and thermal stability of microcapsules were improved, their heat storage capacity decreased. When the pH is 4.5 and the core-to-wall ratio is 2/1, the latent heat of the prepared MUF shell-coated octadecane phase-change microcapsules can still reach 207.3 J g^{-1}.

Author Contributions: L.L. and Z.L. contributed equally to this work. Conceptualization, R.W. and Q.Z.; methodology, Q.Z.; validation, T.M.; investigation, Z.L.; writing—original draft preparation, L.L. and Z.L.; writing—review and editing, J.Z. and J.S.; supervision, J.Z.; funding acquisition, L.L. and J.Z. All authors have read and agreed to the published version of the manuscript.

Funding: This research was funded by the National Natural Science Foundation of China (grant Nos. 32371804 and 12104015), the National Key Research and Development Program of China (grant No. 2023YFD1701600), the Technology Innovation Project of Jilin Province Science and Technology Department (grant No. 20230203046SF), the Open Project of State Key Laboratory of Superhard Materials of Jilin University (grant No. 202423), the Natural Science Foundation of Shandong Province, China (Grant No. ZR2024QC309) and the Special Construction Project Fund for Shandong Province Taishan Scholars.

Institutional Review Board Statement: Not applicable.

Informed Consent Statement: Not applicable.

Data Availability Statement: Data will be made available upon request.

Conflicts of Interest: The authors declare no conflicts of interest.

References

1. Huang, X.; Alva, G.; Liu, L.; Fang, G. Preparation, characterization and thermal properties of fatty acid eutectics/bentonite /expanded graphite composites as novel form–stable thermal energy storage materials. *Sol. Energy Mater. Sol. Cells* **2017**, *166*, 157–166. [CrossRef]
2. Li, Z.; Zhang, J.; Lin, L.; Zhang, X.; Liu, Q.; Shi, J. Layered laser-engraved wood-based composite capable of photothermal conversion and energy storage for indoor thermal management in buildings. *Energy Build.* **2024**, *318*, 114425. [CrossRef]
3. Liu, H.; Wang, X.; Wu, D. Innovative design of microencapsulated phase change materials for thermal energy storage and versatile applications: A review. *Sustain. Energy Fuels* **2019**, *3*, 1091–1149. [CrossRef]
4. Kang, M.; Liu, Y.; Lin, W.; Liang, C.; Cheng, J. The thermal behavior and flame retardant performance of phase change material microcapsules with halloysite nanotube. *J. Energy Storage* **2023**, *60*, 106632. [CrossRef]
5. Németh, B.; Németh, Á.S.; Ujhidy, A.; Tóth, J.; Trif, L.; Gyenis, J.; Feczkó, T. Fully bio-originated latent heat storing calcium alginate microcapsules with high coconut oil loading. *Sol. Energy* **2018**, *170*, 314–322. [CrossRef]
6. Skurkytė-Papievienė, V.; Abraitienė, A.; Sankauskaitė, A.; Rubežienė, V.; Dubinskaitė, K. Enhancement of thermal properties of bio-based microcapsules intended for textile applications. *Open Chem.* **2020**, *18*, 669–680. [CrossRef]
7. Gu, Y.; Li, Y.; Ju, G.; Zheng, T.; Liang, R.; Sun, G. PCM microcapsules applicable foam to improve the properties of thermal insulation and energy storage for cement-based material. *Constr. Build. Mater.* **2023**, *409*, 134144. [CrossRef]
8. Lin, L.; Cao, J.; Zhang, J.; Cui, Q.; Liu, Y. Enhanced Anti-Mold Property and Mechanism Description of Ag/TiO_2 Wood-Based Nanocomposites Formation by Ultrasound-and Vacuum-Impregnation. *Nanomaterials* **2020**, *10*, 682. [CrossRef]
9. Su, W.; Darkwa, J.; Kokogiannakis, G. Development of microencapsulated phase change material for solar thermal energy storage. *Appl. Therm. Eng.* **2017**, *112*, 1205–1212. [CrossRef]
10. Zhang, Q.; Liu, J.; Zhang, J.; Lin, L.; Shi, J. A review of composite phase change materials based on biomass materials. *Polymers* **2022**, *14*, 4089. [CrossRef]
11. Döğüşcü, D.K.; Altıntaş, A.; Sarı, A.; Alkan, C. Polystyrene microcapsules with palmitic-capric acid eutectic mixture as building thermal energy storage materials. *Energy Build.* **2017**, *150*, 376–382. [CrossRef]
12. Sarı, A.; Saleh, T.A.; Hekimoğlu, G.; Tyagi, V.V.; Sharma, R.K. Microencapsulated heptadecane with calcium carbonate as thermal conductivity-enhanced phase change material for thermal energy storage. *J. Mol. Liq.* **2021**, *328*, 115508. [CrossRef]
13. Ji, W.; Cheng, X.; Chen, S.; Wang, X.; Li, Y. Self-assembly fabrication of GO/TiO_2 paraffin microcapsules for enhancement of thermal energy storage. *Powder Technol.* **2021**, *385*, 546–556. [CrossRef]
14. Wan, X.; Zhang, H.; Chen, C.; Wang, R.; Su, L.; Guo, B. Synthesis and characterization of phase change materials microcapsules with paraffin core/cross-linked hybrid polymer shell for thermal energy storage. *J. Energy Storage* **2020**, *32*, 101897. [CrossRef]
15. Qiao, Z.; Mao, J. Multifunctional poly (melamine-urea-formaldehyde)/graphene microcapsules with low infrared emissivity and high thermal conductivity. *Mat. Sci. Eng. B* **2017**, *226*, 86–93. [CrossRef]
16. Konuklu, Y.; Erzin, F. Preparation of pentadecane/poly (melamine-urea-formaldehyde) microcapsules for thermal energy storage applications. *Int. J. Energy Res.* **2019**, *43*, 6322–6326. [CrossRef]
17. Wu, N.; Xu, L.; Zhang, C. The influence of emulsifiers on preparation and properties of microencapsules of melamine–urea–formaldehyde resins with n-dodecanol as phase-change material. *Adv. Polym. Technol.* **2018**, *37*, 3492–3498. [CrossRef]

18. Wang, X.; Lei, W.; Zou, F.; Zhang, C.; Zhu, J. Synthesis and characterization of nano-Fe_3O_4 hybrid phase change microcapsule for infrared stealth and microwave absorption. *J. Mater. Res.* **2022**, *37*, 2335–2346. [CrossRef]
19. Han, S.; Chen, Y.; Lyu, S.; Chen, Z.; Wang, S.; Fu, F. Effects of processing conditions on the properties of paraffin/melamine-urea formaldehyde microcapsules prepared by in situ polymerization. *Colloids Surf. A* **2020**, *585*, 124046. [CrossRef]
20. Yu, F.; Chen, Z.; Zeng, X. Preparation, characterization, and thermal properties of microPCMs containing n-dodecanol by using different types of styrene-maleic anhydride as emulsifier. *Colloid Polym. Sci.* **2009**, *287*, 549–560. [CrossRef]
21. Shi, J.; Wu, X.; Sun, R.; Ban, B.; Li, J.; Chen, J. Synthesis and performance evaluation of paraffin microcapsules with calcium carbonate shell modulated by different anionic surfactants for thermal energy storage. *Colloids Surf. A Physicochem. Eng. Asp* **2019**, *571*, 36–43. [CrossRef]
22. Yan, J.; Ruan, L.; Hu, D.; Liu, W.; Chen, W.; Ma, W. Microencapsulation of phase change materials with a soy oil-based polyurethane shell via Pickering emulsion polymerization. *ACS Appl. Energy Mater.* **2023**, *6*, 6814–6825. [CrossRef]
23. Chang, Y.; Sun, Z. Synthesis and thermal properties of n-tetradecane phase change microcapsules for cold storage. *J. Energ Storage* **2022**, *52*, 104959. [CrossRef]
24. Huang, Z.H.; Yu, X.; Li, W.; Liu, S.X. Preparation of urea-formaldehyde paraffin microcapsules modified by carboxymethy cellulose as a potential phase change material. *J. For. Res.* **2015**, *26*, 253–260. [CrossRef]
25. Babaharra, O.; Choukairy, K.; Faraji, H.; Hamdaoui, S. Improved heating floor thermal performance by adding PCM microcapsule enhanced by single and hybrid nanoparticles. *Heat Transf.* **2023**, *52*, 3817–3838. [CrossRef]
26. Wang, R.; Kang, Y.; Lei, T.; Li, S.; Zhou, Z.; Xiao, Y. Microcapsules composed of stearic acid core and polyethylene glycol-based shell as a microcapsule phase change material. *Int. J. Energy Res.* **2021**, *45*, 9677–9684. [CrossRef]
27. Lv, P.; Liu, L.; Lei, G.; Dong, H. Pore-scale numerical investigation on the thermal storage properties of packed-bed systems with phase change material microcapsules. *J. Clean. Prod.* **2023**, *405*, 137071. [CrossRef]
28. Raeesi, M.; Mirabedini, S.M.; Farnood, R.R. Preparation of microcapsules containing benzoyl peroxide initiator with gelatin-gum arabic/polyurea-formaldehyde shell and evaluating their storage stability. *ACS Appl. Mater. Interfaces* **2017**, *9*, 20818–2082[illegible] [CrossRef]
29. Sun, Y.; Wang, R.; Liu, X.; Dai, E.; Li, B.; Fang, S.; Li, D. Synthesis and performances of phase change microcapsules with a polymer/diatomite hybrid shell for thermal energy storage. *Polymers* **2018**, *10*, 601. [CrossRef]
30. Han, S.; Lyu, S.; Chen, Z.; Fu, F.; Wang, S. Combined stabilizers prepared from cellulose nanocrystals and styrene-maleic anhydride to microencapsulate phase change materials. *Carbohydr. Polym.* **2020**, *234*, 115923. [CrossRef]
31. Di, C.; Jing, L.; Nai-Xun, J. Preparation and thermophysical properties of graphene nanoplatelets-octadecane phase change composite materials. *Acta. Phys. Sin.* **2019**, *68*, 100502-1.
32. Wang, S.; Lei, K.; Wang, Z.; Wang, H.; Zou, D. Metal-based phase change material (PCM) microcapsules/nanocapsules Fabrication, thermophysical characterization and application. *Chem. Eng. J.* **2022**, *438*, 135559. [CrossRef]
33. Su, Y.; Liu, G.; Xie, B.; Fu, D.; Wang, D. Crystallization features of normal alkanes in confined geometry. *Acc. Chem. Res.* **2014** *47*, 192–201. [CrossRef]
34. Giambattista, C.; Sanctuary, R.; Perigo, E.; Baller, J. Relaxations in the metastable rotator phase of n-eicosane. *J. Chem. Phys.* **2015** *143*, 054507. [CrossRef]
35. Marbeuf, A.; Brown, R. Molecular dynamics in n-alkanes: Premelting phenomena and rotator phases. *J. Chem. Phys.* **2006** *124*, 054901. [CrossRef]
36. Shang, Y.; Zhang, D. Preparation and thermal properties of graphene oxide–microencapsulated phase change materials. *Nanoscale Microscale Thermophys. Eng.* **2016**, *20*, 147–157. [CrossRef]
37. Halder, S.; Wang, J.; Fang, Y.; Qian, X.; Imam, M.A. Cenosphere-based PCM microcapsules with bio-inspired coating for thermal energy storage in cementitious materials. *Mater. Chem. Phys.* **2022**, *291*, 126745. [CrossRef]
38. Peng, G.; Dou, G.; Hu, Y.; Sun, Y.; Chen, Z. Phase change material (PCM) microcapsules for thermal energy storage. *Adv. Polym Technol.* **2020**, *2020*, 9490873. [CrossRef]
39. Zhang, Z.; Zhang, Z.; Chang, T.; Wang, J.; Wang, X.; Zhou, G. Phase change material microcapsules with melamine resin shell via cellulose nanocrystal stabilized Pickering emulsion in-situ polymerization. *Chem. Eng. J.* **2022**, *428*, 131164. [CrossRef]
40. Meng, X.; Qin, S.; Fan, H.; Huang, Z.; Hong, J.; Xu, X.; Ouyang, X.; Chen, D.Z. Long alkyl chain-grafted carbon nanotube-decorated binary-core phase-change microcapsules for heat energy storage: Synthesis and thermal properties. *Sol. Energy Mat Sol. C* **2020**, *212*, 110589. [CrossRef]
41. Zhou, J.; Li, S.; Niu, Z. ZnO@ZIF-8@ SiO_2-Modified Microencapsulated Phase Change Material Coating on Cotton Fabric with Radiative Cooling and Thermoregulation Capabilities. *Fiber. Polym.* **2024**, *25*, 2647–2659. [CrossRef]
42. Tang, A.; Pan, J.; Xia, D.; Cai, T.; Zhang, Q.; Tenkolu, G.A.; Jin, Y. Characterization and experimental assessment of hybrid cooling strategy for lithium-ion batteries by integrating microencapsulated phase change materials. *Int. J. Heat. Mass. Tran.* **2024**, *224*, 125389. [CrossRef]
43. Karimi, Y.; Nazar, A.R.S.; Molla-Abbasi, P. Synthesis, characterization and application of a magnetic nanoencapsulated PCM in enhancing the natural convection in a cube cavity. *Therm. Sci. Eng. Prog.* **2024**, *47*, 102342. [CrossRef]
44. Wang, X.; Gao, Y.; Han, N.; Zhang, X.; Li, W. Design and synthesis of microcapsules with cross-linking network supporting core for supercooling degree regulation. *Energy Build.* **2021**, *253*, 111437. [CrossRef]

5. Sarı, A.; Alkan, C.; Döğüşcü, D.K.; Kızıl, Ç. Micro/nano encapsulated n-tetracosane and n-octadecane eutectic mixture with polystyrene shell for low-temperature latent heat thermal energy storage applications. *Sol. Energy* **2015**, *115*, 195–203. [CrossRef]
6. Sarı, A.; Alkan, C.; Bilgin, C. Micro/nano encapsulation of some paraffin eutectic mixtures with poly (methyl methacrylate) shell: Preparation, characterization and latent heat thermal energy storage properties. *Appl. Energy* **2014**, *136*, 217–227. [CrossRef]
7. Xiang, X.; Gao, J.; Wang, F.; Wang, L. Double emulsion templates for efficient production of phase change microcapsules. *Chem. Eng. J.* **2024**, *496*, 154006. [CrossRef]
8. Jiménez-Vázquez, M.; Ramos, F.J.; Garrido, I.; López-Pedrajas, D.; Rodríguez, J.F.; Carmona, M. Production of thermoregulating slurries constituted by nanocapsules from melamine-formaldehyde containing n-octadecane. *J. Energy Storage* **2022**, *51*, 104465. [CrossRef]

molecules

MDPI

Article

Enhancing Electrochemical Performance of Si@CNT Anode by Integrating $SrTiO_3$ Material for High-Capacity Lithium-Ion Batteries

Nischal Oli [1,*], Diana C. Liza Castillo [1], Brad R. Weiner [2], Gerardo Morell [1] and Ram S. Katiyar [1,*]

[1] Department of Physics, University of Puerto Rico-Rio Piedras Campus, San Juan, PR 00925, USA

[2] Department of Chemistry, University of Puerto Rico-Rio Piedras Campus, San Juan, PR 00925, USA

* Correspondence: nischal.oli@upr.edu (N.O.); ram.katiyar@upr.edu (R.S.K.)

Abstract: Silicon (Si) has attracted worldwide attention for its ultrahigh theoretical storage capacity (4200 mA h g^{-1}), low mass density (2.33 g cm^{-3}), low operating potential (0.4 V vs. Li/Li^+), abundant reserves, environmentally benign nature, and low cost. It is a promising high-energy-density anode material for next-generation lithium-ion batteries (LIBs), offering a replacement for graphite anodes owing to the escalating energy demands in booming automobile and energy storage applications. Unfortunately, the commercialization of silicon anodes is stringently hindered by large volume expansion during lithiation–delithiation, the unstable and detrimental growth of electrode/electrolyte interface layers, sluggish Li-ion diffusion, poor rate performance, and inherently low ion/electron conductivity. These present major safety challenges lead to quick capacity degradation in LIBs. Herein, we present the synergistic effects of nanostructured silicon and $SrTiO_3$ (STO) for use as anodes in Li-ion batteries. Si and STO nanoparticles were incorporated into a multiwalled carbon nanotube (CNT) matrix using a planetary ball-milling process. The mechanical stress resulting from the expansion of Si was transferred via the CNT matrix to the STO. We discovered that the introduction of STO can improve the electrochemical performance of Si/CNT nanocomposite anodes. Experimental measurements and electrochemical impedance spectroscopy provide evidence for the enhanced mobility of Li-ions facilitated by STO. Hence, incorporating STO into the Si@CNT anode yields promising results, exhibiting a high initial Coulombic efficiency of approximately 85%, a reversible specific capacity of ~800 mA h g^{-1} after 100 cycles at 100 mA g^{-1}, and a high-rate capability of 1400 mA g^{-1} with a capacity of 800 mA h g^{-1}. Interestingly, it exhibits a capacity of 350 mAh g^{-1} after 1000 lithiation and delithiation cycles at a high rate of 600 mA hg^{-1}. This result unveils and sheds light on the design of a scalable method for manufacturing Si anodes for next-generation LIBs.

Keywords: Si; $SrTiO_3$; CNTs; large volume expansion; ion/electron conductivity; lithium-ion batteries

1. Introduction

Since their commercial debut thirty years ago, lithium-ion batteries (LIBs), a portable, high-density, and Nobel Prize-winning energy storage technology, have become an integral part of modern life and have revolutionized the fields of consumer portable electronics, electric vehicles (EVs), hybrid electric vehicles (HEVs), and large-scale renewable energy storage systems [1–4]. Despite significant advancements in technology, including an impressive thirty-fold price reduction between 1991 and 2018, the most substantial improvements have predominantly occurred on the lithium-metal-oxide cathode side [5–8]. The graphite anodes of LIBs, by contrast, have largely remained unchanged. However, the pursuit of higher energy densities and enhanced battery performance has driven extensive research into advanced alternative energy-dense anode materials [9–12]. The limited specific capacities of commercial graphite anodes (372 mA h g^{-1}) [13] and lithium cobalt oxide cathodes ($LiCoO_2$) (140 mA h g^{-1}) [14,15] are significant barriers to achieving higher performance, driving the exploration of innovative electrode materials. State-of-the-art LIBs utilizing graphite anodes are nearing the upper threshold of their practical specific energy density.

The energy density of LIBs is predominantly influenced by the specific capacity of the electrode materials. Over the past decade, commercial graphite has been the predominant anode material for LIBs due to its low operational potential (~0.1 V vs. Li/Li^+), cost-effectiveness, and robust cycling stability [16]. However, graphite poses several critical limitations. Primarily, it has a moderate specific capacity, which constrains the overall energy density of LIBs. Additionally, the lithium intercalation potential of graphite (~0.1 V vs. Li/Li^+) is close to the lithium metal plating potential (~0 V vs. Li/Li^+), raising significant safety concerns, particularly under conditions of fast charging or low temperatures [16–18]. Consequently, the development of alternative anode materials with higher specific capacities and elevated reaction potentials is imperative for advancing LIB technology [19–21]. Among these, silicon (Si) anodes have attracted worldwide attention due to their remarkably high theoretical specific capacity, $Li_{22}Si_5$ (4200 mA h g^{-1}) [22] or $Li_{15}Si_4$ (3579 mAh g^{-1}) [23], which is nearly ten times that of conventional graphite (372 mA h g^{-1} vs. Li/Li^+), their low mass density of 2.33 g cm^{-3} [24], and favorable operating potential (0.4 V versus Li/Li^+) [25]. When integrated with high capacity and high-voltage cathodes, such as 5 V spinel ($LiMn_xNi_yO_4$, LMNO) [26], Ni-rich $LiNi_xMn_yCo_zO_2$ (NMC) [27], or multi-valent conversion cathodes such as metal fluorides [28] and sulfur [29], Si is expected to achieve specific energy levels of 500 W h kg^{-1} or higher at the cell level [30]. Despite these advantages, the practical application of Si anodes poses significant challenges, including substantial volume expansion (>300%) during lithiation–delithiation, the reactivity of the lithium silicide (Li_xSi) interface, and issues such as delamination and pulverization, which can cause a loss of electrical contact between active materials and current collectors, leading to mechanical degradation and rapid capacity fading [31]. Addressing these challenges is crucial for unlocking the full potential of Si anodes in next-generation LIBs. Hence, to enhance the electrochemical performance of silicon anodes, it is crucial to mitigate the volume expansion effect and boost conductivity.

The full potential of silicon anode materials has yet to be fully realized, indicating a considerable journey ahead. Significant efforts have been dedicated to achieving stable silicon anodes (SiAs), mainly through the design of nanostructures, the synthesis of composite materials, the improvement of polymer binders, and the optimization of electrolytes [32]. Nanostructure design includes zero-dimensional nanoparticles, one-dimensional nanowires (nanotubes), two-dimensional nanofilms, and three-dimensional porous nanostructures [33]. In addition to nanostructure design, significant efforts have been made to optimize electrolyte compositions. Encapsulating the Si interface within non-reactive materials, such as carbon-based substances, polymers, oligomers, and metal oxides, prevents adverse reactions without hindering the lithiation–delithiation processes. Specifically, engineered interfaces, such as 'yolk-shell' and 'pomegranate' structures, effectively manage the 350% volume expansion during lithiation [33].

Polyvinylidene difluoride (PVDF) binders are prevalent in commercial applications due to their chemical and electrochemical stability [34]. However, PVDF binders present several challenges for Si anodes, including insufficient mechanical stretchability to accommodate significant volume changes of Si particles, weak adhesion to Si particles and the copper current collector, and the necessity of toxic organic solvents such as N-methyl pyrrolidone (NMP) in slurry preparation [35,36]. To address these binder issues, research has focused on developing water-based functional binders that can dissipate the stress on Si anodes. Binders with polar groups, such as carboxymethyl cellulose (CMC), alginate, poly(acrylic acid) (PAA), and poly(vinyl alcohol) (PVA), exhibit strong adhesive forces toward Si particles via hydrogen bonding [37–39]. Additionally, combining elastic moieties with polar functional groups in binders imparts the necessary mechanical properties to Si anodes [40–42]. Recently, stress-dissipative elastic waterborne polyurethane (PU) binders for Si anodes have shown promise, possessing both polar functional groups and elasticity [43]. However, the challenge of using toxic solvents, such as NMP, remains. Besides, silicon-based anodes for LIBs, despite innovations, such as stress-dissipating conductive polymer binders [44], accordion frameworks for free-standing, high silicon content [45], carbon fiber-confined yolk-shelled structures [46], and structural tailoring via electrospinning technology [47], continue to be more costly and complex to produce.

Efforts to counteract the volume expansion and potential pulverization of Si have also involved the creation of Si-based composite materials. These composites often include conductive additives (e.g., polymer-based hard carbon, graphene, and carbon nanotubes) or inactive materials (e.g., copper and titanium nitride (TiN)) [48,49]. Additionally, novel architectures have been designed to provide buffering space [50]. Despite significant progress, a trade-off persists between electrochemical performance and processing costs, as achieving excellent properties often requires complex and expensive methods, whereas simpler processes yield unsatisfactory performance [51].

In pursuit of low-cost and commercially viable methods, leveraging the large volume expansion of Si has shown potential. In recent times, advanced energy-generating materials, including ferroelectric, pyroelectric, piezoelectric compounds, as well as their composites, have been increasingly employed in the development of cutting-edge energy conversion technologies [52]. Among these, ferroelectric materials have been integrated into Li-ion batteries to enhance electrochemical performance by generating an electric field in response to mechanical stress [52]. For instance, Xue et al. reported a self-charging Li-ion battery using a piezoelectric poly(vinylidene fluoride) (PVDF) film as a separator, where piezoelectricity was induced by external mechanical stress [53]. In contrast, the current study focuses on internal stress-induced piezoelectricity, where the volume expansion of Si is transferred to a piezoelectric material, generating piezoelectric potential [52]. Furthermore, experiments conducted by Byoung-Sun Lee et al. demonstrated that the piezoelectric potential of the ferroelectric material barium titanate ($BaTiO_3$) significantly enhances the electrochemical performance of Si nanocomposite anodes by increasing Li-ion mobility, resulting in improved discharge capacity and cycle performance [52]. However, Si@CNT anodes exhibit poor cycling performance [52]. Additionally, Xuanmeng et al. utilized ferroelectric STO in lithium–sulfur batteries and claimed that it can adsorb polysulfides, effectively inhibiting the lithium–sulfur shuttle and reducing volume expansion during cycles through the ferroelectric effect, leading to impressive long-cycle performance [54].

Herein, for the first time, we present the integration of high-purity, commercially available ferroelectric STO with silicon nanopowder and multi-walled carbon nanotubes (CNTs) using a facile solid-state planetary ball-milling technique. The electrochemical performance of the resulting STO@Si@CNT composite was systematically examined for Li^+ ion intercalation, demonstrating substantial improvements in battery performance. The CNT matrix functions as a mechanical stress mediator, effectively transferring the volumetric expansion strain of Si to STO while preserving the integrity of conductive pathways key for Li^+ ion transport. STO nanoparticles provide a local ferroelectric potential that is poled by the deformation of nanoparticles during lithiation. It also influences Li^+ mobility [52,54]. Moreover, the optimization of a water-based sodium carboxymethyl cellulose (CMC) binder, combined with a standard carbonate-based electrolyte (1M $LiPF_6$ in EC:DMC, 1:1 wt% + 10% FEC additive), results in an anode exhibiting a remarkably high specific capacity of ~1500 mA h g^{-1}, a superior high-rate capability of ~1400 mA h g^{-1}, and exceptional cycling stability beyond 1000 cycles. These findings underline STO's pivotal role in alleviating the volumetric expansion of silicon anodes and enhancing lithium-ion transport, thereby suggesting that it may hold untapped potential for further advancements in silicon anode battery technologies.

2. Results and Discussion

Structural characterization of the Si/CNT/STO nanocomposite was conducted using X-ray diffraction (XRD) analysis. The XRD spectrum, illustrated in Figure 1a, reveals distinct diffraction peaks corresponding to the (100), (110), (111), (200), (210), and (220) crystallographic planes of strontium titanate (STO), as well as the (111) and (220) planes of crystalline silicon. Notably, near 2θ ≈ 58°, the (311) and (211) planes of Si and STO, respectively, nearly overlap. As shown in Figure 1c, the XRD pattern of STO indicates high purity [55]. It is important to highlight that STO can exhibit either cubic or tetragonal crystal structures (Figure 1d), with the cubic phase predominating when the particle size is below 100 nm. Given that the STO particle diameter in this system was below 100 nm, a cubic structure was anticipated. Although distinct diffraction peaks indicative of the carbon nanotube (CNT) microstructure (Figure 1b), such as C (002), were not sharply resolved, the broad feature around 2θ ≈ 26° was ascribed to the presence of CNTs.

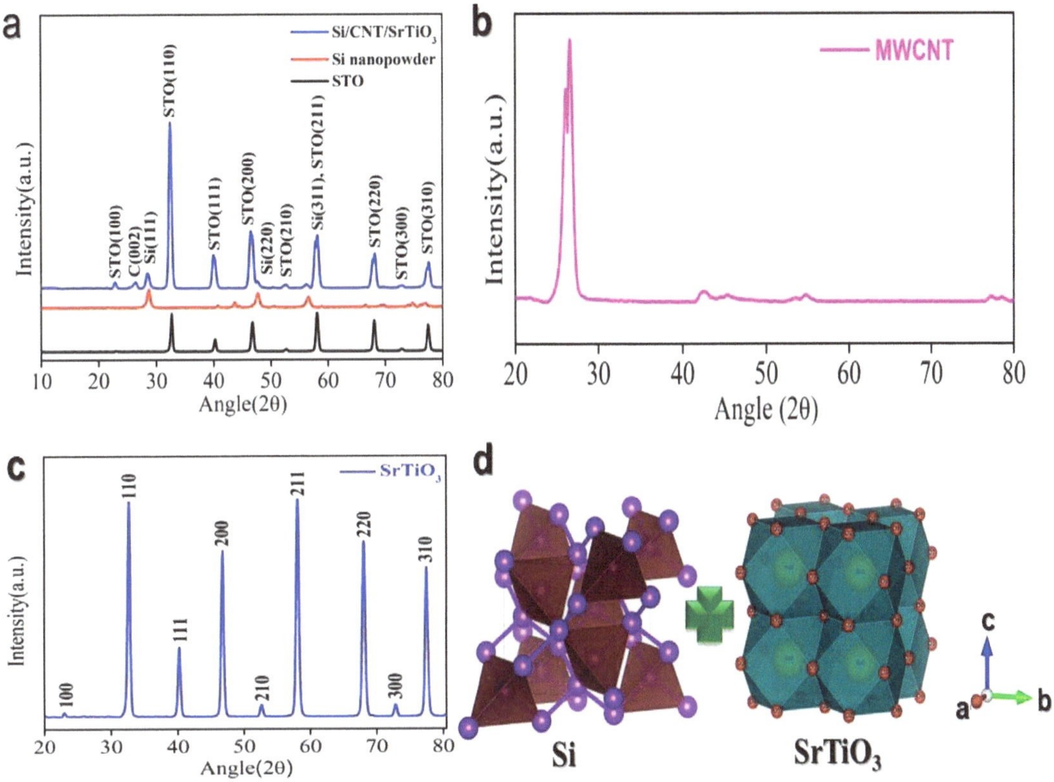

Figure 1. Structural analysis: (**a**) XRD pattern of the Si@STO@CNT composite; (**b**) XRD pattern of CNTs; (**c**) XRD pattern of STO; and (**d**) crystal structure of Si and STO, respectively.

Scanning electron microscopy (SEM) revealed that the nanocomposite surface consists of fragmented CNTs, Si nanoparticles, and STO nanoparticles (Figure 2a,b). The ball-milling process induced CNT breakage, reducing their average length to around 100–150 nm. The SEM image highlights the dense packing of these three components post high-energy ball milling, where the Si and STO regions are distinctly visible and the broken CNTs form a matrix-like framework between them. Furthermore, SEM mapping confirmed a uniform distribution of the Si and STO nanoparticles across the nanocomposite (Figure 2c–f).

The electrochemical performance of the Si/CNT/STO nanocomposite was measured using galvanostatic charge/discharge (GCD) tests. As demonstrated in Figure 3a,e, the voltage curves illustrate stable lithiation and delithiation of Li^+ ions during thew cycles. Characteristic lithiation and delithiation plateaus of silicon were observed at ~0.25 V during discharge and ~0.5 V during charge. Since STO has a low specific capacity, the overall specific capacity was mainly determined by the Si and CNT content. Despite comprising 30 wt%, CNTs contributed less than 5% to the total specific capacity. Thus, the voltage plateaus were primarily linked to the electrochemical activity of the Si nanoparticles.

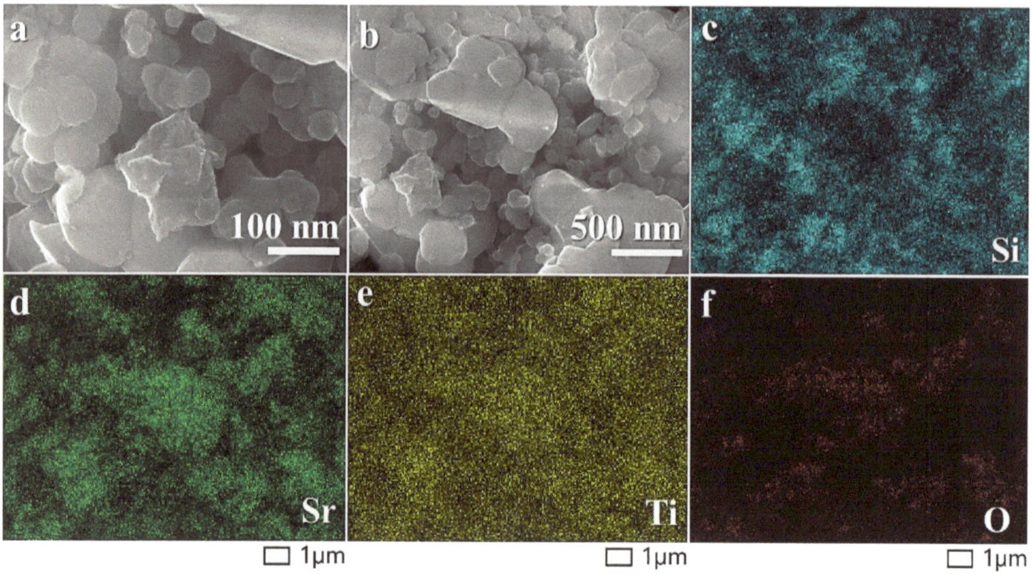

Figure 2. SEM analysis: (**a**) 100 nm; (**b**) 500 nm; and (**c**) SEM mapping of Si; (**d**) Sr; (**e**) Ti; and (**f**) O.

The Si/CNTs/STO nanocomposite exhibited behavior similar to Si/CNT composites in the initial cycle, with a reversible capacity of approximately 1510 mA h g^{-1}, as shown in Figure 3a,e, and an initial Coulombic efficiency of 84.97%, which stabilized to around 98.3% after subsequent cycling. However, the Si/CNT nanocomposite demonstrated cycling performance degradation upon repeated cycling, attributed to the pulverization of the silicon particles and the subsequent loss of conductive pathways, which has been previously reported [52]. In contrast, the Si/CNT/STO nanocomposites demonstrated enhanced cycling stability, which we attribute to enhanced kinetic behavior resulting from the incorporation of STO, as shown in Figure 3b. As depicted in the voltage profiles in Figure 3e, the addition of STO facilitated the mitigation of performance degradation, supporting more stable lithiation and stable electrochemical performance over time.

In addition to its high capacity and consistent performance, the Si@CNTs@STO anode exhibited remarkable rate capabilities and prolonged cycling stability. As illustrated in Figure 3d, the charge capacities recorded at various current densities of 100, 200, 350, 500, 800, 1000, 1200, and 1400 mA g^{-1} were 1238, 1174, 1068, 978, 880, 818, 766, and 729 mA h g^{-1}, respectively. Notably, even at a high current density of 1400 mA g^{-1}, the charge capacity remained at 729 mA h g^{-1}, which is approximately double the theoretical specific capacity of graphite (372 mA h g^{-1}). Upon resuming the original current density of 100 mA g^{-1}, the charge capacity recovered to 896 mA h g^{-1}, reflecting a recovery ratio of approximately 73%. Figure 3c further depicts the charge/discharge profiles corresponding to these current densities. Additionally, Figure 3b demonstrates the cycling performance at 100 mA g^{-1}, where after 150 cycles, the capacity was sustained at 700 mA h g^{-1}, with an average Coulombic efficiency of 98.5%.

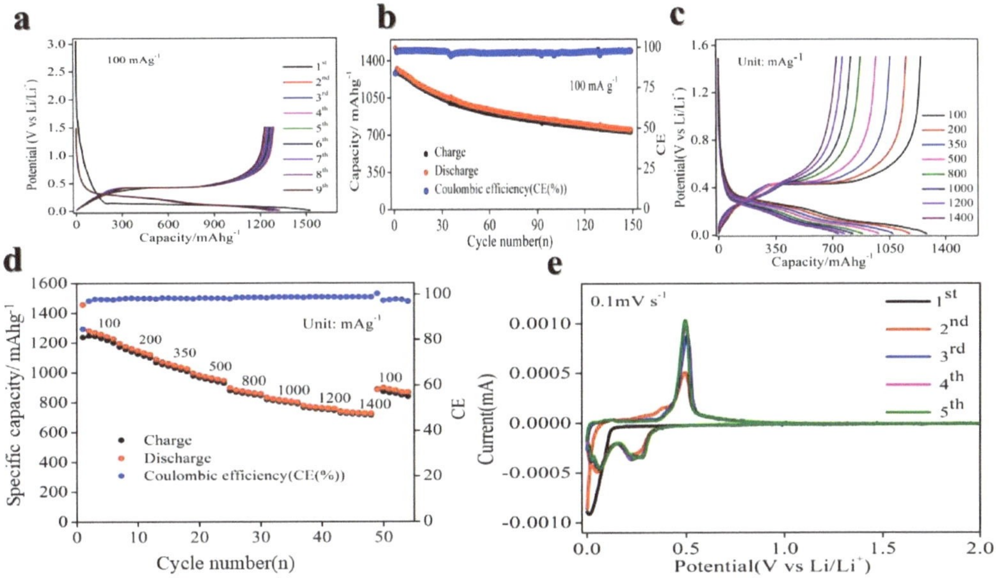

Figure 3. Electrochemical performance of composite Si/STO/CNTs. (a) Galvanostatic charge–discharge (GCD) curve at a low current density of 100 mA g^{-1}; (b) cycling performance at a current density of 100 mA g^{-1}; (c) rate curve at a range of current densities from 100 to 1400 mA g^{-1}; (d) rate performance at a range of current densities from 100 to 1400 mA g^{-1}; (e) cyclic voltammetry at a scan rate of 0.1 mV s^{-1}.

Figure 4a,b presents the cycling performance at current densities of 400 mA g^{-1} and 600 mA g^{-1}, demonstrating consistent stability over more than 1000 cycles. This substantial enhancement in performance can be attributed to the integration of the Si@STO nanocomposite, coupled with the strategic optimization of binders and electrolytes.

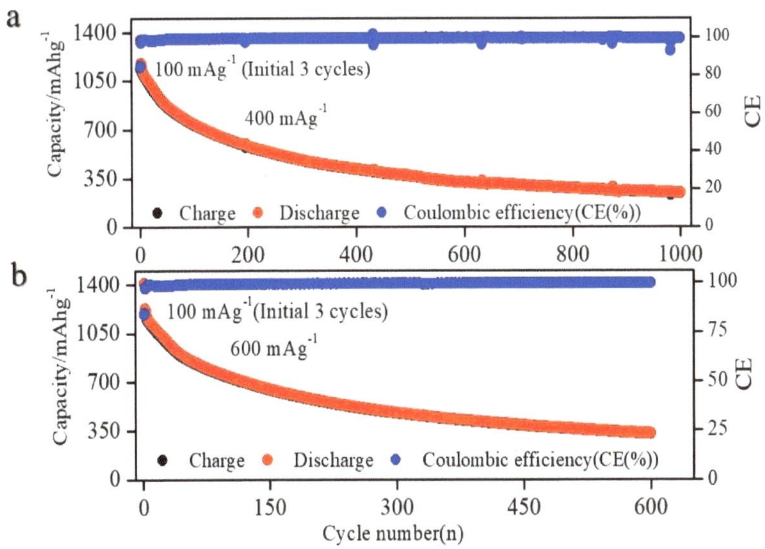

Figure 4. Cyclic performance at current densities of (a) 400 mA g^{-1} and (b) 600 mA g^{-1}.

Electrochemical Impedance Spectroscopy (EIS) is a sophisticated technique for probing the intricate electrochemical dynamics within LIBs. By measuring impedance, which is essentially the resistance to alternating current across a broad frequency range, a detailed analysis of multiple internal processes is provided. This technique is instrumental in elucidating charge transfer mechanisms at the electrode–electrolyte interface, where Li^+ ions shuttle between the anode and cathode during the lithiation–delithiation process. It also offers insights into ion diffusion across the electrodes and electrolyte, particularly in the low-frequency region, which is pivotal for understanding overall battery performance. There are distinct key electrochemical parameters contributing to these processes, mainly ohmic resistance (arising from the bulk properties of the electrode and electrolyte materials), charge transfer resistance (Rct) at the electrode–electrolyte interfaces, and double-layer capacitance at the electrode surface. Through careful interpretation of the impedance plot, performance-limiting factors, such as material degradation, internal resistance buildup, and diffusion inefficiencies, are revealed, thereby guiding the optimization of battery materials, architecture, and operational strategies.

As depicted in Figure 5a, the Rct value is lower compared to that observed after 150 cycles in Figure 5b, which con-firms that the incorporation of STO significantly enhances the electrochemical stability and extends the cycling longevity of the system.

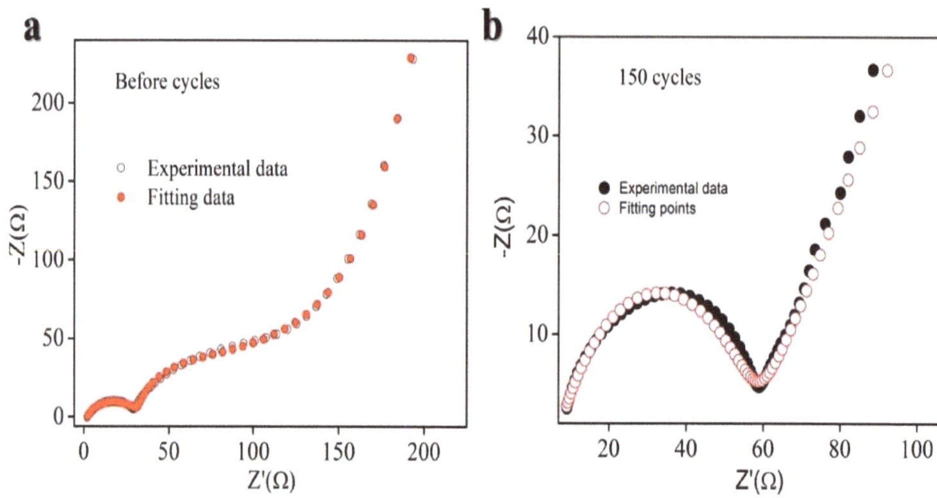

Figure 5. Electrochemical Impedance Spectroscopy analysis of composite Si/STO/CNTs. (**a**) Before lithiation–delithiation and (**b**) after 150 lithiation–delithiation cycles.

3. Experimental Setup

3.1. Synthesis

The Si@@CNT@STO powder was synthesized utilizing a solid-state route. The nanocomposite maintained a Si/CNT/STO mass ratio of 49:21:30. Precise proportions of commercially sourced precursors, including Si nanopowder, $SrTiO_3$, and multiwall carbon nanotubes (Sigma-Aldrich, 99.97% purity, Chicago, IL, USA), were initially ground together using a mortar and pestle for 1 h. Subsequently, these precursors were thoroughly homogenized with a planetary ball mill equipped with zirconia balls (Across International, PQ-N04, Livingston, NJ, USA), operating at 45 Hz (2700 rpm) for 3 h. Finally, the synthesized Si@CNT@STO powder was ground to achieve the desired consistency for further applications.

3.2. Structural Analysis

The structural characterization of the synthesized Si@CNT@STO powder was analyzed using an X-ray diffractometer (Rigaku Smart Lab, The Woodlands, TX, USA) with Cu Kα radiation (λ = 1.5408 Å) over a 2θ range of 20° to 80° at a scan rate of 2° per minute. Additionally, morphological assessment and chemical composition analyses were performed using scanning electron microscopy (SEM) and energy dispersive X-ray spectroscopy (EDS) (JEOL, JSM 6480LV, Tokyo, Japan), respectively.

3.3. Electrochemical Fabrication and Cell Assembly

In order to perform the electrochemical testing, the anode slurry consisted of a Si/CNT/STO nanocomposite and carboxymethyl cellulose (CMC) binder in distilled water, with a nanocomposite-to-binder mass ratio of 9:1. This slurry was coated on a carbon-coated copper foil substrate (thickness of 9 μm) using a doctor blade (MTI Corporation, Richmond, CA, USA) and was subsequently dried at 80 °C overnight. No additional conductive agent was required as the CNTs within the nanocomposites provided sufficient conductivity. The 1 cm diameter anodes were then obtained from the foil, further dried at 60 °C under vacuum for 3 h, and immediately transferred to the glovebox (MBRAUN, Glovebox Workstations, Stratham, NH, USA). For assembling CR2032 coin-type half-cells, lithium foil was used as the counter electrode and polypropylene ethylene as the separator. For the electrolyte, we utilized 1 M LiPF6 in ethylene carbonate (EC)/diethyl carbonate (DEC) [1:1] with 10% FEC additives. All electrochemical performance tests were carried out using a lithium half-cell in the voltage range of 0.01–1.5 V vs. Li^+/Li. A cyclic voltammetry (CV) analysis was conducted using various scan rates between 0.1 and 1 mV s^{-1} with an Arbin instrument, College Station, TX, USA. Furthermore, an electrochemical impedance spectroscopy (EIS) analysis was performed using the Arbin instrument. All charge–discharge tests were performed using a Landt battery tester.

4. Conclusions

Our investigation revealed that the incorporation of STO enhances the electrochemical performance of Si/CNT nanocomposite anodes. Experimental data and electrochemical impedance spectroscopy confirmed that STO facilitates improved Li^+ ion mobility. Consequently, the integration of STO into the Si@CNT anode delivers promising electrochemical performance, with a high initial Coulombic efficiency of approximately 85%, a reversible specific capacity of ~800 mA h g^{-1} after 100 cycles at 100 mA g^{-1}, and an impressive high-rate capability of 1400 mA g^{-1}, maintaining a capacity of 800 mA h g^{-1}. Remarkably, the anode retained a capacity of 350 mA h g^{-1} even after 1000 lithiation–delithiation cycles at a high rate of 600 mA g^{-1}. These findings highlight a scalable pathway for the design and production of Si-based anodes, offering significant potential for next-generation LIBs.

Author Contributions: Conceptualization, N.O.; investigation N.O.; data curation, N.O.; writing—original draft preparation, N.O. and R.S.K.; writing, review and editing, N.O. and D.C.L.C.; supervision, R.S.K., G.M. and B.R.W.; funding acquisition, R.S.K., G.M. and B.R.W. All authors have read and agreed to the published version of the manuscript.

Funding: This research received financial support from the NSF EPSCoR Center for the Advancement of Wearable Technologies (CAWT) under Grant No. OIA-1849243 and the NASA MIRO PR-SPRinT under Grant No. 80NSSC19M0236.

Institutional Review Board Statement: Not applicable.

Informed Consent Statement: Not applicable.

Data Availability Statement: The data presented in this study are available on request from the corresponding author.

Acknowledgments: We would like to extend our sincere appreciation to the Molecular Science Research Center (MSRC) at the University of Puerto Rico for their invaluable support in facilitating the use of Scanning Electron Microscopy (SEM) with Energy Dispersive X-ray (EDX) microanalysis, as well as X-ray Diffraction (XRD).

Conflicts of Interest: The authors declare no conflicts of interest.

References

1. Tarascon, J.M.; Armand, M. Issues and Challenges Facing Rechargeable Lithium Batteries. *Nature* **2001**, *414*, 359–367. [CrossRef] [PubMed]
2. Goodenough, J.B. Evolution of Strategies for Modern Rechargeable Batteries. *Acc. Chem. Res.* **2013**, *46*, 1053–1061. [CrossRef] [PubMed]
3. Goodenough, J.B.; Kim, Y. Challenges for Rechargeable Li Batteries. *Chem. Mater.* **2010**, *22*, 587–603. [CrossRef]
4. Manthiram, A.; Goodenough, J.B. Lithium Insertion into $Fe_2(MO_4)_3$ Frameworks: Comparison of M = W with M = Mo. *J. Solid State Chem.* **1987**, *71*, 349–360. [CrossRef]
5. McNulty, D.; Hennessy, A.; Li, M.; Armstrong, E.; Ryan, K.M. A Review of Li-Ion Batteries for Autonomous Mobile Robots: Perspectives and Outlook for the Future. *J. Power Sources* **2022**, *545*, 231943. [CrossRef]
6. Howard, W.F.; Spotnitz, R.M. Theoretical Evaluation of High-Energy Lithium Metal Phosphate Cathode Materials in Li-Ion Batteries. *J. Power Sources* **2007**, *165*, 887–891. [CrossRef]
7. Betz, J.; Brinkmann, J.P.; Nölle, R.; Lürenbaum, C.; Kolek, M.; Stan, M.C.; Winter, M.; Placke, T. Cross Talk between Transition Metal Cathode and Li Metal Anode: Unraveling Its Influence on the Deposition/Dissolution Behavior and Morphology of Lithium. *Adv. Energy Mater.* **2019**, *9*, 1900574. [CrossRef]
8. Liu, J.; Bao, Z.; Cui, Y.; Dufek, E.J.; Goodenough, J.B.; Khalifah, P.; Li, Q.; Liaw, B.Y.; Liu, P.; Manthiram, A.; et al. Pathways for Practical High-Energy Long-Cycling Lithium Metal Batteries. *Nat. Energy* **2019**, *4*, 180–186. [CrossRef]
9. Oli, N.; Choudhary, S.; Weiner, B.R.; Morell, G.; Katiyar, R.S. Comparative Investigation of Water-Based CMC and LA133 Binders for CuO Anodes in High-Performance Lithium-Ion Batteries. *Molecules* **2024**, *29*, 4114. [CrossRef]
10. Oli, N.; Flórez Gómez, J.F.; Zuluaga Gómez, C.C.; Katiyar, R.K.; Morell, G.; Katiyar, R.S. Revealing Underestimated Performance in the Bismuth Ferrite ($BiFeO_3$) Anode for High-Capacity and Long-Cycling Lithium-Ion Batteries. *ACS Appl. Energy Mater.* **2023**, *6*, 10853–10861. [CrossRef]
11. Pendashteh, A.; Tomey, R.; Vilatela, J.J. Nanotextile 100% Si Anodes for the Next Generation Energy-Dense Li-Ion Batteries. *Adv. Energy Mater.* **2024**, *14*, 2304018. [CrossRef]
12. Kim, N.; Kim, Y.; Sung, J.; Cho, J. Issues Impeding the Commercialization of Laboratory Innovations for Energy-Dense Si Containing Lithium-Ion Batteries. *Nat. Energy* **2023**, *8*, 921–933. [CrossRef]
13. Mo, R.; Li, F.; Tan, X.; Xu, P.; Tao, R.; Shen, G.; Lu, X.; Liu, F.; Shen, L.; Xu, B.; et al. High-Quality Mesoporous Graphene Particles as High-Energy and Fast-Charging Anodes for Lithium-Ion Batteries. *Nat. Commun.* **2019**, *10*, 1474. [CrossRef] [PubMed]
14. Li, J.; Lin, C.; Weng, M.; Qiu, Y.; Chen, P.; Yang, K.; Huang, W.; Hong, Y.; Li, J.; Zhang, M.; et al. Structural Origin of the High-Voltage Instability of Lithium Cobalt Oxide. *Nat. Nanotechnol.* **2021**, *16*, 599–605. [CrossRef]
15. Liu, Z.; Yu, Q.; Oli, N.; Gomez, J.F.F.; Qiu, S.; Tian, H.; Qiu, Q.; Sun, W.; Li, K.; Liu, Z.; et al. A Non-Volatile, Thermo-Reversible and Self-Protective Gel Electrolyte Providing Highly Precise and Reversible Thermal Protection for Lithium Batteries. *Adv. Energy Mater.* **2023**, *13*, 2300143. [CrossRef]
16. Li, S.; Wang, K.; Zhang, G.; Li, S.; Xu, Y.; Zhang, X.; Zhang, X.; Zheng, S.; Sun, X.; Ma, Y. Fast Charging Anode Materials for Lithium-Ion Batteries: Current Status and Perspectives. *Adv. Funct. Mater.* **2022**, *32*, 2200796. [CrossRef]
17. Parekh, M.H.; Sediako, A.D.; Naseri, A.; Thomson, M.J.; Pol, V.G. In Situ Mechanistic Elucidation of Superior Si-C-Graphite Li-Ion Battery Anode Formation with Thermal Safety Aspects. *Adv. Energy Mater.* **2020**, *10*, 1902799. [CrossRef]
18. Xie, L.; Zhang, W.; Chen, X.; Shan, R.; Han, Q.; Qiu, X.; Oli, N.; Florez Gomez, J.F.; Zhu, L.; Wu, X.; et al. Bimetallic Cobalt-Nickel Selenide Nanocubes Embedded in a Nitrogen-Doped Carbon Matrix as an Excellent Li-Ion Battery Anode. *ACS Appl. Mater. Interfaces* **2023**, *15*, 25536–25549. [CrossRef]
19. Placke, T.; Kloepsch, R.; Dühnen, S.; Winter, M. Lithium Ion, Lithium Metal, and Alternative Rechargeable Battery Technologies: The Odyssey for High Energy Density. *J. Solid State Electrochem.* **2017**, *21*, 1939–1964. [CrossRef]
20. Piątek, J.; Afyon, S.; Budnyak, T.M.; Budnyk, S.; Sipponen, M.H.; Slabon, A. Sustainable Li-Ion Batteries: Chemistry and Recycling. *Adv. Energy Mater.* **2021**, *11*, 2003456. [CrossRef]
21. Nzereogu, P.U.; Omah, A.D.; Ezema, F.I.; Iwuoha, E.I.; Nwanya, A.C. Anode Materials for Lithium-Ion Batteries: A Review. *Appl. Surf. Sci. Adv.* **2022**, *9*, 100233. [CrossRef]
22. Bitew, Z.; Tesemma, M.; Beyene, Y.; Amare, M. Nano-Structured Silicon and Silicon Based Composites as Anode Materials for Lithium Ion Batteries: Recent Progress and Perspectives. *Sustain. Energy Fuels* **2022**, *6*, 1014–1050. [CrossRef]
23. Li, P.; Kim, H.; Myung, S.T.; Sun, Y.K. Diverting Exploration of Silicon Anode into Practical Way: A Review Focused on Silicon-Graphite Composite for Lithium Ion Batteries. *Energy Storage Mater.* **2021**, *35*, 550–576. [CrossRef]
24. Kim, S.O.; Manthiram, A. A Facile, Low-Cost Synthesis of High-Performance Silicon-Based Composite Anodes with High Tap Density for Lithium-Ion Batteries. *J. Mater. Chem. A Mater.* **2015**, *3*, 2399–2406. [CrossRef]

25. Pan, H.; Wang, L.; Shi, Y.; Sheng, C.; Yang, S.; He, P.; Zhou, H. A Solid-State Lithium-Ion Battery with Micron-Sized Silicon Anode Operating Free from External Pressure. *Nat. Commun.* **2024**, *15*, 2263. [CrossRef] [PubMed]
26. Xiang, J.; Wei, Y.; Zhong, Y.; Yang, Y.; Cheng, H.; Yuan, L.; Xu, H.; Huang, Y. Building Practical High-Voltage Cathode Materials for Lithium-Ion Batteries. *Adv. Mater.* **2022**, *34*, 2200912. [CrossRef]
27. Gomez-Martin, A.; Reissig, F.; Frankenstein, L.; Heidbüchel, M.; Winter, M.; Placke, T.; Schmuch, R.; Gomez-Martin, A.; Frankenstein, L.; Heidbüchel, M.; et al. Magnesium Substitution in Ni-Rich NMC Layered Cathodes for High-Energy Lithium Ion Batteries. *Adv. Energy Mater.* **2022**, *12*, 2103045. [CrossRef]
28. Clément, R.J.; Lun, Z.; Ceder, G. Cation-Disordered Rocksalt Transition Metal Oxides and Oxyfluorides for High Energy Lithium-Ion Cathodes. *Energy Environ. Sci.* **2020**, *13*, 345–373. [CrossRef]
29. Su, Y.S.; Manthiram, A. Lithium–Sulphur Batteries with a Microporous Carbon Paper as a Bifunctional Interlayer. *Nat. Commun.* **2012**, *3*, 1166. [CrossRef]
30. Wang, Q.; Zhu, M.; Chen, G.; Dudko, N.; Li, Y.; Liu, H.; Shi, L.; Wu, G.; Zhang, D. High-Performance Microsized Si Anodes for Lithium-Ion Batteries: Insights into the Polymer Configuration Conversion Mechanism. *Adv. Mater.* **2022**, *34*, 2109658. [CrossRef]
31. McDowell, M.T.; Woo Lee, S.; Wang, C.; Cui, Y. The Effect of Metallic Coatings and Crystallinity on the Volume Expansion of Silicon during Electrochemical Lithiation/Delithiation. *Nano Energy* **2012**, *1*, 401–410. [CrossRef]
32. Qin, G.; Jia, Z.; Sun, S.; Wu, H.; Hu, K.; Liu, D.; Gao, Y.; Chen, J. Carbon-Coated Si Nanosheets as Anode Materials for High-Performance Lithium-Ion Batteries. *ACS Appl. Nano Mater.* **2024**, *7*, 7595–7604. [CrossRef]
33. Pach, G.F.; Adhikari, P.R.; Quinn, J.; Wang, C.; Singh, A.; Verma, A.; Colclasure, A.; Kim, J.H.; Teeter, G.; Veith, G.M.; et al. Boron-Silicon Alloy Nanoparticles as a Promising New Material in Lithium-Ion Battery Anodes. *ACS Energy Lett.* **2024**, *9*, 2492–2499. [CrossRef] [PubMed]
34. Tabani, Z.; Maghsoudi, H.; Fathollahi Zonouz, A. High Electrochemical Stability of Polyvinylidene Fluoride (PVDF) Porous Membranes Using Phase Inversion Methods for Lithium-Ion Batteries. *J. Solid State Electrochem.* **2021**, *25*, 651–657. [CrossRef]
35. Yuca, N.; Taskin, O.S.; Arici, E. An Overview on Efforts to Enhance the Si Electrode Stability for Lithium Ion Batteries. *Energy Storage* **2020**, *2*, e94. [CrossRef]
36. Eshetu, G.G.; Figgemeier, E. Confronting the Challenges of Next-Generation Silicon Anode-Based Lithium-Ion Batteries: Role of Designer Electrolyte Additives and Polymeric Binders. *ChemSusChem* **2019**, *12*, 2515–2539. [CrossRef]
37. Yang, Y.; Wu, S.; Zhang, Y.; Liu, C.; Wei, X.; Luo, D.; Lin, Z. Towards Efficient Binders for Silicon Based Lithium-Ion Battery Anodes. *Chem. Eng. J.* **2021**, *406*, 126807. [CrossRef]
38. Lai, Y.; Li, H.; Yang, Q.; Li, H.; Liu, Y.; Song, Y.; Zhong, Y.; Zhong, B.; Wu, Z.; Guo, X. Revisit the Progress of Binders for a Silicon-Based Anode from the Perspective of Designed Binder Structure and Special Sized Silicon Nanoparticles. *Ind. Eng. Chem. Res.* **2022**, *61*, 6246–6268. [CrossRef]
39. Zhu, W.; Zhou, J.; Xiang, S.; Bian, X.; Yin, J.; Jiang, J.; Yang, L. Progress of Binder Structures in Silicon-Based Anodes for Advanced Lithium-Ion Batteries: A Mini Review. *Front. Chem.* **2021**, *9*, 712225. [CrossRef] [PubMed]
40. Cao, P.F.; Yang, G.; Li, B.; Zhang, Y.; Zhao, S.; Zhang, S.; Erwin, A.; Zhang, Z.; Sokolov, A.P.; Nanda, J.; et al. Rational Design of a Multifunctional Binder for High-Capacity Silicon-Based Anodes. *ACS Energy Lett.* **2019**, *4*, 1171–1180. [CrossRef]
41. Liu, Z.; Fang, C.; He, X.; Zhao, Y.; Xu, H.; Lei, J.; Liu, G. In Situ-Formed Novel Elastic Network Binder for a Silicon Anode in Lithium-Ion Batteries. *ACS Appl. Mater. Interfaces* **2021**, *13*, 46518–46525. [CrossRef]
42. Jeong, D.; Yook, J.; Kwon, D.S.; Shim, J.; Lee, J.C. Interweaving Elastic and Hydrogen Bond-Forming Polymers into Highly Tough and Stress-Relaxable Binders for High-Performance Silicon Anode in Lithium-Ion Batteries. *Adv. Sci.* **2023**, *10*, 2302027. [CrossRef]
43. Kim, J.O.; Kim, E.; Lim, E.Y.; Kwon, T.; Kim, I.J.; Lee, J.; Ko, J.W.; Lee, J.H. Stress-Dissipative Elastic Waterborne Polyurethane Binders for Silicon Anodes with High Structural Integrity in Lithium-Ion Batteries. *ACS Appl. Energy Mater.* **2024**, *7*, 1629–1639. [CrossRef]
44. Xu, Z.; Chu, X.; Wang, K.; Zhang, H.; He, Z.; Xie, Y.; Yang, W. Stress-Dissipated Conductive Polymer Binders for High-Stability Silicon Anode in Lithium-Ion Batteries. *J. Mater.* **2023**, *9*, 378–386. [CrossRef]
45. Zhao, Z.; Dong, L.; Huang, Q.; Wu, T.; Zhao, F.; Guo, Y.; Hu, B.; Ge, Y.; Zhang, J.; Xie, K.; et al. Accordion Frameworks Enable Free-Standing, High Si Content Anode for Li-Ion Batteries. *Energy Environ. Mater.* **2023**, *6*, e12400. [CrossRef]
46. Jia, D.C.; Feng, Y.Y.; Zhang, C.L.; Li, J.J.; Zhang, B.W.; Dou, Y.H.; Roy, J.C.; Zhu, X.Y.; Zhang, L. Freestanding Carbon Fiber-Confined Yolk-Shelled Silicon-Based Anode for Promoted Lithium Storage Applications. *Rare Metals* **2023**, *42*, 3718–3728. [CrossRef]
47. Huang, A.; Ma, Y.; Peng, J.; Li, L.; Chou, S.; Ramakrishna, S.; Peng, S. Tailoring the Structure of Silicon-Based Materials for Lithium-Ion Batteries via Electrospinning Technology. *eScience* **2021**, *1*, 141–162. [CrossRef]
48. Wang, S.E.; Kim, D.H.; Kim, M.J.; Kim, J.H.; Kang, Y.C.; Roh, K.C.; Choi, J.; Lee, H.W.; Jung, D.S. Achieving Cycling Stability in Anode of Lithium-Ion Batteries with Silicon-Embedded Titanium Oxynitride Microsphere. *Nanomaterials* **2023**, *13*, 132. [CrossRef]
49. Lee, S.S.; Nam, K.H.; Jung, H.; Park, C.M. Si-Based Composite Interconnected by Multiple Matrices for High-Performance Li-Ion Battery Anodes. *Chem. Eng. J.* **2020**, *381*, 122619. [CrossRef]
50. Wu, Z.; Luo, J.; Peng, J.; Liu, H.; Chang, B.; Wang, X. Rational Architecture Design of Yolk/Double-Shells Si-Based Anode Material with Double Buffering Carbon Layers for High Performance Lithium-Ion Battery. *Green Energy Environ.* **2021**, *6*, 517–527. [CrossRef]

51. Martino, A.; Cong, R.; Jo, M.; Park, H.H.; Lee, H.; Lee, C.S. Characteristics and Electrochemical Performance of Hydroxy Functionalized Graphene Quantum Dot-Coated Si Nanoparticles/Reduced Graphene Hybrid Anodes for Advanced Li-Io Batteries. *J. Nanomater.* **2023**, *2023*, 6353894. [CrossRef]
52. Lee, B.S.; Yoon, J.; Jung, C.; Kim, D.Y.; Jeon, S.Y.; Kim, K.H.; Park, J.H.; Park, H.; Lee, K.H.; Kang, Y.S.; et al. Silicon/Carbo Nanotube/BaTiO3 Nanocomposite Anode: Evidence for Enhanced Lithium-Ion Mobility Induced by the Local Piezoelectri Potential. *ACS Nano* **2016**, *10*, 2617–2627. [CrossRef]
53. Wang, Z. Modeling and Simulation of Piezoelectrically Driven Self-Charging Lithium Ion Batteries. *ACS Appl. Mater. Interface* **2017**, *9*, 15893–15897. [CrossRef]
54. He, X.; Lei, Z.; Liu, H.; Wang, S.; Qiao, T.; Wang, X. Synergetic Enhancing Cycling Stability of Li-S Battery by Hollow SrTiO Microspheres Wrapped by Reduced Graphene Oxide. *J. Alloys Compd.* **2022**, *898*, 162987. [CrossRef]
55. Da Silva, L.F.; Avansi, W.; Moreira, M.L.; Mesquita, A.; Maia, L.J.Q.; Andrés, J.; Longo, E.; Mastelaro, V.R. Relationship betwee Crystal Shape, Photoluminescence, and Local Structure in $SrTiO_3$ Synthesized by Microwave-Assisted Hydrothermal Metho *J. Nanomater.* **2012**, *2012*, 890397. [CrossRef]

molecules

MDPI

Article

High-Density Capacitive Energy Storage in Low-Dielectric-Constant Polymer PMMA/2D Mica Nanofillers Heterostructure Composite

Sumit Bera [1,*], Rukshan Thantirige [1], Sujit A. Kadam [1], Anirudha V. Sumant [2] and Nihar R. Pradhan [1,*]

[1] Department of Chemistry, Physics and Atmospheric Science, Jackson State University, 1400 John R. Lynch Street, Jackson, MS 39217, USA
[2] Center for Nanoscale Materials, Argonne National Laboratory, 9700 South Cass Avenue, Lemont, IL 60439, USA
* Correspondence: submit.bera@jsums.edu (S.B.); nihar.r.pradhan@jsums.edu (N.R.P.)

Abstract: The ubiquitous, rising demand for energy storage devices with ultra-high storage capacity and efficiency has drawn tremendous research interest in developing energy storage devices. Dielectric polymers are one of the most suitable materials used to fabricate electrostatic capacitive energy storage devices with thin-film geometry with high power density. In this work, we studied the dielectric properties, electric polarization, and energy density of PMMA/2D Mica nanocomposite capacitors where stratified 2D nanofillers are interfaced between the multiple layers of PMMA thin films using two heterostructure designs of the capacitors, PMMA/2D Mica/PMMA (PMP) and PMMA/2D Mica/PMMA/2D Mica/PMMA (PMPMP). The incorporation of a 2D Mica nanofiller in the low-dielectric-constant PMMA leads to an enhancement in the dielectric constant, with $\Delta\varepsilon \sim 15\%$ and 53% for PMP and PMPMP heterostructures at room temperature. Additionally, a significant improvement in discharged energy density was measured for the PMPMP capacitor ($U_d \sim 38$ J/cm^3 at 825 MV/m) compared to the pristine PMMA ($U_d \sim 9.5$ J/cm^3 at 522 MV/m) and PMP capacitors ($U_d \sim 19$ J/cm^3 at 740 MV/m). This excellent capacitive and energy storage performance of the PMMA/2D Mica heterostructure nanocomposite may inform the fabrication of thin-film, high-density energy storage capacitor devices for potential applications in various platforms.

Keywords: PMMA; 2D materials; mica; polymer composite; energy density; polarization

1. Introduction

The increasing demand for energy storage devices with ultra-high capacity and efficiency has sparked significant research interest in energy storage materials such as lithium-ion batteries, sodium-ion batteries, and dielectric capacitors [1–5]. Dielectric capacitors present compelling advantages over lithium-ion and sodium-ion batteries, including rapid charging, an extended lifespan, and enhanced efficiency, thereby attracting significant attention in electrostatic energy storage research [6]. Polymer dielectrics are the most favored material for the fabrication of electrostatic capacitors due to their efficient energy storage, high dielectric strength, compactness, thermal stability, and cost effectiveness. However, their low energy density, efficiency, and operating voltage limit the application of polymer materials in practical devices [7–9]. The incorporation of various types of 2D filler materials within polymer matrices has been a conventional method for overcoming the low discharged energy density (U_d) and breakdown strength (E_{BD}) of the polymer matrices [9].

Several dielectric polymers, such as polyvinylidene fluoride (PVDF), PVDF-TrFE-CFE, poly(methyl methacrylate) (PMMA), and polyimides (PIs), are commonly used in the fabrication of nanocomposite matrices using various nanofillers, such as hexagonal boron nitride (h-BN), barium titanate nanosheet, Mica, titanium dioxide (TiO_2), aluminum oxide (Al_2O_3), molybdenum disulfide (MoS_2), graphene oxide (GO), reduced graphene oxide (r-GO), MXene, etc. [9–19]. Discharged energy density (U_d), efficiency (η), and breakdown strength (E_{BD}) are the three critical parameters that determine the figure of merit of a capacitor device. These parameters are intrinsically linked with the dielectric polarization at the interface of the polymer and nanofiller entities and how the interface properties change as a function of electric field, temperature, and filler content and configuration. Thus, having a deeper understanding is essential to exploring these composite materials for their potential applications.

The discharged energy density of a dielectric material can be obtained using the area under the electric displacement vs. electric field loop (D-E loop) during discharging. For a linear dielectric material, the maximum discharged energy density of a capacitor $U_{d\_max} = \frac{1}{2}\varepsilon_r\varepsilon_0 E_{BD}{}^2$, where ε_r is the dielectric constant of the material or polymer composites and E_{BD} is the maximum electric field that the capacitor can sustain before the leakage current increases. Thus, ε_r and E_{BD} are the two essential parameters in the optimization of U_d. However, simultaneously enhancing both U_d and E_{BD} is still a challenging task, as E_{BD} degrades with the increasing behavior of ε_r [20]. A concurrent improvement in dielectric properties along with ultra-high energy density was observed with the incorporation of Mica in a high-dielectric-constant PVDF polymer [10]. Thin-layer Mica has a high band gap [3–4 eV] and, along with its two-dimensional structure, offers a significant surface area for interaction with the polymer matrix, resulting in improved mechanical strength, thermal stability, and electrical insulation properties in nanocomposites [10]. Furthermore, the fabrication of multilayer polymer nanocomposites originates interfacial polarization, which contributes significantly to an increase in the value of U_d and E_{BD} [21–24]. This work encompasses both approaches: integrating 2D Mica into the PMMA polymer and fabricating a Mica–PMMA multilayer nanocomposite device using a layer-by-layer method where 2D fillers are aligned in a stratified geometry within the thin film of polymer layers. A significant enhancement in the dielectric constant of up to 55% was observed in the PMPMP devices. Moreover, an enhancement of $U_d \sim 37.5$ J/cm^3 at 825 MV/m with higher $\eta \sim 93\%$ and $E_{BD} \sim 865$ MV/m was estimated in a multilayered stack of polymer and a 2D filler device.

2. Results and Discussion

The optical images after exfoliation onto Si/SiO_2 display 2D Mica flakes of different thicknesses, sizes, and orientations in the 2D plane of the substrate are shown in Figure 1a,b. These flakes were mechanically exfoliated using a scotch-tape method and transferred onto the clean SiO_2 substrate. The distribution of the flakes is non-uniform across the substrate, which is generally observed in the mechanically exfoliated technique. Figure 1c,d depict the optical images of PMMA-coated Mica flakes on Si/SiO_2, confirming that the 2D flakes remain intact on the substrate after spin coating. In Figure 1e,f, the surface of the PMP capacitor reveals the integration of 2D flakes between two layers of polymer films, highlighting their incorporation. The PMP and PMPMP capacitors are shown schematically in Figure 2a,b respectively. The thickness of a single PMMA layer was measured as 200 nm using the AFM tapping mode, recording the cross section of the polymer after making a scratch on the film, as shown in Figure 2c. Figure 2d depicts the height profile of the film thickness measured across the scratch, showing an average of 200 nm thick. The total thickness of the PMP devices is 400 nm, while the PMPMP devices have a total thickness of 600 nm. We measured the dielectric constant of pure PMMA, PMP, and PMPMP capacitors to understand the role of nanofillers within the polymer matrix.

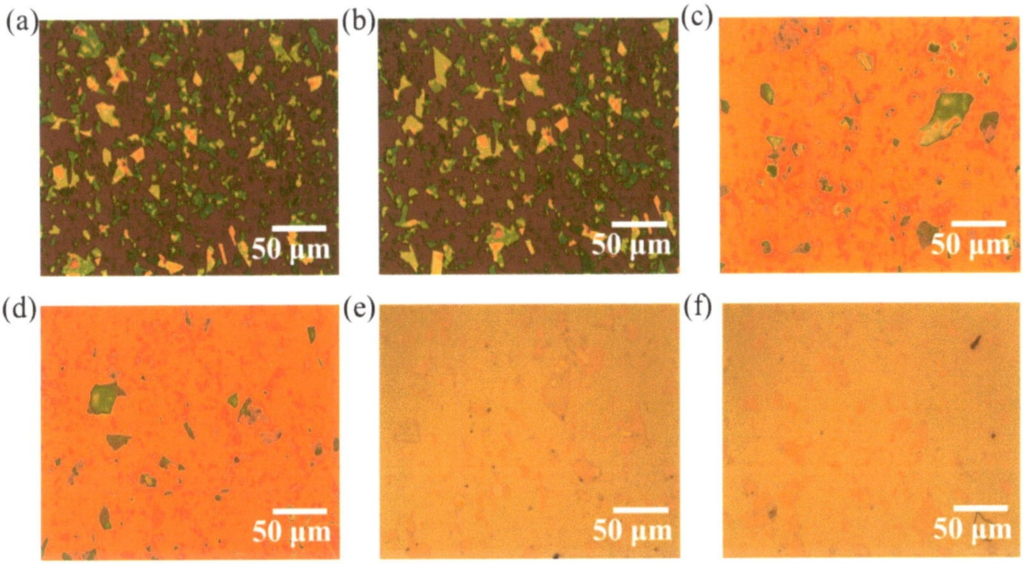

Figure 1. Optical images. (**a**,**b**) Distribution of 2D Mica sheets exfoliated onto Si/SiO_2, (**c**,**d**) PMMA spin coated on the Mica-exfoliated substrates, (**e**,**f**) PMP heterostructure after annealing at 120 °C for 3 h in vacuum.

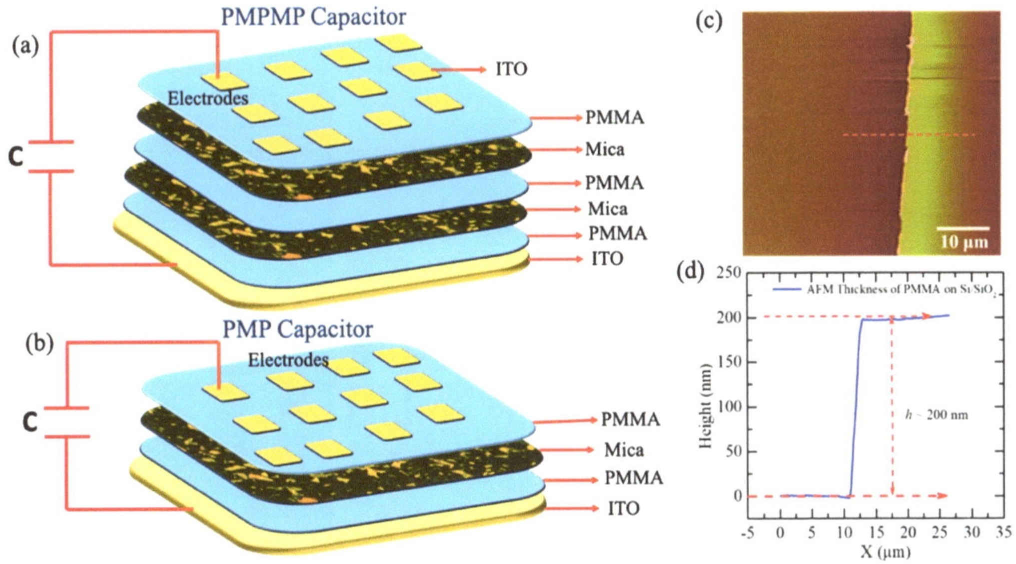

Figure 2. (**a**,**b**) Schematic of PMP and PMPMP heterostructure capacitors. PMP capacitor contained one layer of exfoliated Mica fillers and PMPMP capacitor contained two layers of Mica fillers. (**c**) AFM topography images of PMMA film showing cross-sectional mark. (**d**) Height variation in corresponding PMMA film perpendicular to cross section.

Figure 3a–c show the variation in the dielectric constant and loss tangent of the polymer nanocomposites between 30 and 100 °C, while the frequency varies from 100 Hz to 7 MHz. The dielectric constant gradually decreases with increasing frequency throughout the temperature range. As the frequency increases, the dielectric constant of polymer nanocomposites decreases due to the weakening of the polarization components. At higher frequencies, the dipoles have insufficient time to align with the direction of the electric field, resulting in a lower dielectric constant value [25]. The dielectric constant of the pure PMMA polymer varies between 4.5 at room temperature, 30 °C, and 5.1 at 100 °C. The PMP shows an enhancement of 15% (ε ~ 6.1) and 19% (ε ~ 6.5) at the respective temperatures 30 °C and 100 °C, while the PMPMP exhibits more enhancements of 54% (ε ~ 7.1) and 56% (ε ~ 7.5), shown in Figure 3d. The low enhancement of the PMP is probably due to the low loading of nanofillers compared to the PMPMP capacitors, where the two layers of Mica are stacked in heterostructure geometry. This enhancement is also lower than that of the PVDF/2D Mica heterostructure capacitor reported recently [10,26]. The polarization at the interface between PMMA, a low dielectric polymer, and Mica could be lower, which can be one of the interesting topics to explore using theoretical studies.

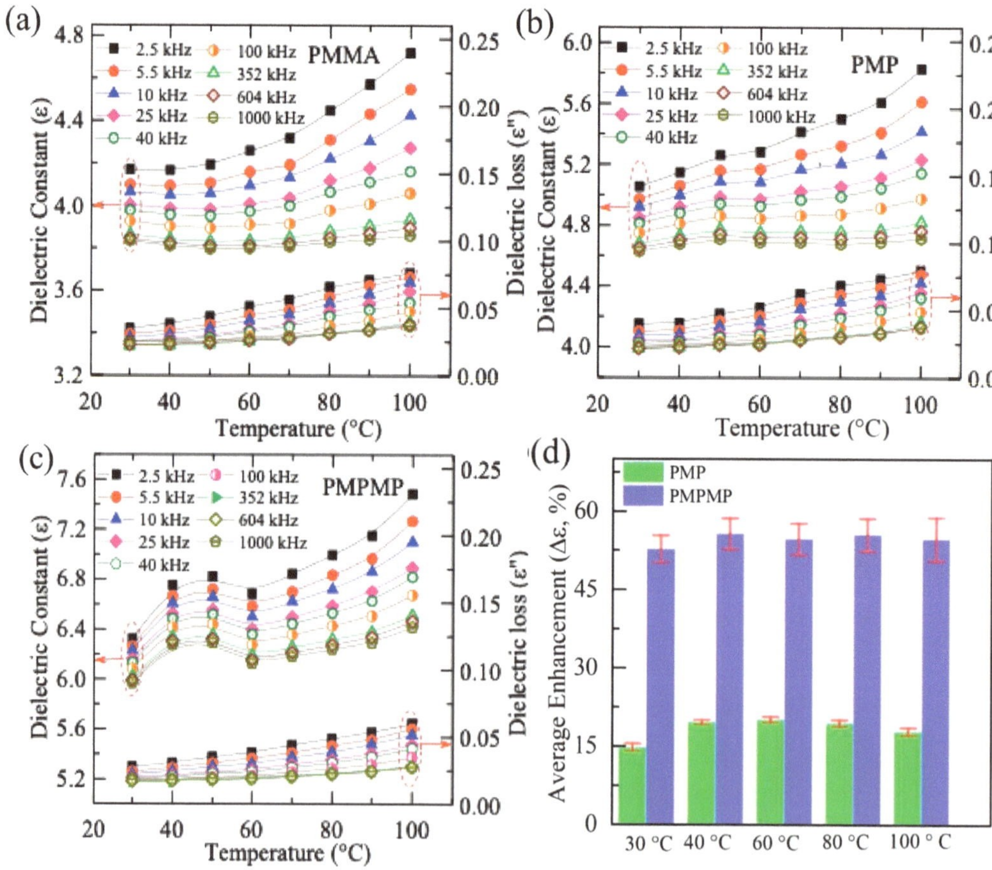

Figure 3. (**a**) Temperature-dependent dielectric constant and dielectric loss of PMMA (**a**), PMP (**b**), and PMPMP (**c**) at different constant frequencies in the range between 2.5 kHz and 1 MHz. (**d**) Histogram showing the enhancement of the dielectric constant (Δε) of PMP and PMPMP with respect to pure PMMA at a constant frequency of 10 kHz and different temperatures.

The enhancement of the dielectric constant in PMP and PMPMP devices can also be attributed to the effect of interfacial polarization occurring at the polymer–polymer interface and the Mica–polymer interface of the multilayer nanocomposite devices [21–24]. The phenomenon of dielectric constant enhancement can be analyzed using series capacitance models of two dissimilar dielectric materials of different thicknesses and dielectric constants. The dielectric constant (ε) of two different dielectric materials (here, PMMA and Mica) of dielectric constants of ε_1, ε_2 and thicknesses of d_1, d_2 with heterojunction can be obtained by the series capacitor model, $\varepsilon = \frac{\varepsilon_1\, \varepsilon_2(d_1+d_2)}{(d_2\varepsilon_1+d_1\varepsilon_2)}$. The average thickness of the 2D Mica flakes is 20 nm, as reported in our previous work [10]. Taking $\varepsilon_1 \sim 4.5$ (dielectric constant of PMMA), $\varepsilon_2 \sim 7$ (dielectric constant of Mica), and $d_1 = 200$ nm, $d_2 = 20$ nm, we obtain $\varepsilon \sim 4.65$, which is higher than the PMMA [27]. Hence, the presence of multiple layers of different dielectric materials justifies the enhancement of the overall dielectric constant in the PMPMP bilayer heterostructure.

At room temperature, the dielectric losses of PMMA, PMP, and PMPMP at 1 kHz are 0.035, 0.038, and 0.028, respectively. Hence, the integration of one layer of Mica in the PMP device does not significantly impact the dielectric loss of the polymer nanocomposite, while the addition of a second layer of PMMA–Mica reduces the loss factor in the PMPMP heterostructure, as shown in Figure 3a–c.

In general, there are two types of losses that play a role in a dielectric material: (a) conduction loss, which arises from the collision, trapping, and recombination of free charges in charged capacitors; and (b) polarization loss, which occurs due to the displacement of bound charges during polarization [28]. The presence of an additional interface and insulating Mica layer in the PMPMP device results in a drop in dielectric loss [24]. The presence of multiple interfaces deviates the electric trees along the interface plane, while the insulating Mica weakens the propagation of the electric tree, restricting the motion of the charge carrier [29–33]. The enhanced dielectric constant and reduced loss factor in the PMP and PMPMP devices point toward improved performance.

Energy density is a key factor in energy storage devices [10,26,34]. We measured the polarization vs. applied electric field to elucidate the energy storage property of the capacitors. In contrast, Figure 4a shows the schematic of the typical P-E loop of a ferroelectric materials capacitor with hysteresis loss (indicated by the pink color), discharge, and charge energy densities. One would expect low hysteresis loss in non-ferroelectric polymers compared to the ferroelectric materials. Figure 4b shows the electric field variation in the electric displacement (D) of PMMA, PMP, and PMPMP heterostructure capacitors at room temperature. The variation in D with the electric field shows a trend of higher polarization and high field sustainability in the nanocomposites with the addition of a 2D Mica filler. This enhancement in field sustainability is also a phenomenon observed in 2D material-based gas sensors [35,36], where the incorporation of layered nanomaterials improves both sensitivity and stability. The energy densities in the charging and discharging cycles are referred to as U_e and U_d, which are basically the areas under the charging and discharging cycles in the D-E loop graph, as shown in the schematic in Figure 4a. The discharged energy density can be calculated as $U_d = \int_{D_r}^{D_{max}} D.dE$, and the area of the D-E loop refers to the hysteresis loss (U_l). The charged energy density $U_e = U_d + U_l$. The ratio of the energy densities at a particular electric field is referred to as efficiency, $\eta\ (\%) = \frac{U_d}{U_e} \times 100$. In optimizing a D-E loop for ultra-high U_d and η, it is necessary to boost the maximum electric displacement (D_{max}) and minimize the remnant electric displacement (D_r) while maintaining a robust electric field sustainability. The PMMA exhibits a low value for both D_r and D_{max}, and high electric field sustainability varies between 400 and 750 MV/m. The capacitors also show minimum hysteresis loss. These combinations result in a moderate U_d of 8 to 15 J/cm^3, while efficiency remains relatively high up to 80% [33,37–40]. Hence, when obtaining a higher D_{max} to elevate the U_d, it is essential to improve the polarization of the polymer. Our capacitor shows the enhanced polarization by introducing the 2D nanofillers between the polymer films. It is observed that the remnant electric displacement (D_r) of the PMMA and PMP capacitor devices is close to ~ 0.5 μC/cm^2, while it is reduced to 0.3 μC/cm^2 for the PMPMP device. The maximum electric displacement (D_{max}) increases significantly with the loading of Mica layers between the polymer layers, as shown in Figure 4c. The $D_{max} - D_r$ increases with the addition of

Mica layers from 5 C/cm^2 in the PMP to 9 C/cm^2 in the PMPMP. This implies a reducing loss feature from PMP to PMPMP capacitors, corroborating the dielectric loss versus temperature variation observed earlier. Figure 4d shows the dynamics of U_d and η of the PMMA, PMP, and PMPMP nanocomposites with the variation in electric field (E). The discharged energy density of pure PMMA was calculated as U_d ~ 9.5 J/cm^3 at 522 MV/m, with η ~ 80.5%. The U_d and η at the same electric field for the PMP capacitor are 9.6 J/cm^3 and 90%, and for the PMPMP they are 14 J/cm^3 and 95%, respectively. The energy density of pure PMMA is similar to the reported values in the literature [39].The highest energy density and efficiency of the PMP device were calculated as U_d ~ 17.7 J/cm^3 and η ~ 81.2% at 720 MV/m. The addition of a second Mica–PMMA layer to the PMP capacitor enhances the energy density and efficiency to 37.35 J/cm^3 and 93% at the highest measured electric field of 825 MV/m, as shown in Figure 4d. Polymer-based dielectrics are widely used as insulation materials for high-voltage applications due to their intrinsic high breakdown strength. Beyond that, these materials, with suitable filler combinations, can possess high charge accumulation but exhibit high resistance to charge percolation by providing ordered traps and scattering centers to mobile charges and by increasing path tortuosity for carrier treeing propagation. We measured the breakdown voltages of the capacitors and used a Weibull probability distribution $P(E) = 1 - exp^{-(\frac{E}{E_{BD}})^{\beta}}$ fitting to obtain the breakdown fields (E_{BD}) of these polymer nanocomposite film capacitors, where P is the cumulative probability of failure, E is the electric field at dielectric failure, E_{BD} is the breakdown strength, which is the dielectric strength at 63.2% probability of failure, and β is a shape parameter. The value of β in a Weibull fit indicates the shape or randomness of data points in the distribution. Having a larger β implies a narrow distribution and higher consistency of breakdown strength of the capacitor. A total of 10 data points were collected from various capacitors to fit the Weibull probability distribution and determine the exact breakdown strength. The PMMA shows high breakdown strength compared to other polymer matrices, but the low polarization limits the discharged energy density (U_d) [16,33,38,41]. Zhang et al. reported a maximum U_d ~ 14.6 J/cm^3 with η ~ 73.47% at E ~ 809 MV/m for the PMMA film [4]. In our study, it is observed that the E_{BD} of the PMMA, PMP, and PMPMP is 535, 672, and 866 MV/m, respectively, as shown in Figure 4e.

Studies indicate that multilayered polymer films consistently exhibit higher electrical breakdown strength compared to their individual-film counterparts [42,43]. There are three important possible mechanisms working behind the ultra-high breakdown field of the PMPMP device: (a) the presence of multiple PMMA and 2D filler layers, (b) the addition of coverage of 2D Mica in a stratified geometry where two-dimensional surfaces of Mica are perpendicular to the applied electric field, and (c) the large surface area interaction between 2D fillers and polymers (interface polarization) between the 2D Mica in the nanocomposite.

During the electrical treeing process in a multilayer dielectric nanocomposite subjected to high electric fields, the propagation of electrical trees diverges at multiple interfaces along the in-plane direction [44,45]. Furthermore, the multilayer configuration restricts the free movement of carriers within the films, which in turn reduces the leakage current and enhances the breakdown strength [37,46–49]. The addition of second layers of Mica–PMMA (MP) onto the PMP device covers more space with 2D fillers by increasing the fillers on the surface while looking perpendicular to the film or along the direction of the applied electric field, resulting in enhanced breakdown strength of the PMPMP device. Therefore, the bilayer 2D Mica filler device (or PMPMP device) is resistant to high electric fields, exhibiting elevated E_{BD} and thereby enhancing U_d. The inclusion of a second layer of 2D Mica also increases the interface between polymer and 2D Mica, leading to the enhancement of the film [10,26].

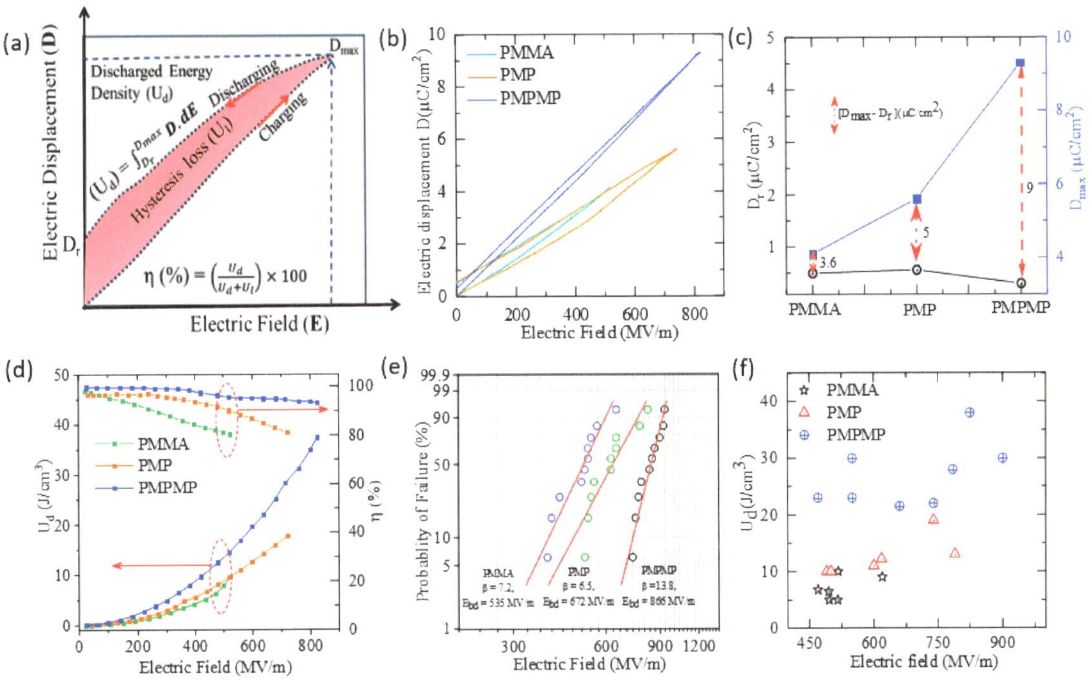

Figure 4. (**a**) Schematic diagram outlining the charging and discharging cycles of a D-E loop, including D_r, D_{max}, and discharged energy density (U_d), (**b**) field-dependent variation in electric displacement (D) of PMMA, PMP, and PMPMP, (**c**) remnant electric displacement (D_r) and maximum electric displacement (D_{max}) of the polymer and polymer nanocomposites, (**d**) variation in discharged energy density (U_d) and efficiency (η) with electric field, (**e**) Weibull probability fit showing the calculated breakdown strength (E_{BD}) and shape parameter (β), (**f**) maximum discharged energy density (U_{dmax}) of different capacitors of the polymer and polymer nanocomposites with electric field.

Our observations showed that 2D fillers oriented perpendicular to the direction of the electric field are more effective at preventing the expansion of electric breakdown paths through the nanocomposite, in contrast to randomly dispersed nanofillers [44,50]. The introduction of in-plane-oriented Mica nanofillers into the PMMA via mechanical exfoliation methods has contributed to improving the breakdown strength of the composite.

The calculated shape parameter (β) for the PMMA, PMP, and PMPMP is 7.2, 6.5, and 13.7, respectively, as shown in Figure 4e. The increase in energy density in multilayer polymer nanocomposites can be attributed to three major factors. (i) Increased dielectric constant: When these 2D nanofillers are integrated within the polymer layers, they can significantly increase the overall dielectric constant of the composite, as discussed earlier in the dielectric constant measurement. The maximum stored energy density (U) in a capacitor is proportional to the dielectric constant (ε) of the material. Therefore, a higher dielectric constant leads to a higher energy density at the same electric field strength. (ii) Improved polarization: The types of polarization that can occur in a polymer nanocomposite include electronic polarization, interfacial polarization, dipolar polarization, and ionic polarization [51]. Due to the substantial difference in dielectric constants between the PMMA and Mica, Maxwell–Wagner–Sillars interfacial polarization can develop at their interface. This phenomenon leads to an additional enhancement of interfacial polarization, and the overall polarization increases in the polymer nanocomposite, leading to an enhancement of discharged energy density. (iii) Improvement of breakdown strength: Multilayer polymer structures incorporated with insulating 2D filler render the electrical treeing process more effective and reduce the leakage current. This leads to an enhancement in breakdown strength and, hence, the energy density, of the nanocomposites.

3. Materials and Methods

In this study, 2D Mica flakes of varying thicknesses (single atomic layers to 30 nm thick) and sizes (1 m to 30 m) were exfoliated onto clean Si/SiO_2. The thin layer of PMMA was spin coated on top of exfoliated Mica at 3000 rpm for 30 s followed by a short time annealing at 100 °C for 5 min. The PMMA-coated 2D Mica layer was etched out using 30 wt % KOH solutions and transferred onto a PMMA-coated ITO wafer to form a PMMA/2D Mica/PMMA (PMP) capacitor. We subsequently transferred a second 2D and PMMA layer using the same procedure onto the 1st PMMA/2D Mica/PMMA capacitor to form a PMMA/2D Mica/PMMA/2D Mica/PPMA (PMPMP) capacitor. Finally, the capacitor devices were annealed at 120 °C inside a vacuum chamber for 2 h. The distribution of the exfoliated 2D Mica flakes was visualized through an optical microscope. The thicknesses of the polymer and the 2D polymer nanocomposite were measured by a Veeco Dimension 3100 atomic force microscope. The temperature-dependent dielectric properties of 2D-nanocomposite films were measured by a HIOKI LCR tester (3522) at frequencies from 1 KHz to 7 MHz. The polarization and breakdown strength measurements were performed using a Radiant ferroelectric precession II instrument with a high-voltage power source (10 kV). Silver electrodes, with diameters ranging between 200 and 300 μm, were placed on top of the devices to serve as the top electrode (and capacitor area) in the experiments. We placed several devices throughout the film to discern the uniformity of the fillers' film quality as well as their effects on dielectric and energy storage properties.

4. Conclusions

This study delved into the dielectric properties and electric polarization of PMMA/Mica nanocomposite fabricating multilayer capacitor devices. The incorporation of a 2D Mica nanofiller led to a remarkable increase in the dielectric constant, with improvements of approximately 15% and 54% observed for PMP and PMPMP structures, respectively, at room temperature. The pursuit of high discharged energy density and efficiency at a high operating electric field was witnessed in the multilayer nanocomposite PMP and PMPMP capacitor devices. Notably, the discharged energy density in the PMPMP reached approximately 38 J/cm^3 at 825 MV/m, surpassing previously reported values for PMMA-based composite capacitors. The presence of interfacial polarization and insulating Mica filler within the polymer contributes to enhanced efficiency and increased breakdown strength. The enhancement of the dielectric constant with the dimensions of loss factor in the PMPMP device also justifies the high energy density and efficiency of the device. The exceptional capacitive and energy storage performance exhibited by these materials opens new avenues for advancing energy storage technologies, paving the way for more efficient and reliable energy storage solutions to meet the growing demands of modern society.

Author Contributions: Conceptualization, N.R.P.; Formal analysis, S.B.; Data curation, S.B.; Writing – original draft, S.B.; Writing – review & editing, R.T., S.A.K., A.V.S. and N.R.P.; Supervision, N.R.P. All authors have read and agreed to the published version of the manuscript.

Funding: This work was supported by the Princeton Alliance for Collaborative Research and Innovation (PACRI) Grant# PACRI-JSU-02. N.R.P. acknowledges NSF HBCU-UP Excellence in Research NSF-DMR-1900692. The authors acknowledge that the use of the Center for Nanoscale Materials, a Department of Energy Office of Science User Facility, was supported by the U.S. DOE, Office of Basic Energy Sciences, under Contract No. DE-AC02-06CH11357. N.R.P. and S.A.K. acknowledged the funding provided by the U.S. Department of Energy, Office of Science, Office of Basic Energy Sciences program under Award Number DE-SC0024072.

Institutional Review Board Statement: Not applicable.

Informed Consent Statement: Not applicable.

Data Availability Statement: The original contributions presented in this study are included in the article; further inquiries can be directed to the corresponding author.

Conflicts of Interest: The authors declare no conflicts of interest.

References

1. Kebede, A.A.; Kalogiannis, T.; Van Mierlo, J.; Berecibar, M. A comprehensive review of stationary energy storage devices for large scale renewable energy sources grid integration. *Renew. Sustain. Energy Rev.* **2022**, *159*, 112213. [CrossRef]
2. Kadam, S.A.; Kadam, K.P.; Pradhan, N.R. Pradhan, Advancements in 2D MXene-based supercapacitor electrodes: Synthesis, mechanisms, electronic structure engineering, flexible wearable energy storage for real-world applications, and future prospects. *J. Mater. Chem. A* **2024**, *12*, 17992–18046. [CrossRef]
3. Kadam, S.A.; Ma, Y.-R.; Chen, Y.-R.; Navale, Y.H.; Salunkhe, A.S.; Patil, V.B.; Ralegankar, S.D.; More, P.D. Mn-Incorporated α-Fe_2O_3 Nanostructured Thin Films: Facile Synthesis and Application as a High-Performance Supercapacitor. *J. Electron. Mater.* **2023**, *52*, 500–513. [CrossRef]
4. Zhang, M.; Du, H.; Wei, Z.; Zhang, X.; Wang, R. Ultrafast Microwave Synthesis of Nickel-Cobalt Sulfide/Graphene Hybrid Electrodes for High-Performance Asymmetrical Supercapacitors. *ACS Appl. Energy Mater.* **2021**, *4*, 8262–8274. [CrossRef]
5. Zhang, M.; Nautiyal, A.; Du, H.; Wei, Z.; Zhang, X.; Wang, R. Electropolymerization of polyaniline as high-performance binder free electrodes for flexible supercapacitor. *Electrochim. Acta* **2021**, *376*, 138037. [CrossRef]
6. Feng, Q.-K.; Zhong, S.-L.; Pei, J.-Y.; Zhao, Y.; Zhang, D.-L.; Liu, D.-F.; Zhang, Y.-X.; Dang, Z.-M. Recent Progress and Future Prospects on All-Organic Polymer Dielectrics for Energy Storage Capacitors. *Chem. Rev.* **2022**, *122*, 3820–3878. [CrossRef]
7. Yang, C.; Wei, H.; Guan, L.; Guo, J.; Wang, Y.; Yan, X.; Zhang, X.; Wei, S.; Guo, Z. Polymer nanocomposites for energy storage, energy saving, and anticorrosion. *J. Mater. Chem. A* **2015**, *3*, 14929–14941. [CrossRef]
8. Arunachalam, P. *6-Polymer-based nanocomposites for energy and environmental applications, In Polymer-Based Nanocomposites for Energy and Environmental Applications*; Jawaid, M., Khan, M.M., Eds.; Woodhead Publishing: Sawston, UK, 2018; pp. 185–203.
9. Bera, S.; Singh, M.; Thantirige, R.; Tiwary, S.K.; Shook, B.T.; Nieves, E.; Raghavan, D.; Karim, A.; Pradhan, N.R. 2D-Nanofiller-Based Polymer Nanocomposites for Capacitive Energy Storage Applications. *Small Sci.* **2023**, *3*, 2300016. [CrossRef]
10. Singh, M.; Das, P.; Samanta, P.N.; Bera, S.; Thantirige, R.; Shook, B.; Nejat, R.; Behera, B.; Zhang, Q.; Dai, Q.; et al. Ultrahigh Capacitive Energy Density in Stratified 2D Nanofiller-Based Polymer Dielectric Films. *ACS Nano* **2023**, *17*, 20262–20272. [CrossRef]
11. Zhu, Y.; Zhu, Y.; Huang, X.; Chen, J.; Li, Q.; He, J.; Jiang, P. High Energy Density Polymer Dielectrics Interlayered by Assembled Boron Nitride Nanosheets. *Adv. Energy Mater.* **2019**, *9*, 1901826. [CrossRef]
12. Li, H.; Ren, L.; Zhou, Y.; Yao, B.; Wang, Q. Recent progress in polymer dielectrics containing boron nitride nanosheets for high energy density capacitors. *High Volt.* **2020**, *5*, 365–376. [CrossRef]
13. Liu, F.; Li, Q.; Li, Z.; Dong, L.; Xiong, C.; Wang, Q. Ternary PVDF-based terpolymer nanocomposites with enhanced energy density and high power density. *Compos. Part A Appl. Sci. Manuf.* **2018**, *109*, 597–603. [CrossRef]
14. Likhi, F.H.; Singh, M.; Chavan, S.V.; Cao, T.; Shanbedi, M.; Karim, A. Effects of Film Confinement on Dielectric and Electrical Properties of Graphene Oxide and Reduced Graphene Oxide-Based Polymer Nanocomposites: Implications for Energy Storage. *ACS Appl. Nano Mater.* **2023**, *6*, 11699–11714. [CrossRef]
15. Ji, M.; Min, D.; Li, Y.; Yang, L.; Wu, Q.; Liu, W.; Li, S. Improved energy storage performance of polyimide nanocomposites by constructing the meso- and macroscopic interfaces. *Mater. Today Energy* **2023**, *31*, 101200. [CrossRef]
16. Xie, B.; Wang, Q.; Zhang, Q.; Liu, Z.; Lu, J.; Zhang, H.; Jiang, S. High Energy Storage Performance of PMMA Nanocomposites Utilizing Hierarchically Structured Nanowires Based on Interface Engineering. *ACS Appl. Mater. Interfaces* **2021**, *13*, 27382–27391. [CrossRef] [PubMed]
17. Yao, L.; Pan, Z.; Liu, S.; Zhai, J.; Chen, H.H.D. Significantly Enhanced Energy Density in Nanocomposite Capacitors Combining the TiO_2 Nanorod Array with Poly(vinylidene fluoride). *ACS Appl. Mater. Interfaces* **2016**, *8*, 26343–26351. [CrossRef]
18. Pan, Z.; Yao, L.; Zhai, J.; Shen, B.; Liu, S.; Wang, H.; Liu, J. Excellent energy density of polymer nanocomposites containing $BaTiO_3@Al_2O_3$ nanofibers induced by moderate interfacial area. *J. Mater. Chem. A* **2016**, *4*, 13259–13264. [CrossRef]
19. Jia, Q.; Huang, X.; Wang, G.; Diao, J.; Jiang, P. MoS_2 Nanosheet Superstructures Based Polymer Composites for High-Dielectric and Electrical Energy Storage Applications. *J. Phys. Chem. C* **2016**, *120*, 10206–10214. [CrossRef]
20. McPherson, J.; Kim, J.; Shanware, A.; Mogul, H.; Rodriguez, J. Proposed universal relationship between dielectric breakdown and dielectric constant. In *Digest. International Electron Devices Meeting*; IEEE: San Francisco, CA, USA, 2002; pp. 633–636.
21. Peng, S.; Du, X.; Liang, Z.; Ma, M.; Guo, Y.; Xiong, L. Multilayer polymer nanocomposites based on interface engineering for high-performance capacitors. *J. Energy Storage* **2023**, *60*, 106636. [CrossRef]
22. Chen, X.; Tseng, J.-K.; Treufeld, I.; Mackey, M.; Schuele, D.E.; Li, R.; Fukuto, M.; Baer, E.; Zhu, L. Enhanced dielectric properties due to space charge-induced interfacial polarization in multilayer polymer films. *J. Mater. Chem. C* **2017**, *5*, 10417–10426. [CrossRef]
23. Rahimabady, M.; Lu, L.; Yao, K. Nanocomposite multilayer capacitors comprising $BaTiO_3@TiO_2$ and poly(vinylidene fluoride-hexafluoropropylene) for dielectric-based energy storage. *J. Adv. Dielectr.* **2014**, *4*, 1450009. [CrossRef]
24. Li, W.; Song, Z.; Zhong, J.; Qian, J.; Tan, Z.; Wu, X.; Chu, H.; Nie, W.; Ran, X. Multilayer-structured transparent MXene/PVDF film with excellent dielectric and energy storage performance. *J. Mater. Chem. C* **2019**, *7*, 10371–10378. [CrossRef]

25. Zulkifli, A. Polymer Dielectric Materials. In *Dielectric Material*; Marius Alexandru, S., Ed.; IntechOpen: Rijeka, Croatia, 201[illegible] p. 314.
26. Bera, S.; Thantirige, R.M.; Wu, J.; Davidson, E.C.; Kadam, S.A.; Sumant, A.V.; Shook, B.T.; Rao, R.; Selhorst, R.; Singh, M.; et a[illegible] Enhanced Energy Density in a Heterostructure Capacitor of Multilayered PVDF and 2D Mica Nanocomposites. *ACS App[illegible] Electron. Mater.* **2024**, *6*, 6582–6590. [CrossRef]
27. Weeks, J.J.R. The Dielectric Constant of Mica. *Phys. Rev.* **1922**, *19*, 319–322. [CrossRef]
28. Pei, J.; Yin, L.; Zhong, S.; Dang, Z. Suppressing the Loss of Polymer-Based Dielectrics for High Power Energy Storage. *Adv. Mate[illegible]* **2022**, *35*, 2203623. [CrossRef]
29. Chen, J.; Wang, Y.; Yuan, Q.; Xu, X.; Niu, Y.; Wang, Q.; Wang, H. Multilayered ferroelectric polymer films incorporating low dielectric-constant components for concurrent enhancement of energy density and charge–discharge efficiency. *Nano Energy* **201[illegible]** *54*, 288–296. [CrossRef]
30. Wang, Y.; Chen, J.; Li, Y.; Niu, Y.; Wang, Q.; Wang, H. Multilayered hierarchical polymer composites for high energy densit[illegible] capacitors. *J. Mater. Chem. A* **2019**, *7*, 2965–2980. [CrossRef]
31. Bai, H.; Zhu, K.; Wang, Z.; Shen, B.; Zhai, J. 2D Fillers Highly Boost the Discharge Energy Density of Polymer-Based Nanocom posites with Trilayered Architecture. *Adv. Funct. Mater.* **2021**, *31*, 2102646. [CrossRef]
32. Liu, Y.; Luo, H.; Xie, H.; Xiao, Z.; Wang, F.; Jiang, X.; Zhou, X.; Zhang, D. Trilayer PVDF nanocomposites with significantl[illegible] enhanced energy density and energy efficiency using $0.55Bi_{0.5}Na_{0.5}TiO_3$-$0.45(Sr_{0.7}Bi_{0.2})$ TiO_3 nanofibers. *Microstructures* **2023**, [illegible] 2023008.
33. Li, Z.; Shen, Z.; Yang, X.; Zhu, X.; Zhou, Y.; Dong, L.; Xiong, C.; Wang, Q. Ultrahigh charge-discharge efficiency and enhance[illegible] energy density of the sandwiched polymer nanocomposites with poly(methyl methacrylate) layer. *Compos. Sci. Technol.* **2020**, *20[illegible]* 108591. [CrossRef]
34. More, P.; Kadam, S.A.; Ma, Y.; Chen, Y.; Tarwal, N.; Navale, Y.; Salunkhe, A.; Patil, V. Spray Synthesized Mn-doped CuC[illegible] Electrodes for High Performance Supercapacitor. *ChemistrySelect* **2022**, *7*, e202202504. [CrossRef]
35. Hingangavkar, G.M.; Kadam, S.A.; Ma, Y.-R.; Sartale, S.D.; Mulik, R.N.; Patil, V.B. Intercalation of two-dimensional graphen[illegible] oxide in WO_3 nanoflowers for NO_2 sensing. *Nano-Struct. Nano-Objects* **2023**, *34*, 100964. [CrossRef]
36. Hingangavkar, G.M.; Kadam, S.A.; Ma, Y.-R.; Bandgar, S.S.; Mulik, R.N.; Patil, V.B. Tailored formation of WO_3-rGO nanohybrid[illegible] for dependable low temperature NO_2 sensing. *Ceram. Int.* **2023**, *49*, 38866–38876. [CrossRef]
37. Zhang, T.; Dan, Z.; Shen, Z.; Jiang, J.; Guo, M.; Chen, B.; Lin, Y.; Nan, C.-W.; Shen, Y. An alternating multilayer architectur[illegible] boosts ultrahigh energy density and high discharge efficiency in polymer composites. *RSC Adv.* **2020**, *10*, 5886–5893. [CrossRef] [PubMed]
38. Zhang, C.; Tong, X.; Liu, Z.; Zhang, Y.; Zhang, T.; Tang, C.; Liu, X.; Chi, Q. Enhancement of Energy Storage Performance o[illegible] PMMA/PVDF Composites by Changing the Crystalline Phase through Heat Treatment. *Polymers* **2023**, *15*, 2486. [CrossRef] [PubMed]
39. Wen, F.; Zhu, C.; Lv, W.; Wang, P.; Zhang, L.; Li, L.; Wang, G.; Wu, W.; Ying, Z.; Zheng, X.; et al. Improving the Energy Density and Efficiency of the Linear Polymer PMMA with a Double-Bond Fluoropolymer at Elevated Temperatures. *ACS Omega* **2021**, *6* 35014–35022. [CrossRef]
40. Liu, Y.; Liu, Z.; Gao, J.; Wu, M.; Lou, X.; Hu, Y.; Li, Y.; Zhong, L. High Energy Density and Temperature Stability in PVDF/PMMA via In Situ Polymerization Blending. *Front. Chem.* **2022**, *10*, 902487. [CrossRef]
41. Wu, Q.; Feng, Z.; Cai, Z.; Lan, C.; Xu, J.; Bi, K.; Hao, Y. Poly(methyl methacrylate)-based ferroelectric/dielectric laminated films with enhanced energy storage performances. *Adv. Compos. Hybrid Mater.* **2022**, *5*, 1137–1144. [CrossRef]
42. Baer, E.; Zhu, L. 50th Anniversary Perspective: Dielectric Phenomena in Polymers and Multilayered Dielectric Films. *Macromolecules* **2017**, *50*, 2239–2256. [CrossRef]
43. Mackey, M.; Hiltner, A.; Baer, E.; Flandin, L.; Wolak, A.; Shirk, M.; Shirk, J.S. Enhanced breakdown strength of multilayered films fabricated by forced assembly microlayer coextrusion. *J. Phys. D Appl. Phys.* **2009**, *42*, 175304. [CrossRef]
44. Jiang, J.; Shen, Z.; Qian, J.; Dan, Z.; Guo, M.; He, Y.; Lin, Y.; Nan, C.-W.; Chen, L.; Shen, Y. Synergy of micro-/mesoscopic interfaces in multilayered polymer nanocomposites induces ultrahigh energy density for capacitive energy storage. *Nano Energy* **2019**, *62* 220–229. [CrossRef]
45. Xie, B.; Zhang, Q.; Zhang, L.; Zhu, Y.; Guo, X.; Fan, P.; Zhang, H. Ultrahigh discharged energy density in polymer nanocomposites by designing linear/ferroelectric bilayer heterostructure. *Nano Energy* **2018**, *54*, 437–446. [CrossRef]
46. Zhang, Y.; Yang, H.; Dang, Z.; Zhan, S.; Sun, C.; Hu, G.; Lin, Y.; Yuan, Q. Multilayer Structured Poly(vinylidene fluoride)-Based Composite Film with Ultrahigh Breakdown Strength and Discharged Energy Density. *ACS Appl. Mater. Interfaces* **2020**, *12* 22137–22145. [CrossRef] [PubMed]
47. Zhang, Y.; Liu, L.; Zhang, T.; Zhang, C.; Chen, Q.; Wang, X.; Lei, Q. PVDF-Based Dielectric Composite Films with Excellent Energy Storage Performances by Design of Nanofibers Composition Gradient Structure. *ACS Appl. Energy Mater.* **2018**, *1*, 6320–6329 [CrossRef]
48. Yang, X.; Zhu, X.; Ji, L.; Hu, P.; Li, Z. Largely enhanced energy storage performance in multilayered ferroelectric polymer nanocomposites with optimized spatial arrangement of ceramic nanofillers. *Compos. Part A Appl. Sci. Manuf.* **2020**, *139*, 106111 [CrossRef]

59. Zhou, L.; Zhao, S.; Xie, P.; Miao, X.; Liu, S.; Sun, N.; Guo, M.; Xu, Z.; Zhong, T.; Shen, Y. Research progress and prospect of polymer dielectrics. *Appl. Phys. Rev.* **2023**, *10*, 031310. [CrossRef]
60. Shen, Z.-H.; Wang, J.-J.; Lin, Y.; Nan, C.-W.; Chen, L.-Q.; Shen, Y. High-Throughput Phase-Field Design of High-Energy-Density Polymer Nanocomposites. *Adv. Mater.* **2018**, *30*, 1704380. [CrossRef]
61. Luo, H.; Zhou, X.; Ellingford, C.; Zhang, Y.; Chen, S.; Zhou, K.; Zhang, D.; Bowen, C.R.; Wan, C. Interface design for high energy density polymer nanocomposites. *Chem. Soc. Rev.* **2019**, *48*, 4424–4465. [CrossRef]

molecules

MDPI

Article

Comparative Investigation of Water-Based CMC and LA133 Binders for CuO Anodes in High-Performance Lithium-Ion Batteries

Nischal Oli [1,*], Sunny Choudhary [1], Brad R. Weiner [2], Gerardo Morell [1] and Ram S. Katiyar [1,*]

[1] Department of Physics, University of Puerto Rico-Rio Piedras Campus, San Juan, PR 00925, USA
[2] Department of Chemistry, University of Puerto Rico-Rio Piedras Campus, San Juan, PR 00925, USA
* Correspondence: nischal.oli@upr.edu (N.O.); ram.katiyar@upr.edu (R.S.K.)

Abstract: Transition metal oxides are considered to be highly promising anode materials for high-energy lithium-ion batteries. While carbon matrices have demonstrated effectiveness in enhancing the electrical conductivity and accommodating the volume expansion of transition metal oxide-based anode materials in lithium-ion batteries (LIBs), achieving an optimized utilization ratio remains a challenging obstacle. In this investigation, we have devised a straightforward synthesis approach to fabricate CuO nano powder integrated with carbon matrix. We found that with the use of a sodium carboxymethyl cellulose (CMC) based binder and fluoroethylene carbonate additives, this anode exhibits enhanced performance compared to acrylonitrile multi-copolymer binder (LA133) based electrodes. CuO@CMC electrodes reveal a notable capacity ~1100 mA h g^{-1} at 100 mA g^{-1} following 170 cycles, and exhibit prolonged cycling stability, with a capacity of 450 mA h g^{-1} at current density 300 mA g^{-1} over 500 cycles. Furthermore, they demonstrated outstanding rate performance and reduced charge transfer resistance. This study offers a viable approach for fabricating electrode materials for next-generation, high energy storage devices.

Keywords: CuO; sodium carboxymethyl cellulose; carbon matrix; lithium-ion batteries; electrochemical reactions

1. Introduction

In the incessant pursuit of greener and more highly efficient energy solutions, lithium-ion batteries (LIBs) have become an indispensable part of modern life and revolutionized energy storage technologies, powering portable electronics to electric vehicles (EVs) or hybrid electric vehicles (HEVs), due to their reasonable energy density (150–200 W h kg^{-1}) and exceptional cycling stability [1–5]. Given the escalating demands for long-range electric vehicles, it is imperative to engineer batteries with superhigh energy densities (350–500 Wh kg^{-1}) [6–10]. Lithium metal is one of the best selections for the anode material in next-generation lithium-air (3862 mA h g^{-1}) [11], lithium-sulfur (1675 mA h g^{-1}) [12], lithium metal batteries and solid-state lithium-based batteries, which have drawn worldwide attention within the battery community, owing to their high theoretical specific capacity of 3860 mA h g^{-1} and low potential of −3.04 V vs. the standard hydrogen electrode (SHE) [13]. Nevertheless, there are critical questions that need to be addressed before lithium metal can be commercially adopted as an anode on a widespread scale. For example, lithium metal is susceptible to dendrite formation, resulting in cell short-circuits and low Coulombic efficiency [14]. The persistent solid–solid interface issue in solid-state electrolytes poses a substantial challenge to contemporary battery manufacturing technology. Thus, the continued development of LIBs with enhanced energy densities remains an appealing option, until the full commercialization of lithium metal or solid-state batteries occurs.

The energy density of a battery primarily depends on the specific capacity of its electrode materials. Presently, existing commercial graphite stands as the dominant anode material for lithium-ion batteries (LIBs), owing to its low operational potential (~0.1 V vs. Li/Li^+), cost-effectiveness, and robust cycling stability [15,16]. However, graphite faces several drawbacks. Mainly, its modest specific capacity of 372 mA h g^{-1} limits the energy density of LIBs. Another drawback is that graphite's Li-insertion potential (~0.1 V vs. Li/Li^+) approximates that of Li metal plating (~0 V vs. Li/Li^+), which raises safety concerns, especially during fast charging and in low-temperature environments [15,17]. Nevertheless, these promising battery systems encounter inherent difficulties that are hard to overcome quickly. Hence, there is pressing demand to innovate high-performance anode materials, possessing a higher capacity, appropriate reaction potential, and environmentally sustainable materials [18,19].

Over the past two decades, there have been tremendous efforts using alloy-based materials in the search for suitable anode materials, notably silicon (Si) (4200 mA h g^{-1}) [20], tin (Sn) (993 mA h g^{-1}) [21], aluminum (Al) (993 mA h g^{-1}) [22], antimony (Sb) (600 mA h g^{-1}) [23], Germanium (Ge) (1624 mA h g^{-1}) [24], and Bismuth (Bi) (385 mA h g^{-1}). [25] Among these, silicon-based anode materials have gained global attention from research communities and manufacturers, owing to their potential for nearly tenfold greater specific (gravimetric) capacity compared to commercial graphite anodes (Si~4200 mA h g^{-1} vs. 372 mA h g^{-1} for graphite), but these silicon-based anode materials undergo a volumetric expansion of up to 300% during lithiation and delithiation processes [26,27]. However, mitigating the substantial capacity degradation of silicon electrodes poses a complex and multifaceted challenge. This capacity fading primarily stems from two distinct factors: primarily, silicon grain disintegration and fracture of the electrode integration, leading to a broken electrical contact with the active material's current collector, and an unstable solid–electrolyte interphase (SEI) that results in electrolyte deformation. Similarly, all other alloy materials often experience significant volume expansion (100–300%), leading to the formation of SEI cracks, thereby compromising cycling stability [28]. There have been extensive efforts made to tackle these issues. Furthermore, there are also several anode materials that undergo a conversion-alloy reaction, involving an initial conversion followed by an alloy reaction. Oxides with high theoretical capacities, namely tin oxide (SnO_2) (1490 mA h g^{-1}), silicon oxide (SiO_x), and bismuth ferrite ($BiFeO_3$) (770 mA h g^{-1}) are common materials undergoing a conversion-alloy reaction [29–31]. Research indicates that these anode materials demonstrate improved cycling performance compared to pure alloy materials, owing to the formation of lithium oxide (Li_2O), which effectively alleviates electrode volume changes [32].

On the other hand, conversion-based transition-metal (TM) oxide materials emerge as promising candidates for anode materials in lithium-ion batteries (LIBs). It is denoted as TM_xO_y (where TM = Ti, V, Cr, Mn, Fe, Co, Ni, Cu, Zn, Mo, etc.). These oxides present high reversible capacities by undergoing a redox mechanism with lithium (TM_xO_y + 2yLi = yLi_2O + xTM) [33–35]. These materials exhibit 2–3 time greater specific capacity than commercial graphite anode. This phenomenon serves as motivation to optimize compounds within the conversion-based material family, aiming to foster the development of high-energy, high-rate capability, and sustainable materials for LIBs [34,36].

Copper oxide (CuO) is an interesting material, with a monoclinic structure and semiconducting properties, but with poor electrical conductivity, which impedes efficient charge transfer [37]. Additionally, CuO-based electrodes suffer from substantial volume expansion and dispersion of copper elements within the Li_2O matrix during charge–discharge process, leading to significant mechanical stress and rapid deterioration in capacity [38]. Enhanced electrochemical performance of CuO-based anodes is crucial to meet the demands of next generation lithium-ion batteries (LIBs). Considerable efforts have been made to bolster the performance of CuO-based electrodes via morphology modification, designing nanostructured configuration and hybridization with composite materials, resulting in materials that exhibit promising electrochemical performance and distinctive structures [38–41]. This material distinguishes itself for its cost-effectiveness and environmentally benign nature. This material is attractive as a high-capacity anode for battery applications, due to its potential for a 2-electron conversion reaction mechanism, resulting in a high theoretical capacity of approximately 670 mA h g^{-1}, nearly doubling the capacity of graphite [39]. The average insertion potential in CuO is approximately ~1.32 V vs. Li^+/Li, higher than that of graphite but lower than that of $Li_4Ti_5O_{12}$ (~1.55 V vs. Li^+/Li), making it suitable for anodes [39,42]. Numerous publications have addressed the lithium-ion battery (LIB) performance of this material, albeit through methods that may be cost-prohibitive. Consequently, it becomes imperative to revisit this material and unveil its complete potential for LIB applications, utilizing commercially viable techniques [38–41,43–46].

In recent years, water-soluble binders have attracted substantial scientific interest as binders for lithium-ion batteries (LIBs), which are imperative for cost-efficient and environmentally sustainable electrode fabrication methodologies [47,48]. Conventionally, *N*-methyl-2-pyrrolidone (NMP) is utilized as a solvent to prepare slurries for cathode or anode materials, a conductive carbon additive, a traditional poly(vinylidene difluoride) (PVDF) binder, and NMP itself. Notably, NMP is classified as a carcinogenic compound with reproductive toxicity, necessitating stringent recycling protocols to mitigate atmospheric contamination [49,50]. Consequently, there is a pronounced shift towards adopting aqueous fabrication processes. Moreover, ongoing research endeavors are extensively investigating the deployment of water-soluble or aqueous binders across both anode and cathode configurations [48,51]. Polymer binders constitute a minor fraction (<5 wt%) of commercial lithium-ion battery (LIB) electrodes, functioning primarily as an adhesive to integrate active materials, conductive agents, and current collectors, thereby preserving both electrical conductivity and mechanical stability within the electrodes [52]. The most commonly employed polymer binders, such as poly(vinylidene difluoride) (PVDF), sodium carboxymethyl cellulose (CMC), and CMC/styrene-butadiene rubber (SBR) blends, have proven effective in graphite anodes due to graphite's limited volumetric expansion (~10%) upon full lithiation, enabling stable long-term cycling [53,54]. However, in the context of Si-based anodes, PVDF-based binder undergoes significant volume expansion, and the inadequate binding capacity of PVDF results in pronounced silicon particle pulverization, huge expansion, and poor cycling performance. Numerous studies indicate that the cycling stability of Si-based anodes is highly contingent on the choice of polymer binder [55–57]. For instance, Si and carbon-coated Si nanoparticle (SiNP) anodes demonstrate markedly improved cycling stability with CMC binders, compared to PVDF. Building on this understanding, our investigation focused on optimizing the performance of a CuO anode by incorporating a CMC binder, enhanced with fluoroethylene carbonate (FEC), as an additive with carbonate-based electrolyte. Comparative analysis was conducted against two binders—CMC and LA133—to evaluate their stabilizing effects on a CuO anode for LIBs.

The molecular structure of PVDF, CMC, and LA133 are depicted in Figure 1a–c. PVDF is characterized as a linear crystalline polymer with a repeating unit of $-CH_2-CF_2-$. The bonding mechanism of PVDF involves the formation of hydrogen bonds with electrode components, via the fluorine atoms along its extended polymer chains; Carboxymethyl CMC features a backbone comprised of D-glucose residues, interconnected through β-1,4-glycosidic linkages. LA133 is a water-soluble polybasic copolymer, primarily based on polyacrylonitrile. The substantial intermolecular forces in LA133 are attributed to the high polarity of the cyano (–CN) groups within its structure [58]. In contrast, this backbone is modified by carboxymethyl groups ($-CH_2-COOH$), attached to some hydroxyl groups of the glucopyranose units, endowing it with high viscosity and pronounced hydrophilicity.

In this study, we utilized high-purity, commercially available CuO compounds, and examined their electrochemical Li-ion insertion performance via a simple solid-state method. Through the optimization of the binder sodium carboxymethyl cellulose (CMC) and electrolyte additives (fluoroethylene carbonate, FEC), we observed a significant enhancement in battery performance. Specifically, the CuO anode demonstrated a remarkable capacity of ~800 mA h g^{-1}, an impressive rate capability ~1200 mA g^{-1}, and exceptional cycling stability with ~500 cycles (~99% retention). Our findings underscore the potential of CuO as a high-capacity and long-lasting anode material for lithium-ion batteries.

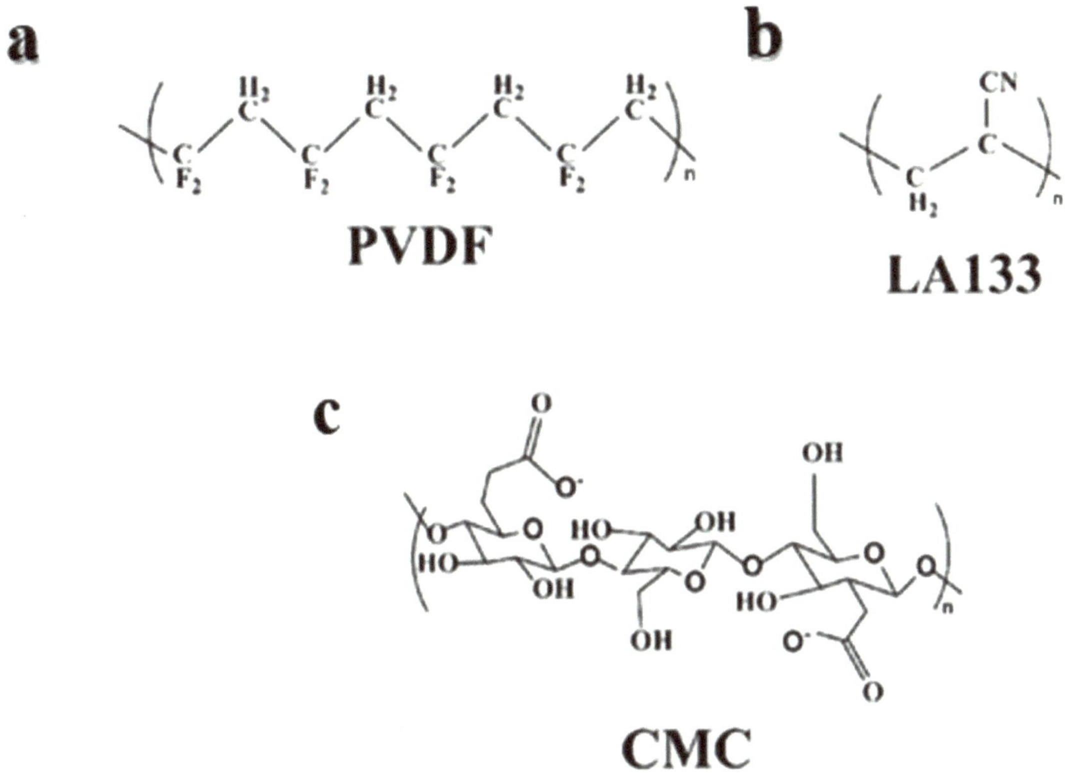

Figure 1. (**a**) PVDF. (**b**) LA133. (**c**) CMC.

2. Result and Discussion

The commercial CuO nano powder presents as a finely powdered material with a black color Figure 2a (inset). To ascertain its quality and crystalline structure, we conducted X-ray diffraction (XRD) analysis at a scan rate of 2° per minute.

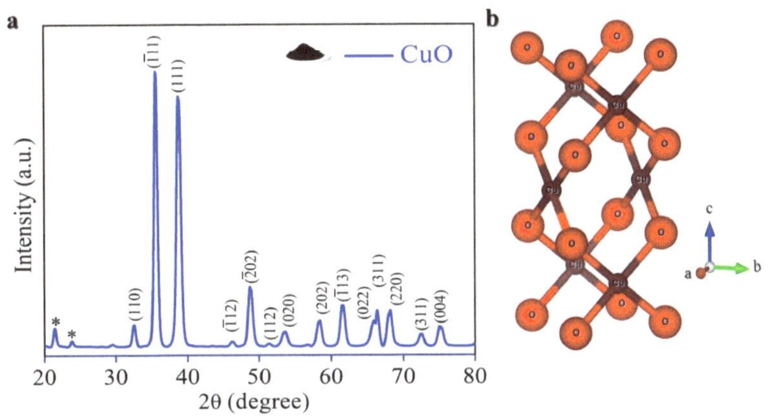

Figure 2. (**a**) XRD pattern; (**b**) Crystal structure.

Figure 2a demonstrates that all XRD patterns align closely with the standard CuO compound (JCPDS No. 48-1548, space group C2/c) [59], with minor characteristic peaks indicating traces of impurities (*) from orthorhombic-phase Cu $(OH)_2$ (JCPDS No. 13-420) [60]. However, these impurities are negligible, suggesting a predominantly pure CuO formation. The material behaves as a typical p-type semiconductor, featuring a narrow band gap, ranging from 1.2–1.8 eV, which is suitable for LIBs [61]. The crystal structure depicted in Figure 2b exhibits monoclinic CuO, characterized by lattice constants a = 4.6837 Å, b = 3.4226 Å, and c = 5.1288 Å, with angles $\beta = 99.54°$ and $\alpha = \gamma = 90°$ [62]. Within this structure, Cu^+ ions coordinate with four coplanar oxygen atoms arranged at the corners of a rectangular parallelogram, forming chains through edge-sharing [63]. The presence of heavy Cu ions contributes to a high crystal density of 6.31 g cm^{-3}, superior to that of graphite, and thus enhancing volumetric energy density [64].

Our investigation extended to the morphology and structural properties of CuO, using scanning electron microscopy (SEM) and Energy dispersive spectroscopy (EDS). Energy dispersive spectroscopy (EDS) analysis at Figure 3a confirms the presence of only Cu and O elements within the sample. Figure 3b–e illustrates the quasi-spherical and mixed morphology of CuO particles, with a size averaging 100–200 nm. Furthermore, elemental mapping results at Figure 3f–h demonstrate homogeneous distribution of Cu and O elements throughout the sample, affirming its purity.

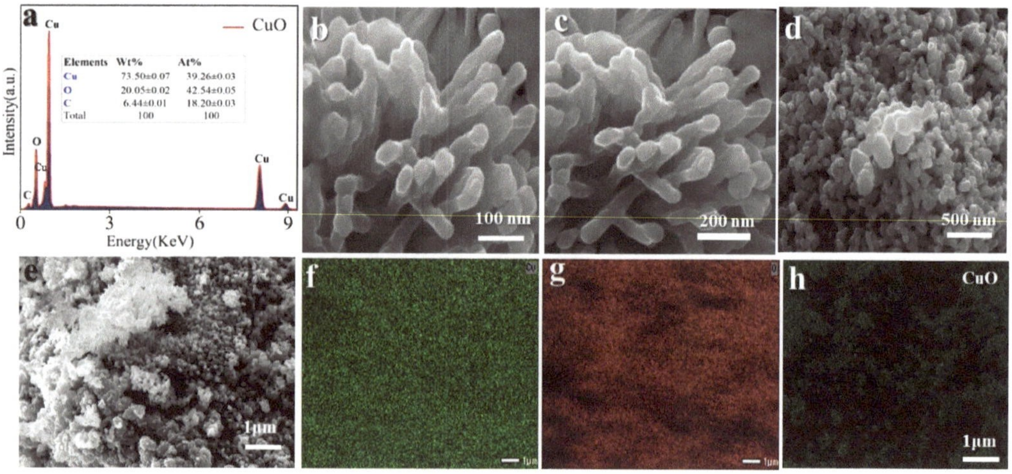

Figure 3. (**a**) EDS patterns; (**b–e**) The SEM images from 100 nm to 500 nm and 1 μm, respectively; (**f–h**) SEM elemental mapping.

To date, investigations have extensively explored the electrochemical performance of CuO-based anodes in various morphologies, including nanowires [65], nanosheets [66], and nanorods [67], as well as distinctive structures such as mesoporous [68], porous [69], hierarchical [70] core/shell architectures [71], and hollow [72]. Additionally, CuO-based composites have been synthesized by incorporating conductive carbonaceous materials and polymers, such as carbon nanotubes [73], graphene [74], and polypyrrole (PPy) [75], aimed at enhancing electronic conductivity cycles. Furthermore, CuO has been combined with high-capacity anode materials, such as TiO_2 [76], ZnO [77], SnO_2 [78], and Fe_2O_3 [79] to achieve superior lithium storage capabilities.

Despite considerable efforts dedicated to synthesizing various morphologies and compounds, researchers still fail to fully meet the commercially viable approach and the desirable performance standards of LIBs. Therefore, there persists a need for more extensive research into developing CuO with novel, cost-effective, and commercially viable synthesis techniques, alongside efforts to enhance electrochemical performance. To gain a clearer insight into the current state of research, we outline the testing conditions and performance metrics of reported CuO in Table 1. It is apparent that each of these reports exhibits some limitations, which are shown in Table 1 with our work.

Table 1. Comparative analysis of this work with previously reported CuO-based anode materials.

Ref.	Samples	Rate	Cycles	Capacity	Notes
[80]	Hollow CuO nanoparticles	100	100	630	Time and energy consuming and expensive synthesis route
[46]	$CuO/Cu_2O/C$ composites	200	600	260	Energy consuming method
[81]	Cu_xO/C anode	100	100	335	Time and energy consuming
[82]	CuO/tube-like carbon	100	100	650	Complicated and expensive synthesis approach
[83]	CuO@Cu microspheres	100	100	876	Sensitive route and time consuming
[65]	CuO nanowire arrays	300	100	550	Complicated synthesis approach
[74]	CuO nanosheets	100	100	600	Poor performance and expensive method
[39]	CuO@C	100 500	100 700	1024	Toxic and energy consuming process
[44]	Peony shaped CuO nanosheets	100 1000	80 100	780 441	Time consuming
This work	Commercial CuO nano powder	100 300	170 500	800 450	Safe, simple, cost-effective, scalable, and financially sustainable strategy for battery production.

It is widely acknowledged that the selection of binders and electrolyte additives significantly affects the performance of anodes, particularly those made of conversion-alloy, conversion, and alloy-based materials prone to substantial volume expansion [84,85]. Previous studies, as summarized in Table 1, predominantly employed poly(vinylidene fluoride) (PVDF) as the binder, which lacks the ability to effectively accommodate the expansion of electrode volume [86]. Additionally, conventional carbonate electrolytes without additives were commonly used, resulting in the failure to establish a robust solid–electrolyte interphase (SEI) [87]. Therefore, our study focuses on investigating two distinct combinations to evaluate the impact of binders and electrolytes on CuO anode performance: the utilization of CMC (carboxymethyl cellulose) as a non-toxic water-based binder in a regular carbonate electrolyte with fluoroethylene carbonate (FEC) additives [1M LiPF6/EC-DMC + 10% FEC]; and the use of LA133, a non-toxic water-based binder, combined with carbonate electrolyte containing FEC additive [1M LiPF6/EC-DMC + 10% FEC].

As depicted in Figure 4a,b, at 100 mA g^{-1} current density, the CuO@CMC electrode initially exhibits a charge capacity of approximately 80 mA h g^{-1} in the first cycle, gradually increasing to ~1100 mA h g^{-1} over 170 cycles. This behavior aligns with prior literature, due to material activation. The average Coulombic efficiency stands at 99%. Notably, the incorporation of the CMC binder with FEC additive demonstrates superior performance, making it the preferred choice for CuO-based anodes. The initial lithiation and delithiation specific capacities are measured at 2138.6 and 807.2 mA h g^{-1}, respectively, corresponding to an initial Coulombic efficiency of 37.7%. The significant irreversible capacity loss during the first cycle is largely due to SEI film formation. Notably, the initial lithiation and subsequent specific capacity exceed the theoretical capacity (674 mA h g^{-1}) [39], a phenomenon observed in a previous study [39]. In this study, the additional capacity is likely attributed to lithium insertion into a considerable quantity of ketjen black conducting agent or nanopores, interfacial lithium storage, and the formation of SEI.

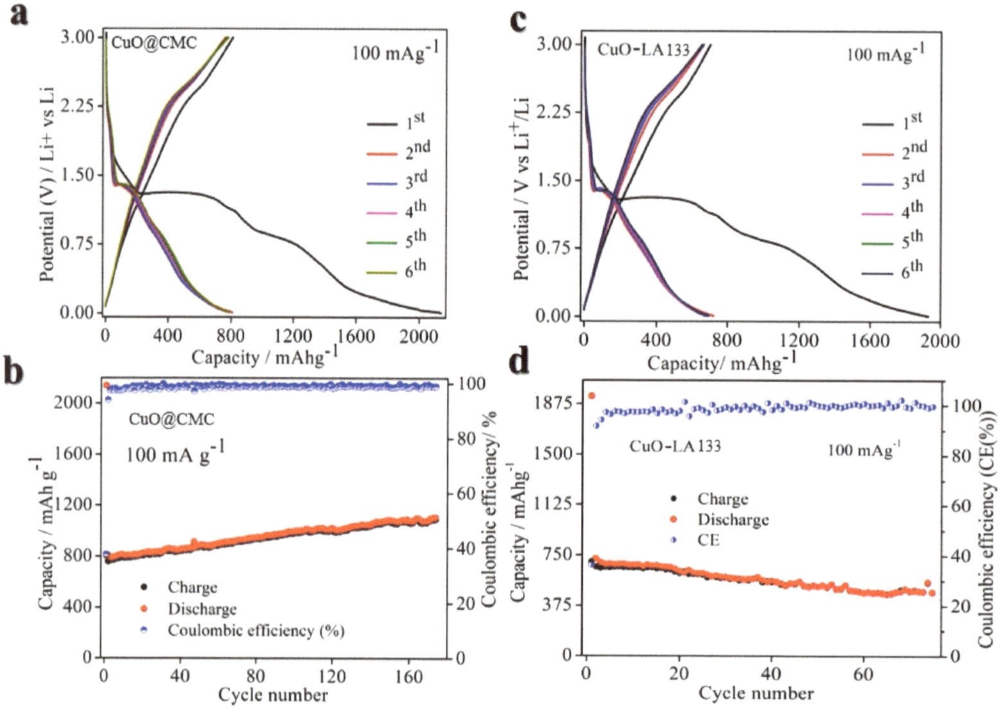

Figure 4. Electrochemical performance of CuO with different binders at a current density of 100 mA g^{-1}. (**a**) Galvanostatic charge-discharge (GCD) curve of CuO@CMC. (**b**) Cycling performance of CuO@CMC. (**c**) Galvanostatic charge-discharge curves of CuO@LA133. (**d**) Cycling performance of CuO@LA132.

On the other hand, the CuO@LA133 electrode demonstrates a comparable capacity, with slightly enhanced cycling performance, as illustrated in Figure 4c,d. However, after 75 cycles, the charge capacity and average Coulombic efficiency stabilize around ~375 mA h g^{-1} and 98%, respectively. The LA133 binder's inadequacy in handling the significant volume changes of CuO anodes leads to particle disintegration, compromising electrical conductivity, and overall performance.

As depicted in Figure 5a, the CuO@LA133 electrode exhibits an initial charge capacity of approximately 700 mA h g^{-1} at low current density of 100 mA g^{-1}, as shown in Figure 4d. This capacity remains stable at 700 mA h g^{-1} for the first 7 cycles. However, when subjected to a higher current density of 300 mA g^{-1}, the capacity rapidly declines to about 300 mA h g^{-1} over 54 cycles. The average Coulombic efficiency observed is 98.31%. This indicates that the LA133 binder is inadequate in accommodating the significant volumetric changes of the CuO anodes, failing to maintain particle cohesion, which is crucial for preserving electrical conductivity and electrochemical stability.

In contrast, as shown in Figure 5d, the CuO@CMC electrode demonstrates an initial charge capacity of around 800 mA h g^{-1}, maintaining this capacity for the first 7 cycles. Upon increasing the current density to 300 mA g^{-1}, the electrode maintains a stable capacity of approximately 450 mA h g^{-1} over ~500 cycles, with an average Coulombic efficiency of 99.10%. This superior performance highlights the effectiveness of the CMC binder in managing the volumetric expansion of CuO anodes, ensuring particle integrity, sustained electrical conductivity, and better electrochemical performance.

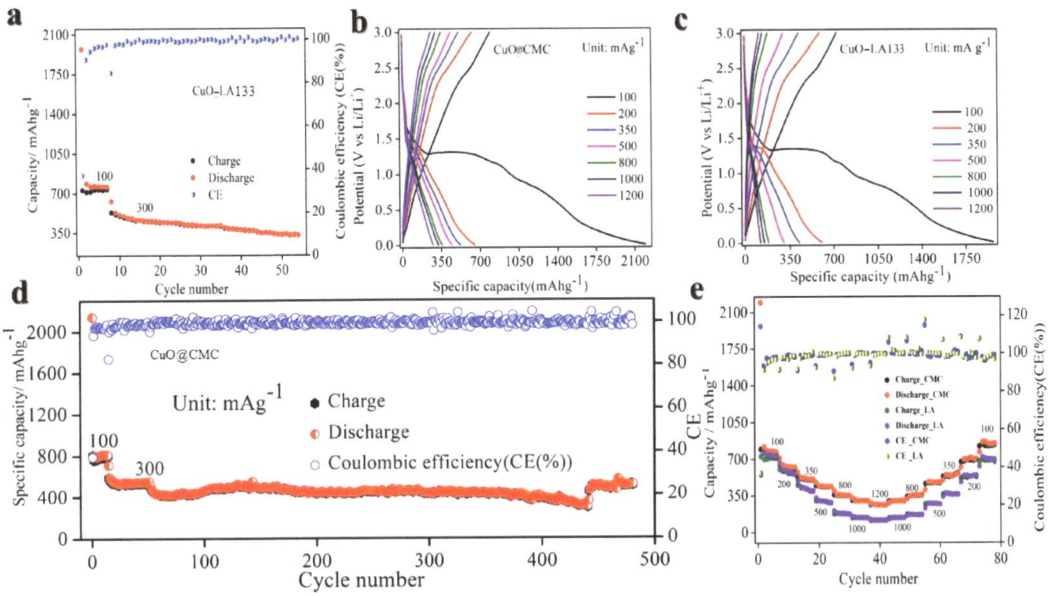

Figure 5. Electrochemical performances of the Li/CuO half-cell in the voltage window of 0.01–3.0 V vs. Li^+/Li. (**a**) Cycling performance of CuO@LA133 at 300 mA g^{-1}. (**b**) Rate curve of CuO@CMC from 100–1200 mA g^{-1} current density. (**c**) Rate curve of CuO@LA132 from 100–1200 mA g^{-1} current density. (**d**) Cycling performance of CuO@CMC at 300 mA g^{-1}. (**e**) Comparison of rate performance of CuO@CMC and CuO@LA133 binders.

In addition to exhibiting high capacity and stable performance, the CuO@CMC anode demonstrates superior rate performance and long-term cycling stability. As illustrated in Figure 5b, the charge capacities of CuO@CMC-based electrodes are 761, 618, 515, 435, 345, 297, and 258 mA h g^{-1} at current densities of 100, 200, 350, 500, 800, 1000, and 1200 mA g^{-1}, respectively. Notably, even at 1200 mA g^{-1} current density, the charge capacity remains at 258 mA h g^{-1}, which is comparable to that of graphite. Upon reverting the current density to 1000, 800, 500, 350, 200, and 100 mA g^{-1}, the charge capacities recover to 294, 350, 477, 544, 693, and 853 mA h g^{-1}, respectively, achieving a recovery ratio of approximately 100%, indicative of excellent rate performance as depicted in Figure 5e, and in separate Figure 6a. Conversely, as shown in Figure 5c, the charge capacities of CuO@LA133-based electrodes are 698, 562, 420, 295, 178, 137, and 111 mA h g^{-1} at current densities of 100, 200, 350, 500, 800, 1000, and 1200 mA g^{-1}, respectively. When the current density is reduced back to 1000, 800, 500, 350, 200, and 100 mA g^{-1}, the charge capacities recover to 135, 164, 266, 355, 530, and 681 mA h g^{-1}, respectively, corresponding to a recovery ratio of approximately 98%. However, this exhibits inferior rate performance, as shown in Figure 5e, and in separate Figure 6b.

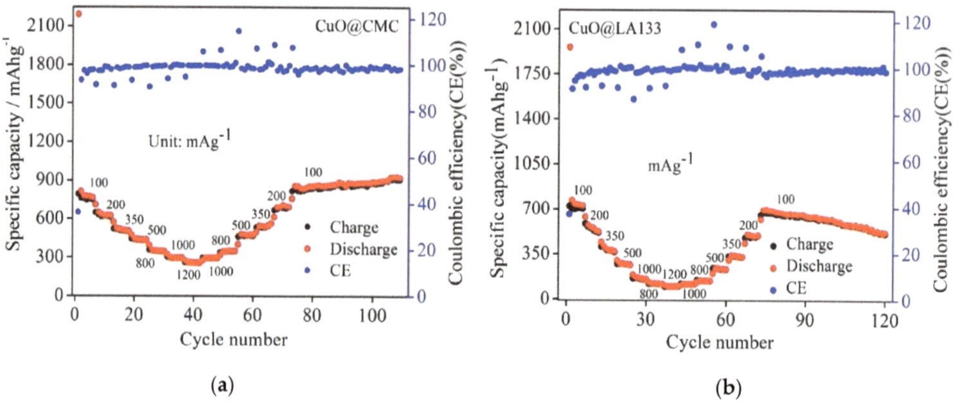

Figure 6. (**a**) CuO with CMC based electrode. (**b**) CuO with LA133 based electrode.

To obtain a more detailed understanding of the Li-insertion mechanism in CuO using both CMC and LA133 binders, we performed a cyclic voltammetry (CV) analysis. Figure 7a illustrates the cyclic voltammetry (CV) profiles obtained at a scan rate of 0.1 mV s^{-1} within a potential window of 0.01–3.0 V. During the initial cathodic sweep, three distinct reduction peaks are observed at 2.08, 1.24, and 0.82 V. These peaks correspond to the stepwise reduction of CuO: the first peak represents the reduction of CuO to an intermediate phase ($Cu^{II}_{1-x}Cu^{I}_{x}O_{1-x/2}$ ($0 \leq x \leq 0.4$), the second peak indicates further reduction to Cu_2O, and the third peak signifies the conversion to metallic Cu and Li_2O. In the subsequent anodic sweep, oxidation peaks appear at 1.36, 2.56, and 2.6 V, which are associated with the oxidation of Cu back to Cu_2O and the reformation of CuO [44]. The reaction mechanisms involved are described by their respective electrochemical equations [44]:

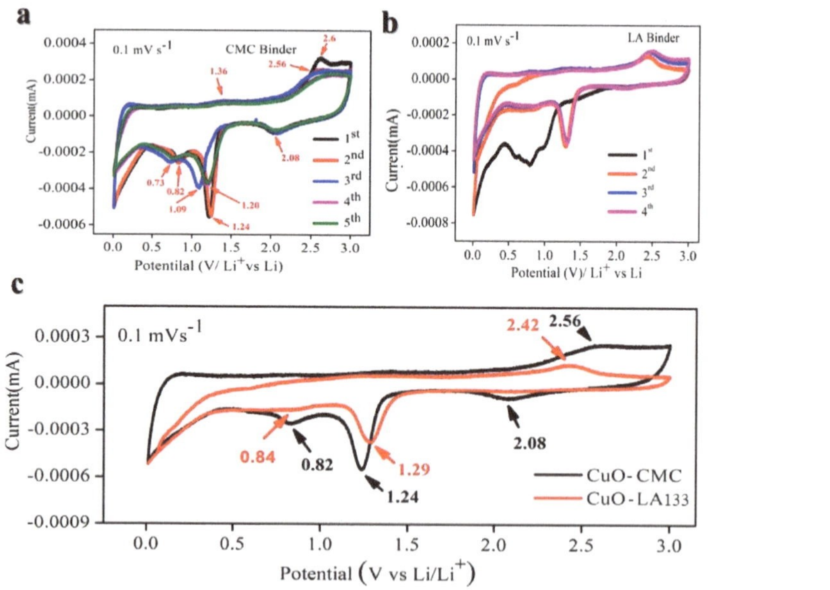

Figure 7. Cyclic voltammetry (CV). (**a**) CuO@CMC based electrode. (**b**) CuO@LA133 based electrode. (**c**) Comparison of CV of CuO@CMC and CuO@LA133 based electrodes.

Discharging

$$CuO + xLi^+ + xe^- \rightarrow Cu^{II}_{1-x}Cu^{I}_{x}O_{1-x/2} + x/2Li_2O\ (0 \leq x \leq 0.4) \quad (1)$$

$$Cu^{II}_{1-x}Cu^{I}_{x}O_{1-x/2} + (1-x)\,Li^+ + (1-x)e^- \rightarrow Cu_2O + (1-x)/2Li_2O\ (0 \leq x \leq 0.4) \quad (2)$$

$$1/2Cu_2O + Li^+ + e^- \rightarrow Cu + 1/2Li_2O \quad (3)$$

Charging

$$Cu + 1/2Li_2O \rightarrow 1/2Cu_2O + Li^+ + e^- \quad (4)$$

$$1/2Cu_2O + 1/2Li_2O \rightarrow CuO^+ + Li^+ + e^- \quad (5)$$

The difference in peak areas between the first cycle and subsequent cycles can be attributed to the formation of the solid-electrolyte interface (SEI) film and electrolyte decomposition [88]. A slight positive shift in potential observed during the second and third cycles indicates structural rearrangements occurring during the initial lithium-ion insertion [39,43]. From the second cycle onwards, the CV curves demonstrate consistent reproducibility, highlighting the superior electrochemical reversibility of the CuO@CMC (Figure 7a) electrode relative to the CuO@LA133 (Figure 7b) electrode, as can be observed in the comparative CV curve in Figure 7c. Figure 4a showcases the charge/discharge voltage profiles for the first six cycles at a current density of 100 mA g^{-1}, revealing three distinct voltage plateaus (2.0–1.28, 1.29–1.26, and 1.27–0.03 V) during the initial discharge, which corroborate the multi-step conversion process of CuO to Cu in the presence of Li, as reflected in the CV results.

From an electrochemical perspective, Electrochemical Impedance Spectroscopy (EIS) was utilized to elucidate the kinetic processes governing the electrode reactions. Figure 8a and 8c present the Nyquist plots for the CuO@CMC and CuO@LA133 electrodes, respectively, prior to cycling at a current density of 100 mA g^{-1}. Both spectra are characterized by the presence of two semicircles and a sloping line. Notably, the CuO@LA133 electrode exhibits a significantly larger semicircle prior to cycling. As depicted in Figure 8c, charge transfer resistance (Rct) is $3.031 \times 10^3\ \Omega$, indicating slower kinetics in the CuO@LA133 electrode. The semicircles observed at high and high-to-medium frequencies correspond to the resistance of the solid electrolyte interphase (SEI) film and the Rct at the electrode/electrolyte interface, respectively. But, as shown in Figure 8a, CuO@CMC electrode Rct value is $2.30 \times 10^2\ \Omega$, which is lower than that of LA133 based electrode. The sloping line at low frequencies is indicative of Li^+ diffusion within the solid state. Importantly, the CuO@CMC electrode (Figure 8b) after 100 cycles demonstrates the smallest semicircle diameter before cycling, signifying the lowest Rct value 21 Ω relative to the CuO@LA133 electrode (Figure 8d) Rct values is 77 Ω. This observation suggests that the CuO@CMC electrode exhibits the lowest polarization and the most rapid reaction kinetics.

Furthermore, SEM analysis was employed to assess the morphological analysis of CuO@CMC and CuO@LA132 electrodes both prior to and following cycling. As depicted in Figure 9a–c, the CuO@CMC electrode exhibits no cracks before cycling, after 40 cycles, and even after 400 cycles. Notably, the SEM image in Figure 9d reveals a smooth, gel-like film on the surface after 400 cycles, indicating increased stability and density.

This improved morphology enhances the electrochemical environment, leading to extended cell lifespan, improved storage performance, and overall superior electrochemical performance. Conversely, Figure 9d–f illustrate the CuO@LA133 electrode before cycling and after 10 cycles, with no visible cracks. However, by 135 cycles, a significant crack is observed, as highlighted by the red circle in Figure 9f, demonstrating that the LA133 binder is not suitable for CuO anode material in LIBs.

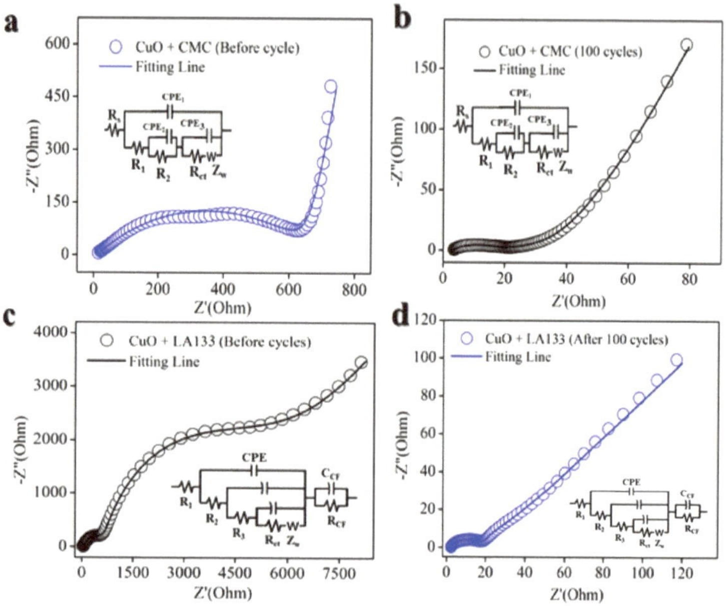

Figure 8. Electrochemical impedance spectroscopy. (**a**) CuO@CMC before cycling. (**b**) CuO@CMC after 100 cycles. (**c**) CuO@LA133 before cycle. (**d**) CuO@LA133 after 100 cycles.

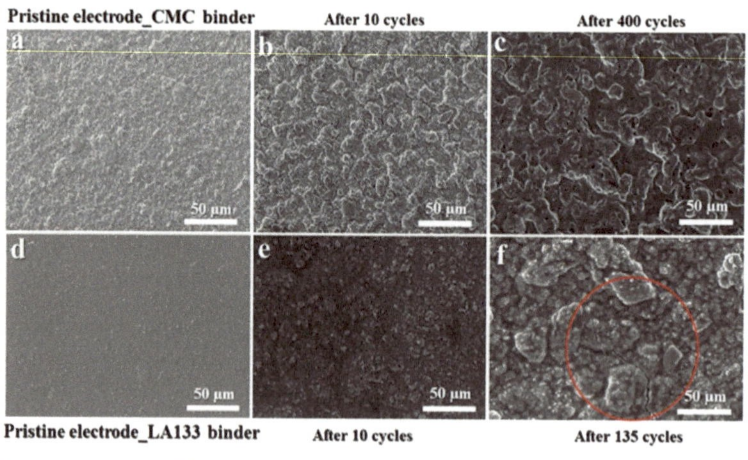

Figure 9. Top-view SEM images Pre- and Post-Cycle Comparisons. (**a**) Prestin CuO@CMC electrode. (**b**) CuO@CMC electrode after 10 cycles. (**c**) CuO@CMC electrode after 400 cycles. (**d**) Prestin CuO@LA133 electrode. (**e**) CuO@LA133 electrode after 10 cycles. (**f**) CuO@LA133 electrode after 135 cycles.

The enhanced performance of CuO@CMC electrodes in lithium-ion batteries (LIBs) can be attributed to several factors. The hydrophilic nature of CMC enhances the electrode's wettability with the electrolyte, resulting in improved ionic conductivity, and more efficient ion transport within the electrode. This reduces internal resistance and boosts the battery's overall performance. CMC's superior mechanical flexibility, compared with PVDF, allows it to accommodate the volume changes and stress associated with CuO's lithiation and delithiation processes [89–91]. This flexibility minimizes electrode cracking and maintains structural integrity, leading to longer cycle life and stable performance. Moreover, CMC ensures a homogeneous distribution of CuO particles within the electrode matrix, preventing agglomeration and promoting consistent and efficient electrochemical reactions throughout the electrode, thereby enhancing capacity and efficiency [48,91]. CMC also forms a conductive network that improves electron transport within the electrode, facilitating efficient electron transfer during charge and discharge processes. This contributes to higher rate capabilities and overall better electrochemical performance [92,93]. The strong adhesion provided by CMC between CuO particles and the current collector, as well as among the active materials, maintains electrode integrity during cycling, reducing the likelihood of active material detachment and capacity loss [47,94]. CMC's compatibility with aqueous processing minimizes side reactions often associated with organic solvents used in PVDF-based electrodes, leading to a more stable electrochemical environment and extending the battery's lifespan. Additionally, the use of water as a solvent in CMC-based electrodes eliminates the need for hazardous organic solvents like NMP, making the manufacturing process safer, more environmentally friendly, and cost-effective. In brief, the scientific reasons for the improved performance of CuO@CMC electrodes in LIBs include enhanced ionic and electron conductivity, superior mechanical flexibility, improved dispersion of active materials, stronger adhesion, and minimized side reactions. These factors collectively contribute to higher capacity, better cycle stability, and overall improved electrochemical performance. Furthermore, the environmental and cost benefits associated with aqueous processing make CMC a highly attractive binder for advanced LIB applications. It is well known that the incorporation of fluoroethylene carbonate (FEC) as an additive in lithium-ion battery (LIB) anodes brings significant advantages, particularly in enhancing performance and extending battery lifespan. A critical benefit is the promotion of a stable and uniform solid electrolyte interphase (SEI) layer, which is especially important for graphite- and silicon-based anodes. This robust SEI layer mitigates continuous electrolyte decomposition, thereby improving battery efficiency. Moreover, the enhanced stability of the SEI with FEC contributes to superior cycling stability by minimizing active lithium loss, a crucial factor for high-capacity anodes like silicon that experience substantial volume expansion. FEC also improves the first-cycle Coulombic efficiency by reducing lithium consumption during initial SEI formation, leading to better overall battery performance. Additionally, it reduces capacity fade over extended cycles by stabilizing the SEI and limiting electrolyte degradation, thus helping the battery maintain its capacity across numerous charge–discharge cycles. Another key advantage is the suppression of lithium dendrite formation, a critical concern in high-energy-density anodes, thereby lowering the risk of short circuits, and enhancing battery safety. Finally, FEC is compatible with high-voltage electrolyte systems, making it well-suited for advanced LIB chemistries that operate at elevated voltages, which in turn increases energy density.

3. Materials and Methods

3.1. Synthesis

Commercially available CuO nano powder (Sigma-Aldrich, 99.99% purity, Rockville, MD, USA) was utilized as the anode material for this experimental investigation, without any additional modifications.

3.2. Characterization

Structural examination of the CuO nano powder was conducted using an X-ray diffractometer (Rigaku SmartLab, The Woodlands, TX, USA), equipped with Cu Kα radiation (λ = 1.5408 Å), over a range from 20–80° at a scan rate of 2° per minute. Additionally, scanning electron microscopy (SEM) and energy-dispersive X-ray spectroscopy (EDS) were employed, equipped with a backscattered secondary electron detector (SEI) operating at an accelerating voltage of 20 kV and Dry SD_{30} detector for morphological examination and chemical composition identification, respectively, utilizing a JEOL JSM 6480LV instrument (JEOL, Akishima, Tokyo).

3.3. Cell Fabrication & Electrochemical Measurement

To conduct the electrochemical testing, the anode was prepared by mixing 50% (*w*/*w*) of the active material, CuO, with 25% (*w*/*w*) Ketjen black as a conducting agent, and 25% (*w*/*w*) binder [either LA133 or CMC]. Both slurries were prepared using water as a solvent. The resulting slurries were spread onto a copper foil (9 μm thickness, MTI Corporation, Richmond, CA, USA), using a doctor blade (MTI Corporation, Richmond, CA, USA). The coated foil was then dried overnight at 60 °C for 16 h. Anodes with a diameter of 1 cm were punched (MTI Corporation, Richmond, CA, USA) from the dried coated foil, and further dried at 60 °C under vacuum for 12 h before being transferred into the glovebox (MBRAUN Glovebox Workstations, Stratham, NH, USA). The average active mass loading of electrodes is in the range of 1.2–1.7 mg. For assembling CR2032 coin-type half-cells, lithium chips (MSE Supplies LLC, Tucson, AZ, USA) were used as the counter electrode, and polypropylene ethylene was used as the separator. Regarding the electrolyte, a commercial solution of 1 M $LiPF_6$ in ethylene carbonate (EC) and diethyl carbonate (DEC) [1:1] with 10% fluoroethylene carbonate (FEC) additives was prepared. Subsequently, the electrochemical performance of CuO for both LA133 and CMC-based electrodes was evaluated. We employed the Landt battery tester (LANDT Instruments, Vestal, NY, USA) to conduct electrochemical measurements, setting the voltage range at 0.01–3.0 V. Cyclic voltammetry (CV) was performed within the same range, while electrochemical impedance spectroscopy (EIS) was measured using the Arbin tester (Arbin Instruments, College Station, TX, USA) across a frequency range from 0.01 Hz to 100 kHz.

4. Conclusions

In this study, we introduce an efficient and cost-effective method for producing CuO electrode material. CuO@CMC based electrodes demonstrate remarkable electrochemical performance as compared to CuO@LA133 based electrodes, revealing a high specific capacity of approximately ~1100 mA h g^{-1} at a current density of 100 mA g^{-1} after 170 cycles, impressive rate performance of around 325 mA h g^{-1} at high current density 1200 mA g^{-1}, and excellent long-term stability of about 450 mA h g^{-1} at 300 mA g^{-1} for over ~500 cycles. By employing CuO nanoparticles in combination with a suitable amount of high surface area conducting ketjen black carbon, and an appropriate amount of CMC binder, along with the addition of 10% FEC additive to a regular carbonate-based electrolyte, we effectively address the volume expansion issue encountered in CuO anodes for lithium-ion batteries (LIBs). Additionally, this strategy enhances both electron and ion conductivity. Our findings offer a novel and facile approach for manufacturing transition metal oxide (TMO) anode materials with enhanced energy density, thereby facilitating their widespread adoption in advanced energy storage technologies.

Author Contributions: Conceptualization, N.O. and R.S.K.; investigation, N.O.; data curation, N.O. and S.C.; writing—original draft preparation, N.O. and R.S.K.; writing, review and editing, N.O. and R.S.K.; supervision, R.S.K., G.M. and B.R.W.; funding acquisition, R.S.K., G.M. and B.R.W. All authors have read and agreed to the published version of the manuscript.

Funding: This research received financial support from the NSF EPSCoR Center for the Advancement of Wearable Technolo-gies (CAWT) under Grant No. OIA-1849243, the PR NASA EPSCoR program through Grant No. 80NSSC22M0025, and the NASA MIRO PR SPRinT initiative via Grant No. 80NSSC19M02346.

Institutional Review Board Statement: Not applicable.

Informed Consent Statement: Not applicable.

Data Availability Statement: The data presented in this study are available on request from the corresponding author.

Acknowledgments: We gratefully acknowledge financial support from the National Science Foundation—Center for the Advancement of Wearable Technologies (NSF-CAWT) (Grant No. 1849243) and IRG2 NASA MIRO (Grant No. 80NSSC19M02346). We wish to express our gratitude to Molecular Science Research Center (MSRC) at the university of Puerto Rico, for their help during the experimental process and fruitful discussions.

Conflicts of Interest: The authors declare no conflicts of interest.

References

1. Tarascon, J.M.; Armand, M. Issues and Challenges Facing Rechargeable Lithium Batteries. *Nature* **2001**, *414*, 359–367. [CrossRef]
2. Goodenough, J.B. Evolution of Strategies for Modern Rechargeable Batteries. *Acc. Chem. Res.* **2013**, *46*, 1053–1061. [CrossRef] [PubMed]
3. Goodenough, J.B.; Kim, Y. Challenges for Rechargeable Li Batteries. *Chem. Mater.* **2010**, *22*, 587–603. [CrossRef]
4. Goodenough, J.B.; Park, K.S. The Li-Ion Rechargeable Battery: A Perspective. *J. Am. Chem. Soc.* **2013**, *135*, 1167–1176. [CrossRef]
5. Mauger, A.; Julien, C.M.; Goodenough, J.B.; Zaghib, K. Tribute to Michel Armand: From Rocking Chair—Li-Ion to Solid-State Lithium Batteries. *J. Electrochem. Soc.* **2020**, *167*, 070507. [CrossRef]
6. Fang, C.; Wang, X.; Meng, Y.S. Key Issues Hindering a Practical Lithium-Metal Anode. *Trends Chem.* **2019**, *1*, 152–158. [CrossRef]
7. Khan, F.M.N.U.; Rasul, M.G.; Sayem, A.S.M.; Mandal, N.K. Design and Optimization of Lithium-Ion Battery as an Efficient Energy Storage Device for Electric Vehicles: A Comprehensive Review. *J. Energy Storage* **2023**, *71*, 108033. [CrossRef]
8. Zou, F.; Manthiram, A. A Review of the Design of Advanced Binders for High-Performance Batteries. *Adv. Energy Mater.* **2020**, *10*, 2002508. [CrossRef]
9. Liu, Z.; Yu, Q.; Oli, N.; Gomez, J.F.F.; Qiu, S.; Tian, H.; Qiu, Q.; Sun, W.; Li, K.; Liu, Z.; et al. A Non-Volatile, Thermo-Reversible, and Self-Protective Gel Electrolyte Providing Highly Precise and Reversible Thermal Protection for Lithium Batteries. *Adv. Energy Mater.* **2023**, *13*, 2300143. [CrossRef]
10. Xie, L.; Zhang, W.; Chen, X.; Shan, R.; Han, Q.; Qiu, X.; Oli, N.; Florez Gomez, J.F.; Zhu, L.; Wu, X.; et al. Bimetallic Cobalt-Nickel Selenide Nanocubes Embedded in a Nitrogen-Doped Carbon Matrix as an Excellent Li-Ion Battery Anode. *ACS Appl. Mater. Interfaces* **2023**, *15*, 25536–25549. [CrossRef]
11. Wang, Q.; Liu, B.; Shen, Y.; Wu, J.; Zhao, Z.; Zhong, C.; Hu, W. Confronting the Challenges in Lithium Anodes for Lithium Metal Batteries. *Adv. Sci.* **2021**, *8*, 2101111. [CrossRef] [PubMed]
12. Sapkota, N.; Chiluwal, S.; Parajuli, P.; Rowland, A.; Podila, R. Insights into the Pseudocapacitive Behavior of Sulfurized Polymer Electrodes for Li–S Batteries. *Adv. Sci.* **2023**, *10*, 2206901. [CrossRef]
13. Qian, J.; Henderson, W.A.; Xu, W.; Bhattacharya, P.; Engelhard, M.; Borodin, O.; Zhang, J.G. High Rate and Stable Cycling of Lithium Metal Anode. *Nat. Commun.* **2015**, *6*, 6362. [CrossRef] [PubMed]
14. Aslam, M.K.; Niu, Y.; Hussain, T.; Tabassum, H.; Tang, W.; Xu, M.; Ahuja, R. How to Avoid Dendrite Formation in Metal Batteries: Innovative Strategies for Dendrite Suppression. *Nano Energy* **2021**, *86*, 106142. [CrossRef]
15. Peled, E.; Menachem, C.; Bar-Tow, D.; Melman, A. Improved Graphite Anode for Lithium-Ion Batteries Chemically: Bonded Solid Electrolyte Interface and Nanochannel Formation. *J. Electrochem. Soc.* **1996**, *143*, L4–L7. [CrossRef]
16. Zhang, H.; Yang, Y.; Ren, D.; Wang, L.; He, X. Graphite as Anode Materials: Fundamental Mechanism, Recent Progress and Advances. *Energy Storage Mater.* **2021**, *36*, 147–170. [CrossRef]
17. Liu, X.; Yin, L.; Ren, D.; Wang, L.; Ren, Y.; Xu, W.; Lapidus, S.; Wang, H.; He, X.; Chen, Z.; et al. In Situ Observation of Thermal-Driven Degradation and Safety Concerns of Lithiated Graphite Anode. *Nat. Commun.* **2021**, *12*, 4235. [CrossRef]
18. Aurbach, D.; Zinigrad, E.; Cohen, Y.; Teller, H. A Short Review of Failure Mechanisms of Lithium Metal and Lithiated Graphite Anodes in Liquid Electrolyte Solutions. *Solid. State Ion.* **2002**, *148*, 405–416. [CrossRef]
19. Chen, G.; Yan, L.; Luo, H.; Guo, S. Nanoscale Engineering of Heterostructured Anode Materials for Boosting Lithium-Ion Storage. *Adv. Mater.* **2016**, *28*, 7580–7602. [CrossRef]
20. Zhang, Y.; Wu, B.; Mu, G.; Ma, C.; Mu, D.; Wu, F. Recent Progress and Perspectives on Silicon Anode: Synthesis and Prelithiation for LIBs Energy Storage. *J. Energy Chem.* **2022**, *64*, 615–650. [CrossRef]
21. Ying, H.; Han, W.Q. Metallic Sn-Based Anode Materials: Application in High-Performance Lithium-Ion and Sodium-Ion Batteries. *Adv. Sci.* **2017**, *4*, 1700298. [CrossRef] [PubMed]
22. Chang, X.; Xie, Z.; Liu, Z.; Zheng, X.; Zheng, J.; Li, X. Aluminum: An Underappreciated Anode Material for Lithium-Ion Batteries. *Energy Storage Mater.* **2020**, *25*, 93–99. [CrossRef]
23. Yang, X.; Zhang, R. High-Capacity Graphene-Confined Antimony Nanoparticles as a Promising Anode Material for Potassium-Ion Batteries. *J. Alloys Compd.* **2020**, *834*, 155191. [CrossRef]
24. Fugattini, S.; Gulzar, U.; Andreoli, A.; Carbone, L.; Boschetti, M.; Bernardoni, P.; Gjestila, M.; Mangherini, G.; Camattari, R.; Li, T.; et al. Binder-Free Nanostructured Germanium Anode for High Resilience Lithium-Ion Battery. *Electrochim. Acta* **2022**, *411*, 139832. [CrossRef]
25. Devina, W.; Setiadi Cahyadi, H.; Albertina, I.; Chandra, C.; Park, J.H.; Yoon Chung, K.; Chang, W.; Kyu Kwak, S.; Kim, J. High-Energy–Density Carbon-Coated Bismuth Nanodots on Hierarchically Porous Molybdenum Carbide for Superior Lithium Storage. *Chem. Eng. J.* **2022**, *432*, 134276. [CrossRef]
26. Eshetu, G.G.; Figgemeier, E. Confronting the Challenges of Next-Generation Silicon Anode-Based Lithium-Ion Batteries: Role of Designer Electrolyte Additives and Polymeric Binders. *ChemSusChem* **2019**, *12*, 2515–2539. [CrossRef]

27. Imashuku, S.; Taguchi, H.; Fujieda, S.; Suzuki, S.; Wagatsuma, K. Three-Dimensional Lithium Mapping of Graphite Anode Usin Laser-Induced Breakdown Spectroscopy. *Electrochim. Acta* **2019**, *293*, 78–83. [CrossRef]
28. He, Y.; Jiang, L.; Chen, T.; Xu, Y.; Jia, H.; Yi, R.; Xue, D.; Song, M.; Genc, A.; Bouchet-Marquis, C.; et al. Progressive Growth c the Solid–Electrolyte Interphase towards the Si Anode Interior Causes Capacity Fading. *Nat. Nanotechnol.* **2021**, *16*, 1113–112 [CrossRef]
29. Wang, J.; Wang, H.; Yao, T.; Liu, T.; Tian, Y.; Li, C.; Li, F.; Meng, L.; Cheng, Y. Porous N-Doped Carbon Nanoflakes Supporte Hybridized SnO2/Co3O4 Nanocomposites as High-Performance Anode for Lithium-Ion Batteries. *J. Colloid. Interface Sci.* **202** *560*, 546–554. [CrossRef]
30. Revathi, J.; Jyothirmayi, A.; Rao, T.N.; Deshpande, A.S.; Revathi, J.; Jyothirmayi, A.; Rao, T.N.; Deshpande, A.S. Wood-Derive Carbon Fibers Embedded with SnOx Nanoparticles as Anode Material for Lithium-Ion Batteries. *Glob. Chall.* **2020**, *4*, 190004 [CrossRef]
31. Shin, Y.; Ding, H.; Persson, K.A. Revealing the Intrinsic Li Mobility in the Li2MnO3 Lithium-Excess Material. *Chem. Mater.* **201** *28*, 2081–2088. [CrossRef]
32. Abouimrane, A.; Cui, Y.; Chen, Z.; Belharouak, I.; Yahia, H.B.; Wu, H.; Assary, R.; Curtiss, L.A.; Amine, K. Enabling High Energ Density Li-Ion Batteries through Li2O Activation. *Nano Energy* **2016**, *27*, 196–201. [CrossRef]
33. Zhu, J.; Ding, Y.; Ma, Z.; Tang, W.; Chen, X.; Lu, Y. Recent Progress on Nanostructured Transition Metal Oxides As Anod Materials for Lithium-Ion Batteries. *J. Electron. Mater.* **2022**, *51*, 3391–3417. [CrossRef]
34. Muchuweni, E.; Mombeshora, E.T.; Muiva, C.M.; Sathiaraj, T.S. Lithium-Ion Batteries: Recent Progress in Improving the Cyclin and Rate Performances of Transition Metal Oxide Anodes by Incorporating Graphene-Based Materials. *J. Energy Storage* **202** *73*, 109013. [CrossRef]
35. Ortiz, W.; Malca, C.; Barrionuevo, D.; Aldalbahi, A.; Pacheco, E.; Oli, N.; Feng, P. Two-Dimensional Tungsten Disulfide Nanosheet and Their Application in Self-Powered Photodetectors with Ultra-High Sensitivity and Stability. *Vacuum* **2022**, *201*, 11109 [CrossRef]
36. Poizot, P.; Laruelle, S.; Grugeon, S.; Dupont, L.; Tarascon, J.M. Nano-Sized Transition-Metal Oxides as Negative-Electrod Materials for Lithium-Ion Batteries. *Nature* **2000**, *407*, 496–499. [CrossRef] [PubMed]
37. Zoolfakar, A.S.; Rani, R.A.; Morfa, A.J.; O'Mullane, A.P.; Kalantar-Zadeh, K. Nanostructured Copper Oxide Semiconductors: A Perspective on Materials, Synthesis Methods and Applications. *J. Mater. Chem. C Mater.* **2014**, *2*, 5247–5270. [CrossRef]
38. Mai, Y.J.; Wang, X.L.; Xiang, J.Y.; Qiao, Y.Q.; Zhang, D.; Gu, C.D.; Tu, J.P. CuO/Graphene Composite as Anode Materials fo Lithium-Ion Batteries. *Electrochim. Acta* **2011**, *56*, 2306–2311. [CrossRef]
39. Xu, Y.; Chu, K.; Li, Z.; Xu, S.; Yao, G.; Niu, P.; Zheng, F. Porous CuO@C Composite as High-Performance Anode Materials fo Lithium-Ion Batteries. *Dalton Trans.* **2020**, *49*, 11597–11604. [CrossRef]
40. Chai, J.; Wang, K.; Li, Q.; Du, J.; Jiang, L.; Han, N.; Zhang, W.; Tang, B.; Rui, Y. Highly Dispersed CuO Nanoparticle on ZIF-4 Framework as Anode Material for LIBs. *J. Alloys Compd.* **2022**, *914*, 165316. [CrossRef]
41. Pu, F.; Bai, Y.; Lv, J.; Zhao, X.; Wu, G.; Kong, C.; Lei, B.; Zhang, X.; Jin, H.; Yang, Z. Yolk–Shell Cu2O@CuO-Decorated RGO fo High-Performance Lithium-Ion Battery Anode. *Energy Environ. Mater.* **2022**, *5*, 253–260. [CrossRef]
42. Ezhyeh, Z.N.; Khodaei, M.; Torabi, F. Review on Doping Strategy in Li4Ti5O12 as an Anode Material for Lithium-Ion Batteries *Ceram. Int.* **2023**, *49*, 7105–7141. [CrossRef]
43. Wang, B.; Wu, X.L.; Shu, C.Y.; Guo, Y.G.; Wang, C.R. Synthesis of CuO/Graphene Nanocomposite as a High-Performance Anode Material for Lithium-Ion Batteries. *J. Mater. Chem.* **2010**, *20*, 10661–10664. [CrossRef]
44. Hu, L.; Yang, Z.; Li, H.; Yang, X.; Li, J.; Wang, P.; Jin, C.; Zhang, C. Three-Dimensional Clusters of Peony-Shaped CuO Nanosheets as a High-Rate Anode for Li-Ion Batteries. *RSC Adv.* **2021**, *11*, 10760–10766. [CrossRef]
45. Zhao, J.; Zhao, Y.; Yue, W.C.; Zheng, S.M.; Li, X.; Gao, N.; Zhu, T.; Zhang, Y.J.; Xia, G.M.; Wang, B. Facile Fabrication of Hollow CuO Nanocubes for Enhanced Lithium/Sodium Storage Performance. *CrystEngComm* **2021**, *23*, 6107–6116. [CrossRef]
46. Xu, Q.; Jiu, H.; Zhang, L.; Song, W.; Wei, H.; Wang, C.; Yang, J.; Guo, F.; Gao, T. Structure Design of CuO/Cu2O@C Heterostructure Polyhedron Accumulated by Hollow Microspheres for High-Performance Lithium Storage. *J. Alloys Compd.* **2021**, *887*, 161417 [CrossRef]
47. Lingappan, N.; Kong, L.; Pecht, M. The Significance of Aqueous Binders in Lithium-Ion Batteries. *Renew. Sustain. Energy Rev* **2021**, *147*, 111227. [CrossRef]
48. Trivedi, S.; Pamidi, V.; Pinto Bautista, S.; Nur, F.; Shamsudin, A.; Weil, M.; Barpanda, P.; Bresser, D.; Fichtner, M.; Trivedi, S.; et al Water-Soluble Inorganic Binders for Lithium-Ion and Sodium-Ion Batteries. *Adv. Energy Mater.* **2024**, *14*, 2303338. [CrossRef]
49. Kukay, A.; Polizos, G.; Bott, E.; Ielvev, A.; Tao, R.; Sharma, J.; Li, J. Mechanical and Electrochemical Implications of Drying Temperature on Lithium-Ion Battery Electrodes. *Batter. Supercaps* **2024**, *7*, e202400113. [CrossRef]
50. David, M.; Gerofke, A.; Lange, R.; Kolossa-Gehring, M.; Apel, P. The European Human Biomonitoring Initiative (HBM4EU) Human Biomonitoring Guidance Values (HBM-GVs) for the Aprotic Solvents N-Methyl-2-Pyrrolidone (NMP) and N-Ethyl-2- Pyrrolidone (NEP). *Int. J. Hyg. Environ. Health* **2021**, *238*, 113856. [CrossRef]
51. Salini, P.S.; Gopinadh, S.V.; Kalpakasseri, A.; John, B.; Thelakkattu Devassy, M. Toward Greener and Sustainable Li-Ion Cells: An Overview of Aqueous-Based Binder Systems. *ACS Sustain. Chem. Eng.* **2020**, *8*, 4003–4025. [CrossRef]
52. Chou, S.L.; Pan, Y.; Wang, J.Z.; Liu, H.K.; Dou, S.X. Small Things Make a Big Difference: Binder Effects on the Performance of Li and Na Batteries. *Phys. Chem. Chem. Phys.* **2014**, *16*, 20347–20359. [CrossRef]

53. Li, J.-T.; Wu, Z.-Y.; Lu, Y.-Q.; Zhou, Y.; Huang, Q.-S.; Huang, L.; Sun, S.-G.; Li, J.-T.; Wu, Z.-Y.; Lu, Y.-Q.; et al. Water Soluble Binder, an Electrochemical Performance Booster for Electrode Materials with High Energy Density. *Adv. Energy Mater.* **2017**, *7*, 1701185. [CrossRef]
54. Lestriez, B. Functions of Polymers in Composite Electrodes of Lithium Ion Batteries. *Comptes Rendus Chim.* **2010**, *13*, 1341–1350. [CrossRef]
55. Mazouzi, D.; Karkar, Z.; Hernandez, C.R.; Manero, P.J.; Guyomard, D.; Roué, L.; Lestriez, B. Critical Roles of Binders and Formulation at Multiscales of Silicon-Based Composite Electrodes. *J. Power Sources* **2015**, *280*, 533–549. [CrossRef]
56. Magasinski, A.; Zdyrko, B.; Kovalenko, I.; Hertzberg, B.; Burtovyy, R.; Huebner, C.F.; Fuller, T.F.; Luzinov, I.; Yushin, G. Toward Efficient Binders for Li-Ion Battery Si-Based Anodes: Polyacrylic Acid. *ACS Appl. Mater. Interfaces* **2010**, *2*, 3004–3010. [CrossRef]
57. Parikh, P.; Sina, M.; Banerjee, A.; Wang, X.; D'Souza, M.S.; Doux, J.M.; Wu, E.A.; Trieu, O.Y.; Gong, Y.; Zhou, Q.; et al. Role of Polyacrylic Acid (PAA) Binder on the Solid Electrolyte Interphase in Silicon Anodes. *Chem. Mater.* **2019**, *31*, 2535–2544. [CrossRef]
58. Wang, R.; Feng, L.; Yang, W.; Zhang, Y.; Zhang, Y.; Bai, W.; Liu, B.; Zhang, W.; Chuan, Y.; Zheng, Z.; et al. Effect of Different Binders on the Electrochemical Performance of Metal Oxide Anode for Lithium-Ion Batteries. *Nanoscale Res. Lett.* **2017**, *12*, 575. [CrossRef]
59. Parekh, Z.R.; Chaki, S.H.; Hirpara, A.B.; Patel, G.H.; Kannaujiya, R.M.; Khimani, A.J.; Deshpande, M.P. CuO Nanoparticles – Synthesis by Wet Precipitation Technique and Its Characterization. *Physica B Condens. Matter* **2021**, *610*, 412950. [CrossRef]
60. Li, H.; Zhang, Z.; Ren, Z.; Chen, Y.; Huang, J.; Lei, Z.; Qian, X.; Lai, Y.; Zhang, S. A Quadruple Biomimetic Hydrophilic/Hydrophobic Janus Composite Material Integrating $Cu(OH)_2$ Micro-Needles and Embedded Bead-on-String Nanofiber Membrane for Efficient Fog Harvesting. *Chem. Eng. J.* **2023**, *455*, 140863. [CrossRef]
61. Iqbal, T.; ur Rehman, A.; Khan, M.A.; Shafique, M.; Ahmad, P.; Mahmood, H.; Naeem, M.; Iqbal, J. Copper Oxide Nanosheets Prepared by Facile Microplasma Electrochemical Technique with Photocatalytic and Bactericidal Activities. *J. Mater. Sci. Mater. Electron.* **2020**, *31*, 16649–16660. [CrossRef]
62. Musa, A.M.M.; Farhad, S.F.U.; Gafur, M.A.; Jamil, A.T.M.K. Effects of Withdrawal Speed on the Structural, Morphological, Electrical, and Optical Properties of CuO Thin Films Synthesized by Dip-Coating for CO_2 Gas Sensing. *AIP Adv.* **2021**, *11*, 115004. [CrossRef]
63. Asbrink, S.; Waskowska, A. CuO: X-Ray Single-Crystal Structure Determination at 196 K and Room Temperature. *J. Phys. Condens. Matter* **1991**, *3*, 8173. [CrossRef]
64. Kaushik, H.K.; Kaur, A.; Garg, V.; Abida, K.; Kumar, S.; Singh, K.; Singh, S.P.; Khan, S. Effect of CuO on Physical, Structural and Optical Properties of Lithium Borosilicate Glasses. *Mater. Today Commun.* **2023**, *35*, 106208. [CrossRef]
65. Zhang, R.; Liu, J.; Guo, H.; Tong, X. Synthesis of CuO Nanowire Arrays as High-Performance Electrode for Lithium Ion Batteries. *Mater. Lett.* **2015**, *139*, 55–58. [CrossRef]
66. Zhang, C.; Lv, W.; Zhou, G.; Huang, Z.; Zhang, Y.; Lyu, R.; Wu, H.; Yun, Q.; Kang, F.; Yang, Q.-H.; et al. Vertically Aligned Lithiophilic CuO Nanosheets on a Cu Collector to Stabilize Lithium Deposition for Lithium Metal Batteries. *Adv. Energy Mater.* **2018**, *8*, 1703404. [CrossRef]
67. Zhang, W.; Ma, G.; Gu, H.; Yang, Z.; Cheng, H. A New Lithium-Ion Battery: CuO Nanorod Array Anode versus Spinel LiNi0.5Mn1.5O4 Cathode. *J. Power Sources* **2015**, *273*, 561–565. [CrossRef]
68. Chen, X.; Zhang, N.; Sun, K. Facile Fabrication of CuO Mesoporous Nanosheet Cluster Array Electrodes with Super Lithium-Storage Properties. *J. Mater. Chem.* **2012**, *22*, 13637–13642. [CrossRef]
69. Yuan, S.; Huang, X.-L.; Ma, D.-L.; Wang, H.-G.; Meng, F.-Z.; Zhang, X.-B.; Yuan, S.; Huang, X.-L.; Ma, D.-L.; Wang, H.-G.; et al. Engraving Copper Foil to Give Large-Scale Binder-Free Porous CuO Arrays for a High-Performance Sodium-Ion Battery Anode. *Adv. Mater.* **2014**, *26*, 2273–2279. [CrossRef]
70. Guo, Z.; Chen, X.; Li, J.; Liu, J.H.; Huang, X.J. ZnO/CuO Hetero-Hierarchical Nanotrees Array: Hydrothermal Preparation and Self-Cleaning Properties. *Langmuir* **2011**, *27*, 6193–6200. [CrossRef]
71. Yuan, W.; Luo, J.; Pan, B.; Qiu, Z.; Huang, S.; Tang, Y. Hierarchical Shell/Core CuO Nanowire/Carbon Fiber Composites as Binder-Free Anodes for Lithium-Ion Batteries. *Electrochim. Acta* **2017**, *241*, 261–271. [CrossRef]
72. Deng, C.; Hu, H.; Ge, X.; Han, C.; Zhao, D.; Shao, G. One-Pot Sonochemical Fabrication of Hierarchical Hollow CuO Submicrospheres. *Ultrason. Sonochem* **2011**, *18*, 932–937. [CrossRef] [PubMed]
73. Liu, Y.; Cai, X.; Shi, W. Free-Standing Graphene/Carbon Nanotubes/CuO Aerogel Paper Anode for Lithium Ion Batteries. *Mater. Lett.* **2016**, *172*, 72–75. [CrossRef]
74. Pu, F.; Kong, C.; Lv, J.; BoMa; Zhang, W.; Zhang, X.; Yang, S.; Jin, H.; Yang, Z. CuO Ultrathin Nanosheets Decorated Reduced Graphene Oxide as a High Performance Anode for Lithium-Ion Batteries. *J. Alloys Compd.* **2019**, *805*, 355–362. [CrossRef]
75. Zhou, Y.; Jin, X.; Ni, J.; Zhang, S.; Yang, J.; Liu, P.; Wang, Z.; Lei, J. Evaporation Induced Uniform Polypyrrole Coating on CuO Arrays for Free-Standing High Lithium Storage Anode. *J. Solid. State Electrochem.* **2019**, *23*, 1829–1836. [CrossRef]
76. Rahman, M.M.; Khan, S.B.; Marwani, H.M.; Asiri, A.M.; Alamry, K.A. Selective Iron(III) Ion Uptake Using CuO-TiO2 Nanostructure by Inductively Coupled Plasma-Optical Emission Spectrometry. *Chem. Cent. J.* **2012**, *6*, 158. [CrossRef] [PubMed]
77. Xie, Q.S.; Lin, L.; Ma, Y.T.; Yang, J.R.; Huang, J.; Wang, L.S.; Peng, D.L. Facile Fabrication of ZnO–CuO Porous Hybrid Microspheres as Lithium Ion Battery Anodes with Enhanced Cyclability. *Rare Met.* **2017**, *36*, 403–410. [CrossRef]
78. Zhu, X.; Shi, H.; Yin, J.; Zhu, H.; Zhou, Y.; Tang, Y.; Wu, P.; Lu, T. Facile Preparation of CuO@SnO_2 Nanobelts as a High-Capacity and Long-Life Anode for Lithium-Ion Batteries. *RSC Adv.* **2014**, *4*, 34417–34420. [CrossRef]

79. Di Lecce, D.; Verrelli, R.; Campanella, D.; Marangon, V.; Hassoun, J. A New CuO-Fe_2O_3-Mesocarbon Microbeads Conversion Anode in a High-Performance Lithium-Ion Battery with a $Li_{1.35}Ni_{0.48}Fe_{0.1}Mn_{1.72}O_4$ Spinel Cathode. *ChemSusChem* **2017**, *10*, 1607–1615. [CrossRef]
80. Dong, Y.; Jiang, X.; Mo, J.; Zhou, Y.; Zhou, J. Hollow CuO Nanoparticles in Carbon Microspheres Prepared from Cellulose-Cuprammonium Solution as Anode Materials for Li-Ion Batteries. *Chem. Eng. J.* **2020**, *381*, 122614. [CrossRef]
81. Zhang, H.; Chen, R.; Zhang, X.; Zong, P.; Bai, Y.; Jin, H.; Xu, H.; Lian, K.; Ma, F. Cotton as a Sustainable Source of CuxO/C Anode for High-Performance Li-Ion Battery. *Ionics* **2019**, *25*, 2519–2524. [CrossRef]
82. Chen, Z.; Hou, Z.; Xu, W.; Chen, Y.; Li, Z.; Chen, L.; Wang, W. Ultrafine CuO Nanoparticles Decorated Activated Tube-like Carbon as Advanced Anode for Lithium-Ion Batteries. *Electrochim. Acta* **2019**, *296*, 206–213. [CrossRef]
83. Li, Z.; Xu, Y.; Chen, Y.; Zhang, W.; Li, K.; Zhang, H. In Situ Fabrication of Hierarchical CuO@Cu Microspheres Composed of Nanosheets as High-Performance Anode Materials for Lithium-Ion Batteries. *ChemistrySelect* **2019**, *4*, 13569–13575. [CrossRef]
84. Su, J.; Gao, Z.; Xie, Y.; Zhang, Z.; Wang, H. Boosting Li-Storage Properties of Conversion-Type Anodes for Lithium-Ion Batteries via Steric Effect of Intercalation-Type Materials: A Case of $MnCO_3$. *Compos. B Eng.* **2021**, *212*, 108733. [CrossRef]
85. Oli, N.; Ortiz Lago, W.; Tripathi, B.; Bhattarai, M.; Weiner, B.R.; Morell, G.; Katiyar, R.S. Enhanced Electrochemical Performance of Bi_2O_3 via Facile Synthesis as Anode Material for Ultra-Long Cycle Lifespan Lithium-Ion Batteries. *Electrochem. Commun.* **2024**, *159*, 107656. [CrossRef]
86. Song, Z.; Wang, L.; Yang, K.; Gong, Y.; Yang, L.; Liu, X.; Pan, F. Intermolecular Chemistry for Designing Functional Binders in Silicon/Carbon Composite Anodes. *Mater. Today Energy* **2022**, *30*, 101153. [CrossRef]
87. Chung, G.J.; Tran, Y.H.T.; Han, J.; Kim, K.; Lee, Y.S.; Song, S.W. Novel Additives-Package to Mitigate the Failure Modes of High-Capacity $LiNi_{0.82}Co_{0.11}Mn_{0.07}O_2$-Based Lithium-Ion Battery. *Chem. Eng. J.* **2022**, *446*, 137288. [CrossRef]
88. Su, Y.; Liu, T.; Zhang, P.; Zheng, P. CuO Nanowire Arrays Synthesized at Room Temperature as a High-Performance Anode Material for Li/Na-Ion Batteries. *Thin Solid. Films* **2019**, *690*, 137522. [CrossRef]
89. Yoon, D.H.; Marinaro, M.; Axmann, P.; Wohlfahrt-Mehrens, M. Study of the Binder Influence on Expansion/Contraction Behavior of Silicon Alloy Negative Electrodes for Lithium-Ion Batteries. *J. Electrochem. Soc.* **2020**, *167*, 160537. [CrossRef]
90. Huang, W.; Wang, W.; Wang, Y.; Qu, Q.; Jin, C.; Zheng, H. Overcoming the Fundamental Challenge of PVDF Binder Use with Silicon Anodes with a Super-Molecular Nano-Layer. *J. Mater. Chem. A Mater.* **2021**, *9*, 1541–1551. [CrossRef]
91. Park, J.H.; Kim, S.H.; Ahn, K.H. Role of Carboxymethyl Cellulose Binder and Its Effect on the Preparation Process of Anode Slurries for Li-Ion Batteries. *Colloids Surf. A Physicochem. Eng. Asp.* **2023**, *664*, 131130. [CrossRef]
92. Akhlaq, M.; Mushtaq, U.; Naz, S.; Uroos, M. Carboxymethyl Cellulose-Based Materials as an Alternative Source for Sustainable Electrochemical Devices: A Review. *RSC Adv.* **2023**, *13*, 5723–5743. [CrossRef] [PubMed]
93. Xie, L.; Zhao, L.; Wan, J.; Shao, Z.; Wang, F.; Lv, S. The Electrochemical Performance of Carboxymethyl Cellulose Lithium as a Binding Material for Anthraquinone Cathodes in Lithium Batteries. *J. Electrochem. Soc.* **2012**, *159*, A499–A505. [CrossRef]
94. Lestriez, B.; Bahri, S.; Sandu, I.; Roué, L.; Guyomard, D. On the Binding Mechanism of CMC in Si Negative Electrodes for Li-Ion Batteries. *Electrochem. Commun.* **2007**, *9*, 2801–2806. [CrossRef]

molecules

MDPI

Article

The Preparation of N, P-Doped NiSe Nanorod Electrode Materials on Nickel Foam Using the Microwave Method for High-Performance Supercapacitors

Zhen Lu [1], Hongjie Kang [1], Qianwen Duan [1], Chao Lv [2], Rui Liu [1], Feng Feng [1,*] and Haidong Zhao [1,*]

1 School of Chemistry and Chemical Engineering, Shanxi Datong University, Datong 037009, China; luzhen2024@sxdtdx.edu.cn (Z.L.); 15603450662@163.com (H.K.); 18834839312@163.com (Q.D.); liurui198689@163.com (R.L.)
2 School of Coal Engineering, Shanxi Datong University, Datong 037009, China; lvchao0711@126.com
* Correspondence: feng-feng64@263.net (F.F.); zhaohd@sxdtdx.edu.cn (H.Z.)

Abstract: Transition metal selenides have the leading position in the field of energy storage and conversion due to their high theoretical capacity, good electrical conductivity, and cycling stability. Nickel is widely used for the construction of positive electrodes in devices due to its good conductivity, variable valence state, and ideal redox activity. NiSe materials have high internal resistance and are prone to volume change during charging and discharging, thus affecting the practical application of this electrode material, and the reported NiSe materials have not achieved a more desirable capacity value. Therefore, in this study, N, P-NiSe nanoelectrode materials were prepared using nickel foam as the nickel source and hexachlorocyclotriphonitrile as the nitrogen and phosphorus dopant using an efficient, energy-saving, and simple microwave method. It was also characterised by XRD and XPS to confirm the successful preparation of N, P-NiSe materials. In addition, the material yielded a high capacitance value (3184 F g^{-1}) and good cycling stability (72% of the initial capacitance value was retained after 4000 cycles) in electrochemical tests. To demonstrate its excellent suitability for practical applications, an asymmetric supercapacitor was assembled using N, P-NiSe as the anode and activated carbon as the cathode. At an operating voltage of 1.6 V, the device achieved an energy density of 289.06 Wh kg^{-1} and a power density of 799.26 W kg^{-1} and retained 80% of its initial capacity after 20,000 cycles.

Keywords: N, P doped; microwave method; supercapacitors

1. Introduction

Energy storage technology represents one of the most effective solutions to the growing energy crisis and environmental protection problems facing the world today [1]. Among the commonly used green energy storage technologies, supercapacitors have been widely noticed and researched due to their excellent capacitance storage capacity and cycle stability [2,3]. The key technology for the development of supercapacitors is the development of high-performance electrode materials [4–7]. Currently, the most commonly used electrode materials include carbon materials [8–10], polyaniline (PANI) [11], polypyrrole (PPy) [12], transition metal oxides [13], manganese dioxide and other active electrode materials [14]. However, the electrochemical properties of these electrode materials are not ideal. Therefore, the design and development of high-performance supercapacitor electrode materials is still a crucial topic in the field of green energy storage.

Nickel selenide, a semiconductor material, has received considerable attention in the fields of catalysis and energy storage due to its significant electronic and magnetic properties [15–17]. In 2012, Gao et al. prepared NiSe using a hydrothermal method and used it as a catalyst for electrochemical hydrogen production [18]. In 2017, Tian et al. prepared NiSe nanorods using a hydrothermal method, obtaining a specific capacitance of 6.81 F cm^{-2}

(current density 5 mA cm^{-2}) and demonstrating their good cycling stability. Furthermore, the capacitance retention was 90.09% after 3000 cycles (current density 3.6 A g^{-1}) [19]. Younas et al. successfully prepared $Ni_{0.85}Se$ nanosheets. The hexagonal-phase structure of $Ni_{0.85}Se$ nanosheets exhibited excellent electrochemical performance in supercapacitors, as demonstrated with a fast microwave method. The nanosheet morphology shortened the diffusion of the ion transfer paths, thereby increasing the reaction kinetics. The assembled device exhibited a high energy density of 63.5 Wh kg^{-1} and excellent cycling performance (95% of the starting capacitance value after 8000 cycles) [20].

However, single nickel selenide materials often do not exhibit desirable capacitance values, and researchers have attempted to make greater progress by introducing heteroatoms to modify their performance [21–23]. Liu et al. prepared P-$NiSe_2$ using a hydrothermal method. The results of electrochemical tests demonstrated that P doping could markedly enhance the HER-catalysed activity of $NiSe_2$ nanomaterials and reduce their hydrogen precipitation reaction overpotentials [24]. Yan et al. demonstrated that N doping disrupted the octahedral coordination environment of the Ni atoms in NiSe, resulting in a decrease in the simplicity of orbitals and an upward band centre shift. This improved interfacial charge transfer kinetics and enhanced redox kinetics. The assembled Li-S batteries exhibited high performance (682.6 mA h g^{-1} at 5 C) and a low capacity decay rate, with the inclusion of N-NiSe [25]. In 2022, Wang prepared NiSe-RGo using a hydrothermal method. This method resulted in the homogeneous distribution of NiSe on graphene, which was beneficial for the electrochemical performance of the battery. This resulted in faster charge transport and the diffusion of NiSe-RGO nanohybrids, as well as an abundance of NiSe-RGO with a high capacitance value (781 C g^{-1} at 1 A g^{-1}) and good cycling stability (90% of the initial capacitance after 5000 cycles) [26]. It can be concluded that doping with heteroatoms can alter the electrochemical properties of composites, demonstrating great potential for energy storage. The incorporation of heteroatoms not only alters the lattice structure to a greater extent, thereby exposing more active sites, but also induces stronger electronic interactions and optimises its free energy. Secondly, doping other elements can regulate the microstructure of the electrode material and prevent the structure from collapsing and affecting its performance during the electrolysis process.

In this study, nickel foam (NF) served as the nickel source, while hexachlorocyclotriphosphonitrile (HCCP) was employed as the nitrogen–phosphorus source. These components were utilised in the synthesis of nitrogen–phosphorus-doped nickel selenide (N, P-NiSe) electrode materials with self-supported structures. The self-supported structure reduced the loss of electron transport in the bonding agent, and the reticulated structure of nickel foam provided a larger surface area for electrochemical reactions, thus increasing the availability of active sites to enhance electrochemical performance. In this study, the properties were optimised through the introduction of N and P elements, and the preparation method was optimised to improve the quality and performance of the resulting materials. In comparison to the conventional synthesis method characterised by a lengthy reaction time and a significant energy expenditure, the (N, P-NiSe) rod-shaped nanocomposite electrode materials were successfully synthesised through the precise control of microwave power and microwave time. The specific capacitance of the N, P-NiSe electrode reached 3184 F g^{-1} at 1 A g^{-1} and retained 72% of its initial capacity after 4000 (10 A g^{-1}) cycles. Furthermore, an asymmetric supercapacitor (ASC) was prepared with N, P-NiSe as the anode and activated carbon (AC) as the cathode. The ASC demonstrated a power density of 799.26 W kg^{-1} and an energy density of 289.06 Wh kg^{-1}. Additionally, 80% of the initial capacity was retained after 20,000 cycles.

2. Results

2.1. Morphology and Microstructure

Figure 1a shows the SEM images of the product prepared under the following conditions: 800 W, 90 s (W1). As the microwave heating time was relatively short, the N, P-NiSe nanorods were formed less and were distributed unevenly. As illustrated in Figure 1b, with the increase in reaction time (800 W, 120 s) (W2), the nanorod particles of N, P-NiSe exhibit a uniform distribution. As shown in Figure 1c,d, when the microwave heating time is further increased or the power is increased, the rod particles aggregate on the nickel foam and eventually agglomerate into irregular shapes. The microscopic morphology of N, P-NiSe particles can be regulated by microwave power and time, with this morphology having a significant impact on electrochemical performance. The formation of agglomerates or uneven dispersion can lead to a deterioration in electrochemical performance. This is consistent with the subsequent electrochemical performance tests, where the highest capacitance values were also obtained for uniform rod-like nanomorphology observed at 800 W, 120 s (W2). The results demonstrated that the microstructure could be modulated by microwave power and time and directly affected electrochemical properties. As illustrated in Figure 1e, nitrogen (N) and phosphorus (P) were evenly distributed throughout the product, indicating successful doping. In addition, the presence of a large amount of Au elements in the EDX spectra is due to the fact that the samples were sprayed with Au in order to obtain a clearer imaging effect when performing electron microscopy. The presence of potassium is due to the use of a small amount of KOH solution to dissolve the Se powder during sample preparation, resulting in a small amount of K in the sample. Oxygen was detected because the air itself contains a large amount of oxygen.

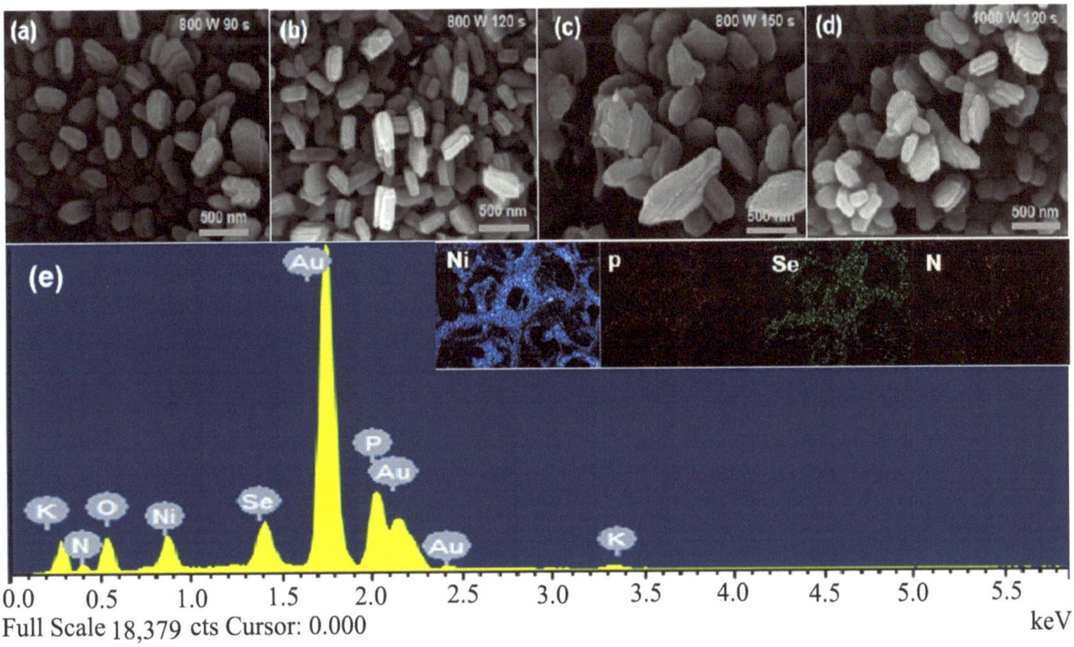

Figure 1. SEM images and EDS pattern of N, P-NiSe electrode material: (**a–d**) products prepared at 800 W 90 s (W1), 800 W 120 s (W2), 800W 150 s (W3), and 1000 W 120 s (W4), respectively; (**e**) the electronic energy pattern of the product prepared at 800 W 120 s (W2).

The nanostructures within the N, P-NiSe samples (W2) were examined using TEM. As illustrated in Figure 2a, the N, P-NiSe nanorods (W2) (approximately 30 nm in width and 150–200 nm in length) are uniformly distributed on the nickel foam. The peculiar morphology and structure are further confirmed with the HRTEM images in Figure 2b,c. The lattice fringes of NiSe were 2.72 nm (101) and those of Ni were 2.02 nm (111), which is consistent with the strongest crystalline surface results of XRD. This provides further evidence of the formation of the N, P-NiSe nanorods on nickel foam. Figure 2d illustrates the result of the selected area electron diffraction (SAED) analysis, revealing the presence of dispersed bright crystalline spots, thereby confirming that N, P-NiSe is a single-crystal structure.

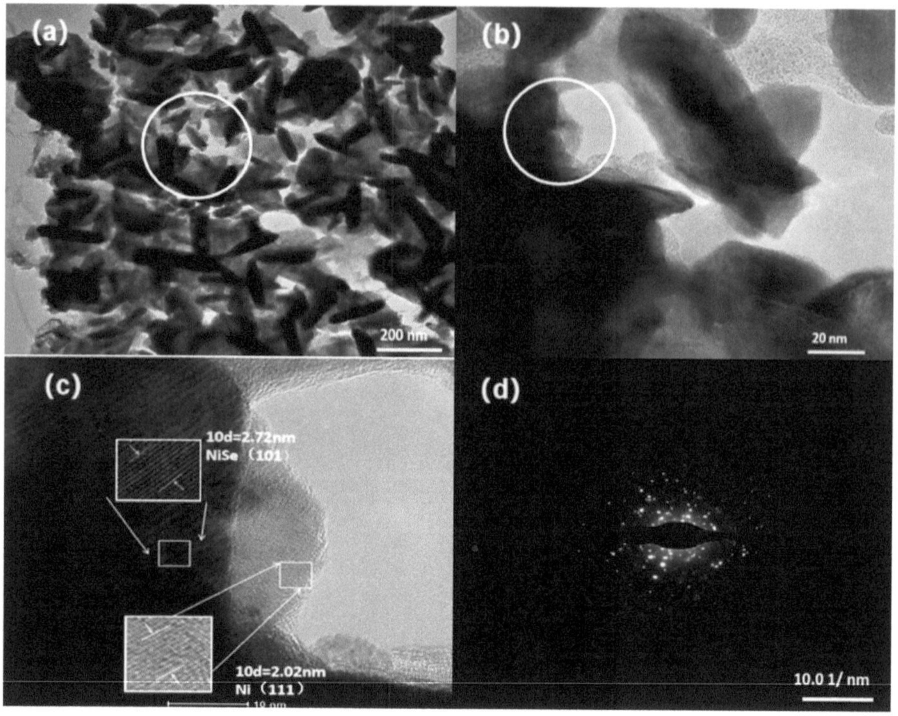

Figure 2. TEM images of N, P-NiSe electrode material (W2): (**a**) TEM image; (**b**,**c**) HRTEM images, (**b**) is the enlarge image labelled with a circle in (**a**), and (**c**) is the enlarge image labelled with a circle in (**b**), the bigger square in (**c**) is the enlarge image of smaller square; (**d**) a selected area electron diffraction (SAED) image.

The composition and structure of the N, P-NiSe electrode material (NiSe, W2, and W4) were investigated by X-ray powder diffraction (XRD), and the results are presented in Figure 3. The diffraction peaks of NiSe were observed at 2θ angles of 28.13°, 32.85°, 44.37°, 49.79°, 59.61°, 60.93°, and 68.87°, and could be attributed to the (100), (101), (102), (110), (103), (201), and (202) crystal faces of NiSe. The other diffraction peaks labelled with an asterisk (*) were located at 35.58°, 40.92°, 42.58°, and 53.96°, all of which can be attributed to Ni_5Se_5, another crystalline form of NiSe. The XRD spectra of W2 and W4 samples are also shown in Figure 3. The results show that most of the diffraction peaks can be attributed to NiSe and Ni_5Se_5. In addition, the impurity peaks appearing in the XRD spectra of the W2 and W4 samples (diffraction angles of 51.51° and 76.20°, respectively) were most likely owing to N and P doping.

Further analysis of the elemental valence and elemental composition of N, P-NiSe (W2) was conducted using XPS. Figure 4a shows the full spectrum of the N, P-NiSe electrode material, which revealed the presence of Ni, Se, N, P, and C elements. Figure 4b depicts the Se 3D spectrum. The binding energies of SeOx and Se $3d_{5/2}$ correspond to 58.91 eV and 54.32 eV, representing the metal–selenide bond (Ni-Se Ni^{2+}, Se^{2-}) [27,28]. Figure 4c presents the P 2p spectrum, with binding energies at 133.4 eV and 136.8 eV. Figure 4d displays the N 1s spectrum, with the binding energy of pyrrole N corresponding to 399.5 eV, indicating the presence of both N and P elements. Figure 4e shows the spectra of Ni 2p, with binding energies at 855.83 eV and 873.35 eV for Ni $2p_{3/2}$ and Ni $2p_{1/2}$, representing the characteristics of Ni^{2+} for NiSe [27,28]. Additionally, binding energies at the satellite peaks of Ni 2p were observed at 879.16 eV and 861.74 eV, respectively. The XPS analyses indicated the successful preparation of nickel foam self-supported N, P-NiSe electrode materials.

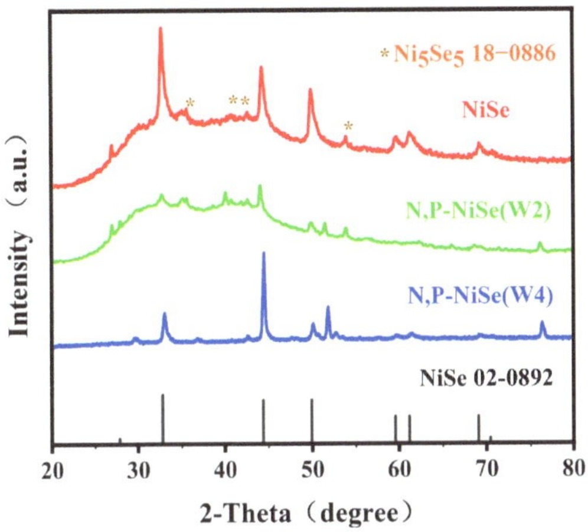

Figure 3. XRD patterns of N, P-NiSe electrode material (NiSe, W2, and W4).

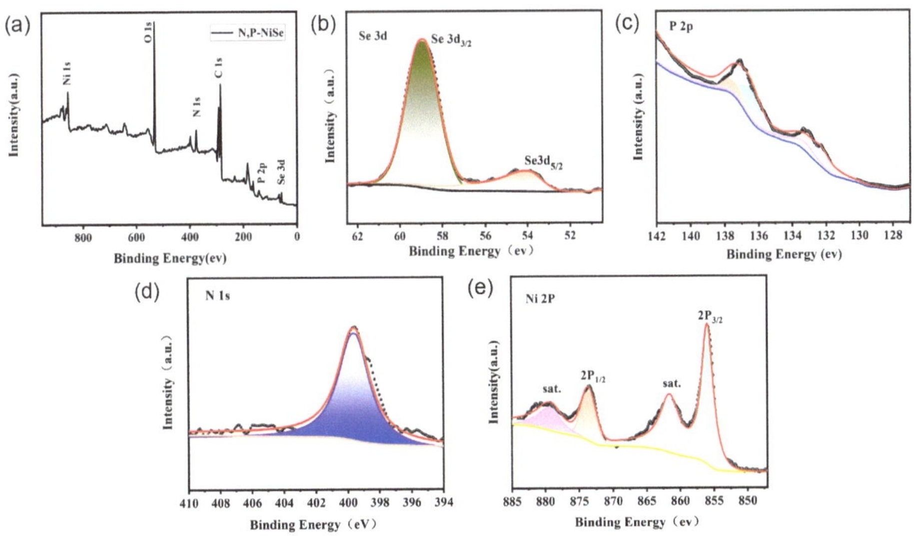

Figure 4. XPS patterns of N, P-NiSe electrode material (W2): (**a**) the full spectrum; (**b**–**e**) the spectra of Se, P, N, and Ni, respectively.

2.2. Electrochemical Performance Analysis

2.2.1. Electrochemical Properties of the Three-Electrode System

The electrochemical performance of the N, P-NiSe/NF electrode was tested in the three-electrode system with 6 M KOH under different conditions. The variation in the cyclic voltammetry (CV) curves for sample W2 at different scan rates is shown in Figure 5a. It can be seen in Figure 5a that the oxidation peak shifts to the right (from 0.44 to 0.56 V) and the reduction peak shifts to the left (from 0.22 to 0.17 V) as the scan rate increases.

This is due to the fact that the electrode material exhibits increased resistance as the scan rate increases, which hinders the diffusion of ions and electrolytes between the electrodes. Figure 5b shows a GCD curve of sample W2. The capacity of the N, P-NiSe material was calculated to be 3184 F g^{-1} at a current density of 1 A g^{-1}. A clear charge–discharge plateau can be observed in this curve, indicating that there was pseudocapacitance in the N, P-NiSe material and that it exhibited a pronounced cell-type behaviour [29]. The N, P-NiSe electrode presents a pseudocapacitive nature, and its charge storage mechanism is purely a faradaic redox reaction. Its proposed reaction mechanism is explained in [27–29].

$$\text{NiSe} + \text{OH}^{-1} \rightarrow \text{NiSeOH} + \text{e}^{-1} \quad (1)$$

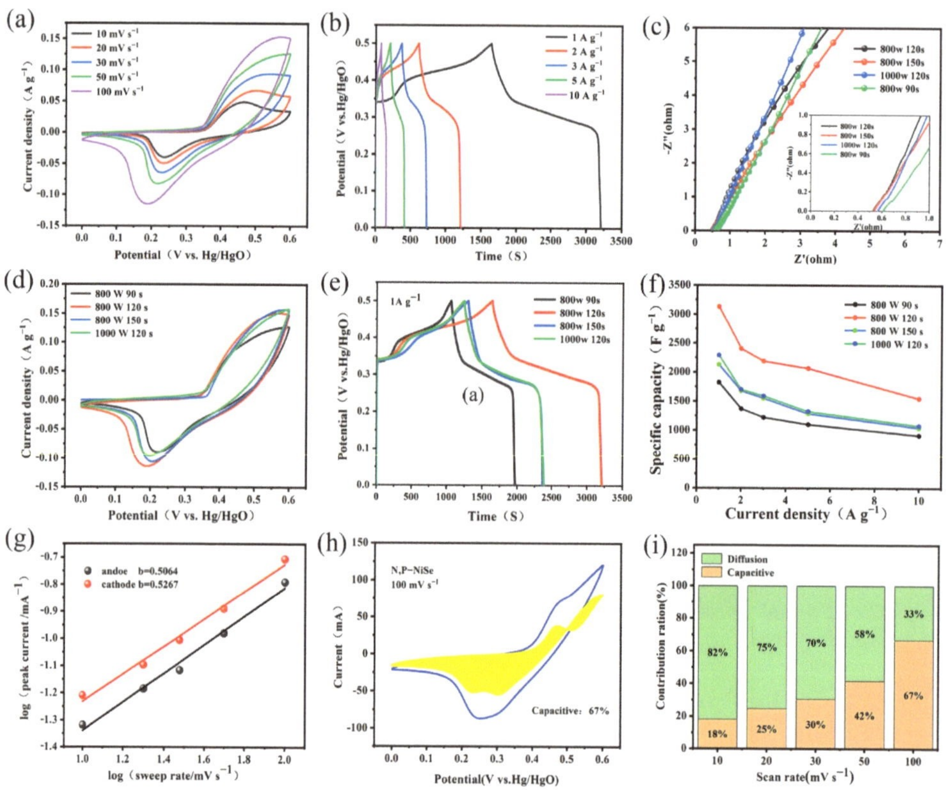

Figure 5. (**a**) CV curve of N, P-NiSe under optimal conditions (800 W/120 S) (W2); (**b**) GCD curve of (sample W2); (**c**) Nernquist diagram of N, P-NiSe under different reaction parameters (W1, W2, W3, and W4); (**d**) CV curves of N, P-NiSe under different microwave conditions (W1, W2, W3, and W4); (**e**) GCD curves of N, P-NiSe under different reaction parameters (W1, W2, W3, and W4); (**f**) the specific capacity of N, P-NiSe samples under different preparation conditions (W1, W2, W3, and W4); (**g**) N, P-NiSe charge storage dynamics calculation (W2); (**h**) capacity share of N, P-NiSe energy storage at 100 mV^{-1} (W2); (**i**) relative contribution of capacitance and diffusion-controlled charge storage (W2).

The differences in the internal resistance of the samples under different reaction conditions are shown in Figure 5c. Sample W2 exhibited a smaller solution resistance in the low-frequency region compared with the material under other conditions. Figure 5d shows the CV curves of N, P-NiSe materials under a variety of reaction conditions, from which it can be seen that the mathematical integral area of the CV curve is greatest for sample W2. Figure 5e illustrates the GCD curves of N, P-NiSe under different conditions. This figure also confirms that sample W2 exhibits a longer discharge time and superior Coulombic efficiency compared to the other samples. The capacitance values of samples W1, W3, and W4 were 1816 F g^{-1}, 2130 F g^{-1}, and 2242 F g^{-1}, respectively, which were in agreement with the results of CV tests. Figure 5f presents the line graphs of the capacitance values of N, P-NiSe at different current densities for different conditions. It can be observed that 120 s has the highest capacitance value regardless of the condition of 800 W. The charge storage mechanism of N, P-NiSe material was investigated by fitting and calculating cyclic voltammetry curves [30]. The relationship between the peak current (I) and the scan rate (v) was determined using the following formula:

$$I = a \times v^{b} \quad (2)$$

where b is the slope of the fitted straight line. If b = 1, it can be concluded that a non-Faraday process controls charge storage. Conversely, if b = 0.5, it can be inferred that a Faraday process controls charge storage. As shown in Figure 5g, the fitted b values for the anodic oxidation and cathodic reduction peaks of N, P-NiSe were 0.50554 and 0.5267, respectively. This finding suggests that diffusion-controlled Faraday processes (cell behaviour) dominate charge storage. Furthermore, the total charge storage contribution can be determined using the following equation:

$$i\,(v) = k_1 v + k_2 v^{1/2} \quad (3)$$

where k_1v represents the surface control capacitance, and $k_2v^{1/2}$ is the diffusion control capacitance [31]. As shown in Figure 5h, the surface control capacitance accounted for 67% of the total capacity at 100 mV s^{-1}. Figure 5i illustrates the capacitance occupancy at varying sweep rates, demonstrating that the capacitive behaviour dominates the charge storage mechanism of N, P-NiSe/NF as the sweep rate increases. The doping of the N, P elements provides a plethora of active sites for electron storage, endowing the N, P-NiSe/NF with excellent reversible pseudocapacitive charge storage capability.

In addition, with the preparation condition of 800 W and 120 S, we performed comparison experiments with the dosage of 5 mg, 10 mg, and 15 mg of hexachlorocyclotriphosphonitrile as the dopant precursor and found that the total amount of N and P doping could be adjusted, but due to the small amount, it was difficult to determine the accurate total amount of N and P doping directly. In order to explore the effect of N and p doping on the electrochemical performance of the electrodes, cyclic voltammetry and galvanostatic charge–discharge tests were carried out on the samples with different dopant precursor amounts (as shown in Figure S1, Supplementary Materials). The specific capacitance values for precursor doping amounts of 5 mg, 10 mg, and 15 mg were 2340, 3184, and 2280 F g^{-1}, respectively, indicating that the optimal precursor doping amount was 10 mg.

The electrochemical properties of N, P-NiSe electrode materials and the undoped single NiSe were compared. The cyclic voltammetry (CV), galvanostatic charge–discharge (GCD), and electrochemical impedance spectroscopy (EIS) plots of the NiSe and N, P-NiSe (W2) materials are shown in Figure 6a, Figure 6b, and Figure 6c, respectively. There is a significant difference between the two CV plots at 100 mV s^{-1}. The N, P-NiSe integral area is larger than that of NiSe, indicating that N, P doping improves the electrochemical performance. In comparison to the maximum specific capacitance value of NiSe (514 F g^{-1} at 1 A g^{-1}), the capacitance value of N, P-doped NiSe reached 3184 F g^{-1} at the same current density. This can be attributed to the distortion of the lattice of the electrode material caused by the doping of N, P elements, which generates more electrochemically active sites and thus increases the capacitance value. The introduction of electrochemically active sites, which increased the capacitance value dramatically, is evident in the electrochemical impedance spectroscopy (EIS) impedance plots. The resistance (Rs) of the N, P-doped NiSe electrode material and that of the NiSe electrode material were 0.53 Ω and 0.58 Ω, respectively. This indicates an increase in conductivity after doping, which is consistent with the large capacitance values obtained for N, P-doped NiSe. As illustrated in Figure 6d, the N, P-NiSe electrode material exhibits excellent cycling stability, with a capacitance retention of above 72% following a 4000-cycle performance test at 15 A g^{-1}. Figure 6e illustrates that, despite the notable enhancement in stability, it is still possible that the internal volume of N, P-NiSe underwent expansion due to repeated charging and discharging cycles. This may have resulted in the irreversible oxidation of some electrodes, which led to a decline in performance. Furthermore, a slight increase in the internal resistance was observed. The multiplicative performance of N, P-NiSe is demonstrated in Figure 6f, which illustrates that after multiple charging and discharging cycles at different current densities, the capacitance retention of N, P-NiSe remains at more than 72% after 4000 cycles. The capacitance value of sample W2 was maintained at 3184 F g^{-1} at 1 A g^{-1} after several charge–discharge cycles, indicating that the material has good multiplicative properties. A comparison of N, P-NiSe with the same type of material is presented in Table 1.

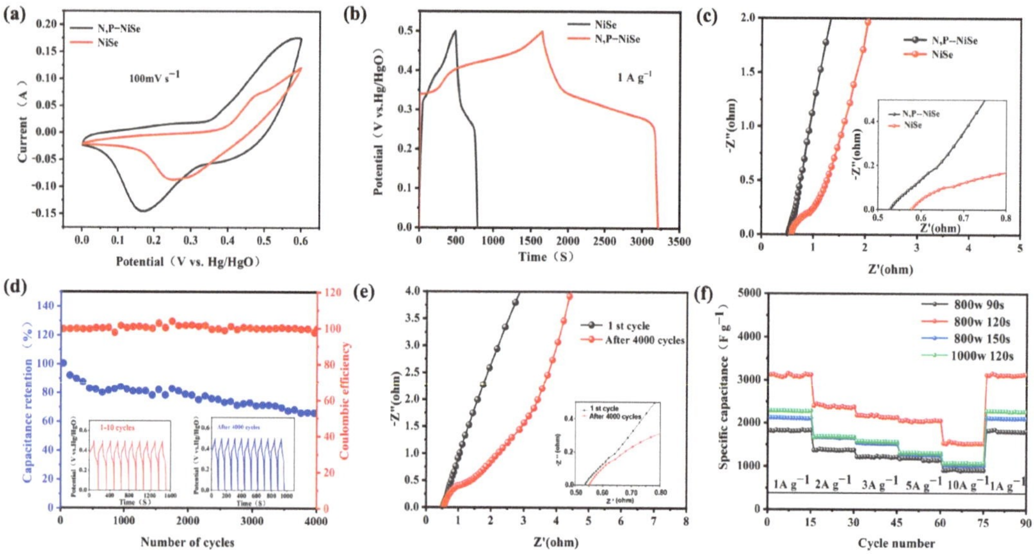

Figure 6. (**a**) CV curves of N, P-NiSe (W2) and NiSe; (**b**) GCD curves of N, P-NiSe (W2) and NiSe; (**c**) EIS plots of N, P-NiSe (W2) and NiSe; (**d**) capacitance retention and Coulomb efficiency; (**e**) EIS plots before and after cycling (W2); (**f**) multiplicity performance of sample (W1, W2, W3, and W4).

Table 1. Comparison of electrochemical properties of N, P-NiSe (W2) with other materials.

Material	Method	Specific Capacity	Retention Rate	Ref.
NiSe microspheres	Hydrothermal	492 F g^{-1} (0.5 A g^{-1})	200 cycles, 84.6%	[32]
$NiSe_2$@rGO	Microwave	580 F g^{-1} (1 A g^{-1})	5000 cycles, 78.13%	[33]
NiSe@$MoSe_2$	Hydrothermal	223 Fg^{-1} (1 g^{-11})	1000 cycles, 93.7%	[34]
$Ni_{0.85}Se$@$MoSe_2$	Hydrothermal	774 g^{-1} (1 A g^{-1})	5000 cycles, 88%	[35]
NiSe@RGO	Hydrothermal	781 C g^{-1} (1 A g^{-1})	5000 cycles, 90%	[26]
N, P-NiSe	Microwave	3184 F g^{-1} (1 Ag^{-1})	4000 cycles, 72%	This Work

2.2.2. Electrochemical Properties of the Two-Electrode System

In order to explore the performance of the N, P-NiSe electrode material in greater depth, sample W2 and activated carbon were combined to form an asymmetric supercapacitor. Figure 7a–d illustrate the electrochemical performance of the N, P-NiSe//AC supercapacitor, which comprises sample W2 and activated carbon in a liquid electrolyte.

The mass ratios of the AC electrode and the N, P-NiSe (sample W2) electrode were adjusted according to Equation (4). Figure 7a illustrates the cyclic voltammetry (CV) curves of the AC electrode (operating at −1 to 0 V) and the N, P-NiSe electrode (operating at 0 to 0.6 V) in the N, P-NiSe//AC configuration. This configuration was based on the complementary relationship between the working electrodes, which comprise a bilayer and exhibit pseudocapacitance. Figure 7b shows the CV curves of N, P-NiSe//AC, with no discernible distortion, indicating that it has good current responsiveness. N, P-NiSe//AC was capable of functioning properly at 1.6 V. Figure 7c illustrates the CV curves of N, P-NiSe//AC at 10–100 mV s^{-1}. Figure 7d shows the constant current GCD curves at different voltages (1.3–1.7 V). It can be observed that the shape of the curve changes at 1.7 V, which may be attributed to electrode polarisation. Consequently, the optimum operating voltage window for this device was 1.6 V. Figure 7e illustrates the constant current GCD curves at varying current densities. Figure 7f presents a line graph of the capacitance values at different current densities. Figure 7g depicts the cycling performance of N, P-NiSe//AC, demonstrating that the capacitance retention remains high even after 20,000 consecutive charge–discharge cycles. Figure 7h illustrates the change in the internal resistance of the solution following 20,000 cycles. The p-value of N, P-NiSe//AC was 289.06 W kg^{-1}, and the corresponding E-value was 799.26 Wh kg^{-1}, indicating that the N, P-NiSe//AC has a high energy density. The Ragone diagrams of N, P-NiSe//AC with other electrode materials are shown in Figure 7i. Furthermore, the E-value of N, P-NiSe//AC is demonstrably higher than that of other related materials reported in the literature [32–34,36,37], thereby indicating that the electrochemical performance of N, P-NiSe//AC is superior.

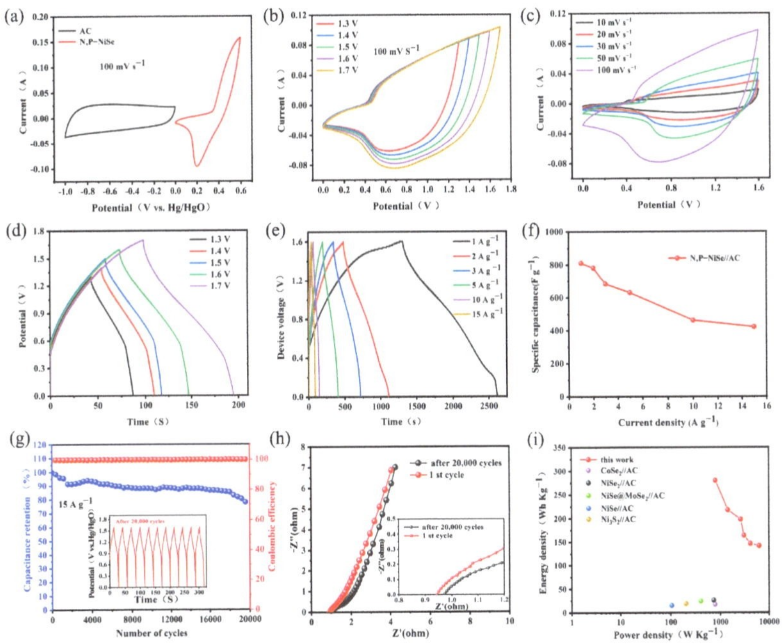

Figure 7. (**a**) CV curves of N, P-NiSe (W2) and AC; (**b**) CV curves at different voltages; (**c**) CV curves of N, P-NiSe//AC at different scanning rates at 1.6 V; (**d**) variable current GCD curves of N, P-NiSe//AC at different voltages; (**e**) GCD curves of N, P-NiSe//AC AC at different current currents; (**f**) line chart of capacitance value of N, P-NiSe//AC at different current densities; (**g**) cycle stability and coulomb efficiency after 20,000 cycles; (**h**) EIS image before and after 20,000 cycles; (**i**) energy–power density diagram of different devices.

3. Materials and Methods

3.1. Material Preparation

Nickel foam (NF) was purchased from Long Sheng Bao Products Company (Guangzhou, China). Prior to the experiment, NF was ultrasonically cleaned with hydrochloric acid, ethanol, and ultrapure water to ensure that it was free of impurities. It was then vacuum-dried at 80 °C. Selenium powder was purchased from Aladdin Industries. The hexachlorocyclotriphosphonitrile (HCCP) was sourced from Leyan (Shanghai, China). Anhydrous ethanol, ethylene glycol (EG), ethylenediamine (EDA), and potassium hydroxide (KOH) were procured from the Tianjin Damao Chemical Company (Tianjin, China).

3.2. Preparation of N, P-NiSe/Ni

Briefly, 20 mg of selenium powder, 10 mg of hexachlorocyclotriphosphonitrile powder, 300 μL of potassium hydroxide (6 M), 2 mL of $C_2H_8N_2$, and 1 mL of C_2H_6O were dissolved by stirring to obtain a purplish-black liquid. Subsequently, nickel foam was immersed in the solution and sonicated for 15 min. The foam was then transferred to a heat-resistant crucible (10 mL). After being kept in a microwave oven at a certain power and time, the nickel foam was washed with deionised water and anhydrous ethanol and further dried (80 °C, 12 h). The preparation process is illustrated in Figure 8, and the different reaction conditions are presented in Table 2.

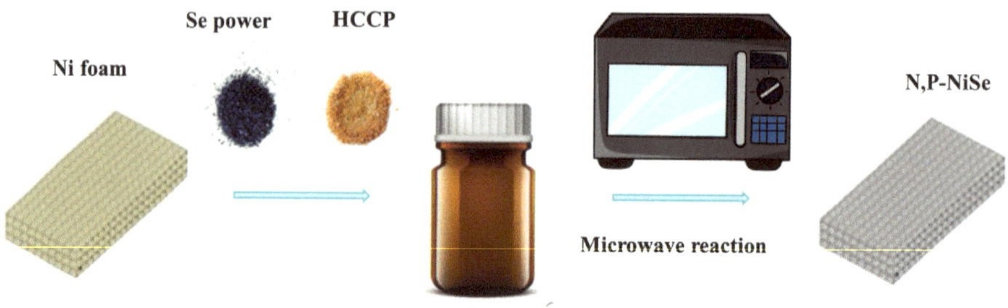

Figure 8. The preparation process of N, P-NiSe/Ni.

Table 2. Specific capacitance values of N, P-NiSe/Ni under different reaction conditions.

Sample	Microwave Watts/W	Microwave Duration/s	Duration/s Capacitance/F g^{-1} (1 A g^{-1})
W1	800 W	90 S	1816 F g^{-1}
W2	800 W	120 S	3184 F g^{-1}
W3	800 W	150 S	2130 F g^{-1}
W4	1000 W	120 S	2242 F g^{-1}

3.3. Preparation of N, P-NiSe/Ni//AC

N, P-NiSe/Ni was employed as the positive electrode, while a 1 × 1 cm^2 commercial AC electrode served as the negative electrode, and the resulting device was tested in 6 M potassium hydroxide. The negative electrode was prepared as follows: Activated carbon, acetylene black, and PVDF were mixed into a homogeneous slurry in a mass ratio of 80:10:10. The slurry was uniformly applied to 1 cm^2 NF, dried at 100 °C for 10 h, and finally pressed into the electrode sheet.

3.4. Characterisation of N, P-NiSe

The crystal structure of the resulting samples was characterised by powder X-ray diffraction (XRD). The surface morphology and microscopic internal structure of the samples were determined by scanning electron microscopy (SEM) and transmission electron microscopy (TEM). X-ray photoelectron spectroscopy (XPS) and energy-dispersive spectroscopy (EDS) were used to identify the elemental composition and distribution of the samples.

3.5. Electrochemical Testing

The electrochemical properties of N, P-NiSe composites were tested by cyclic voltammetry (CV), constant current charge–discharge (GCD), and electrochemical impedance spectroscopy (EIS) in 6 M potassium hydroxide. A platinum sheet was used as the counter electrode, and a mercuric oxide electrode was used as the reference electrode. An electrochemical workstation (CHI660E) was utilised for electrochemical data testing. The cycling stability of the materials was evaluated using the Landt battery test system (CT2001A). The capacitance values were calculated using the following equation:

$$C\ (F\ g^{-1}) = I\ \Delta t/(m\ \Delta V) \tag{4}$$

A two-electrode test system was assembled with AC as the negative electrode and the synthetic material as the positive electrode. In this case, the mass ratio between the positive and negative electrodes was determined with the following equation:

$$m^{+}/m^{-} = C^{-}\ \Delta V^{-}/C^{+}\Delta V^{+} \tag{5}$$

The energy density (E) and power density (P) were calculated using the following equations:

$$E = C\ \Delta V^{2}/7.2 \tag{6}$$

$$P = 3600\ E/\Delta t \tag{7}$$

4. Conclusions

In summary, N, P-NiSe composites were prepared using the microwave method, exhibiting excellent capacitance performance and remarkable capacitance retention at high current density and long operation time. The synergistic interaction between N, P and NiSe resulted in the excellent electrochemical properties of N, P-NiSe. The capacitance values of the material were up to 3184 F g^{-1} (1 A g^{-1}), and it was observed that 72% of the original capacitance value was maintained through 4,000 charge–discharge cycles. Furthermore, the practical application of the electrode was demonstrated through the assembly of devices comprising N, P-NiSe//AC, which exhibited an energy density of 289.06 W h kg^{-1} and a power density of 799.26 W kg^{-1} and retained 80% of the initial capacitance after 20,000 cycles. These outcomes not only offer novel insights into the optimisation of NiSe but also facilitate the advancement of novel supercapacitors.

Supplementary Materials: The following supporting information can be downloaded at: https://www.mdpi.com/article/10.3390/molecules29133224/s1. Figure S1 (a) cyclic voltammetry curve and (b) galvanostatic charge-discharge curve of different N, P dopant precursor contents.

Author Contributions: Conceptualisation, Z.L. and H.Z.; methodology, C.L. and R.L.; formal analysis, Z.L.; data curation, H.K. and Q.D.; writing—original draft preparation, H.K. and Z.L.; writing—review and editing, Z.L. and H.Z.; project administration, Z.L.; funding acquisition, F.F. All authors have read and agreed to the published version of the manuscript.

Funding: This research was funded by the program of the Scientific Activities of Selected Returned Overseas Professionals in Shanxi Province (20220032), Da Tong Key Research Project of Science and Technology Planning (2023025), and Key Research and Development Project of Shanxi Province (202102090301008).

Institutional Review Board Statement: Not applicable.

Informed Consent Statement: Not applicable.

Data Availability Statement: The data presented in this study are available on request from the corresponding author.

Acknowledgments: The authors express their gratitude to the Shanxi Province Union Laboratory of Clean Energy Materials and Shanxi Province Key Laboratory of Chemical Biosensing, for providing the laboratories used in this study.

Conflicts of Interest: The authors declare no conflicts of interest.

References

1. Zhang, R.; Sharma, R.; Tan, Z.; Kautish, P. Do export diversification and stock market development drive carbon intensity? Th role of renewable energy solutions in top carbon emitter countries. *Renew. Energy* **2022**, *185*, 1318–1328. [CrossRef]
2. Kebede, A.A.; Kalogiannis, T.; Mierlo, J.M.; Berecibar, M. A comprehensive review of stationary energy storage devices for larg scale renewable energy sources grid integration. *Renew. Sustain. Energy Rev.* **2022**, *159*, 112213. [CrossRef]
3. Usman, F.O.; Ani, E.C.; Ebirim, W. Danny Jose Portillo Montero, Kehinde Andrew Olu-lawal, Nwakamma Ninduwezuor-Ehiobu Integrating renewable energy solutions in the manufacturing industry: Challenges and opportunities: A review. *Eng. Sci. Techno J.* **2024**, *5*, 674–703.
4. Huang, J.; Xie, Y.; You, Y.; Yuan, J.; Xu, Q.; Xie, H.; Chen, Y. Rational Design of Electrode Materials for Advanced Supercapacitors From Lab Research to Commercialization. *Adv. Funct. Mater.* **2023**, *33*, 2213095. [CrossRef]
5. Zhao, H.; Hu, X.; Kang, H.; Feng, F.; Guo, Y.; Lu, Z. Microwave Construction of NiSb/NiTe Composites on Ni-Foam fo High-Performance Supercapacitors. *ACS Omega* **2023**, *9*, 2597–2605. [CrossRef] [PubMed]
6. Shahedi, M.; Hadi, A.R.; Salehghadimi, L.; Tabrizi, A.G.; Farhoudian, S.; Babapoor, A.; Pahlevani, M. Flexible all-solid-stat supercapacitors with high capacitance, long cycle life, and wide operational potential window: Recent progress and futur perspectives. *J. Energy Storage* **2022**, *50*, 104223.
7. Lu, Z.; Zhao, H.; Luo, J.; Bi, R. Self-supporting lignin-based carbon material flexible supercapacitor prepared by the microwav method. *Energy Fuels* **2022**, *36*, 8480–8487. [CrossRef]
8. Richard, L. McCreery. Advanced Carbon Electrode Materials for Molecular Electrochemistry. *Chem. Rev.* **2008**, *108*, 2646–2687.
9. Mandal, S.; Thangarasu, S.; Thong, P.T.; Kim, S.-C.; Shim, J.-Y.; Jung, H.-Y. Positive electrode active material developmen opportunities through carbon addition in the lead-acid batteries: A recent progress. *J. Power Sources* **2021**, *485*, 229336. [CrossRef
10. Fang, B.; Binder, L. A Novel Carbon Electrode Material for Highly Improved EDLC Performance. *J. Phys. Chem. B* **2006**, *110*, 7877–7882. [CrossRef]
11. Elumalai, P.; Charles, J. Investigation of structural and optical properties of ternary polyaniline-polypyrrole-nickel oxide (PANI PPy-NiO) nanocomposite for optoelectronic devices. *Polym. Int.* **2023**, *72*, 176–188. [CrossRef]
12. Kumar, A.; Ibraheem, S.; Ali, S.; Maiyalagan, T.; Javed, M.S.; Saad, R.K.G.A.; Yasin, G. Polypyrrole and polyaniline-based membranes for fuel cell devices: A review. *Surf. Interfaces* **2022**, *29*, 101738. [CrossRef]
13. Cao, Y.; He, Y.; Gang, H.; Wu, B.; Yan, L.; Wei, D.; Wan, H. Stability study of transition metal oxide electrode materials. *J. Power Sources* **2023**, *560*, 232710. [CrossRef]
14. Ahmed, F.; Umar, A.; Kumar, S.; Shaalan, N.M. Manganese dioxide nanoparticles/reduced graphene oxide nanocomposites for hybrid capacitive desalination. *Adv. Compos. Hybrid Mater.* **2023**, *6*, 19. [CrossRef]
15. Hussain, I.; Sahoo, S.; Lamiel, C.; Nguyenc, T.T.; Ahmeda, M.; Xi, C.; Iqbal, S.; Ali, A.; Abbas, N.; Javedf, M.S.; et al. Research progress and future aspects: Metal selenides as effective electrodes. *Energy Storage Mater.* **2022**, *47*, 13–43. [CrossRef]
16. Tang, G.; Liang, J.; Wu, W. Transition Metal Selenides for Supercapacitors. *Adv. Funct. Mater.* **2024**, *34*, 2310399. [CrossRef]
17. Khan, S.; Ullah, N.; Mahmood, A.; Saad, M.; Ullah, Z.; Ahmad, W.; Ullah, S. Recent Advancements in the Synthetic Mechanism and Surface Engineering of Transition Metal Selenides for Energy Storage and Conversion Applications. *Energy Technol* **2023**, *11*, 2201416. [CrossRef]
18. Gao, M.-R.; Lin, Z.Y.; Zhuang, T.T.; Jiang, J.; Xu, Y.F.; Zheng, Y.R.; Yu, S.H. Mixed-solution synthesis of sea urchin-like NiSe nanofiber assemblies as economical Pt-free catalysts for electrochemical H_2 production. *J. Mater. Chem.* **2012**, *22*, 13662–13668 [CrossRef]
19. Tian, Y.; Ruan, Y.; Zhang, J.; Yang, Z.; Jiang, J.; Wang, C. Controllable growth of NiSe nanorod arrays via one-pot hydrothermal method for high areal-capacitance supercapacitors. *Electrochim. Acta* **2017**, *250*, 327–334. [CrossRef]
20. Younas, W.; Naveed, M.; Cao, C.; Zhu, Y.; Du, C.; Ma, X.; Mushtaq, N.; Tahir, M.; Naeem, M. Facile One-Step Microwave-Assisted Method to Synthesize Nickel Selenide Nanosheets for High-Performance Hybrid Supercapacitor. *J. Colloid Interface Sci.* **2022**, *608*, 1005–1014. [CrossRef]
21. Meenal Gupta1, P.K.; Singh Bhattacharya1, Y.M.; Shulga, N.Y.; Shulga, Y.K. Progress, status and prospects of non-porous, heteroatom-doped carbons for supercapacitors and other electrochemical applications. *Appl. Phys. A* **2019**, *125*, 1–15.
22. Li, Z.; Lin, J.; Li, B.; Yu, C.; Wang, H.; Li, Q. Construction of heteroatom-doped and three-dimensional graphene materials for the applications in supercapacitors: A review. *J. Energy Storage* **2021**, *44*, 103437. [CrossRef]
23. Feng, X.; Bai, Y.; Liu, M.; Li, Y.; Yang, H.; Wang, X.; Wu, C. Untangling the respective effects of heteroatom-doped carbon materials in batteries, supercapacitors and the ORR to design high performance materials. *Energy Environ. Sci.* **2021**, *14*, 2036–2089 [CrossRef]

24. Liu, H.; Wang, K.; He, W.J.; Zheng, X.; Gong, T.; Li, Y.; Zhao, J.; Zhang, J.; Liang, L. Phosphorus-doped nickel selenides nanosheet arrays as highly efficient electrocatalysts for alkaline hydrogen evolution. *Int. J. Hydrogen Energy* **2021**, *46*, 1967–1975. [CrossRef]
25. Yan, T.; Feng, J.; Zeng, P.; Zhao, G.; Wang, L.; Yuan, C.; Cheng, C.; Li, Y.; Zhang, L. Modulating eg orbitals through ligand engineering to boost the electrocatalytic activity of NiSe for advanced lithium-sulfur batteries. *J. Energy Chem.* **2022**, *74*, 317–323. [CrossRef]
26. Wang, J.; Li, S.; Zhu, Y.; Zhai, S.; Liu, C.; Fu, N.; Hou, S.; Niu, Y. Metal-organic frameworks-derived NiSe@RGO composites for high-performance asymmetric supercapacitors. *J. Electroanal. Chem.* **2022**, *919*, 116548. [CrossRef]
27. Chen, H.; Chen, S.; Fan, M.; Li, C.; Chen, D.; Tian, G.; Shu, K. Bimetallic nickel cobalt selenides: A new kind of electroactive material for high-power energy storage. *J. Mater. Chem. A* **2015**, *3*, 23653. [CrossRef]
28. Kirubasankar, B.; Murugadoss, V.; Lin, J. In situ grown nickel selenide on graphene nanohybrid electrodes for high energy density asymmetric supercapacitors. *Nanoscale* **2018**, *10*, 20414–20425. [CrossRef] [PubMed]
29. Gogotsi, Y.; Penne, R.M. Energy Storage in Nanomaterials-Capacitive, Pseudocapacitive, or Battery-like? *ACS Nano* **2018**, *12*, 2081–2083. [CrossRef]
30. Girard, H.-L.; Wang, H.; d'Entremont, A.; Pilon, L. Physical Interpretation of Cyclic Voltammetry for Hybrid Pseudocapacitors. *J. Phys. Chem. C* **2015**, *119*, 11349–11361. [CrossRef]
31. Eftekhari, A. Surface Diffusion and Adsorption in Supercapacitors. *ACS Sustain. Chem. Eng.* **2019**, *7*, 3692–3701. [CrossRef]
32. Guo, K.L.; Yang, F.F.; Cui, S.Z.; Chen, W.; Mi, L. Controlled synthesis of 3D hierarchical NiSe microspheres for high-performance supercapacitor design. *RSC Adv.* **2016**, *6*, 46523–46530. [CrossRef]
33. Lu, Z.; Zhao, H.; Luo, J.; Wang, J. Reduced-graphene-oxide-modified selfsupported $NiSe_2$ nanospheres on nickel foam as a battery-type electrode material for high efficiency supercapacitors. *J. Phys. Chem. Solids* **2022**, *163*, 110593. [CrossRef]
34. Peng, H.; Zhou, J.Z.; Sun, K.J.; Ma, G.; Zhang, Z.; Feng, E.; Lei, Z. High-performance asymmetric supercapacitor designed with a novel NiSe@$MoSe_2$ nanosheet array and nitrogen-doped carbon nanosheet. *ACS Sustain. Chem. Eng.* **2017**, *5*, 5951–5963. [CrossRef]
35. Peng, H.; Wei, C.; Wang, K.; Meng, T.; Ma, G.; Lei, Z.; Gong, X. $Ni_{0.85}Se$@$MoSe_2$ Nanosheet Arrays as the Electrode for High Performance Supercapacitors. *ACS Appl. Mater. Interfaces* **2017**, *9*, 17067–17075. [CrossRef] [PubMed]
36. Dai, C.S.; Chien, P.Y.; Lin, J.Y.; Chou, S.W.; Wu, W.K.; Li, P.H.; Wu, K.Y.; Lin, T.W. Hierarchically Structured Ni_3S_2/Carbon Nanotube Composites as High-Performance Cathode Materials for Asymmetric Supercapacitors. *ACS Appl. Mater. Interfaces* **2013**, *5*, 12168–12174. [CrossRef]
37. Yun, X.; Lu, T.; Zhou, R.; Lu, Z.; Li, J.; Zhu, Y. Heterostructured $NiSe_2$/$CoSe_2$ hollow microspheres as battery-type cathode for hybrid supercapacitors: Electrochemical kinetics and energy storage mechanism. *Chem. Eng. J.* **2021**, *426*, 131328. [CrossRef]

molecules

MDPI

Article

The Effect of Expanded Graphite Content on the Thermal Properties of Fatty Acid Composite Materials for Thermal Energy Storage

Dongyi Zhou [1,*], Shuaizhe Xiao [2] and Yicai Liu [3,*]

1. School of Energy and Mechanical Engineering, Hunan University of Humanities, Science and Technology, Loudi 417000, China
2. School of Mechanical and Energy Engineering, Shaoyang University, Shaoyang 422000, China; 19375383902@163.com
3. School of Energy Science and Engineering, Central South University, Changsha 410083, China

* Correspondence: zhoudongyi@hnsyu.edu.cn (D.Z.); lyc0300@163.com (Y.L.); Tel.: +86-0739-5236-6600 (D.Z.)

Abstract: The mass content of expanded graphite (EG) in fatty acid/expanded graphite composite phase-change materials (CPCMs) affects their thermal properties. In this study, a series of capric–myristic acid/expanded graphite CPCMs with different EG mass content (1%, 3%, 5%, 8%, 12%, 16%, and 20%) were prepared. The adsorption performance effect of EG on the PCMs was observed and analyzed. The structure and thermal properties of the prepared CPCMs were characterized via scanning electron microscopy, differential scanning calorimetry, thermal conductivity measurements, and heat energy storage/release experiments. The results show that the minimum mass content of EG in the CPCMs is 7.6%. The phase-change temperature of the CPCMs is close to that of the PCMs, at around 19 °C. The latent heat of phase change is equivalent to that of the PCM at the corresponding mass content, and that of phase change with an EG mass content of 8% is 138.0 J/g. The CPCMs exhibit a large increase in thermal conductivity and a significant decrease in storage/release time as the expanded graphite mass content increases. The thermal conductivity of the CPCM with a mass content of 20% is 418.5% higher than that with a mass content of 5%. With an increase in the EG mass content in CPCMs, the heat transfer mainly transitions from phase-change heat transfer to thermal conductivity.

Keywords: expanded graphite; thermal properties; composite phase-change materials; thermal energy storage

1. Introduction

Energy scarcity and environmental pollution are global challenges. The main means of solving the current energy crisis and environmental pollution problems are to save energy and rapidly develop renewable and clean energy, and thermal energy storage technology cannot be ignored in these measures [1]. Sensible energy storage, latent energy storage, and chemical energy storage are the three main types of thermal energy storage techniques [2]. Latent heat storage is the utilization of heat absorbed or released by energy storage materials during phase changes, and the amount of heat stored mainly depends on the latent heat of the phase change of the energy storage materials. A high energy storage density, minimal temperature fluctuations, strong material stability, and increased safety are some of the attributes of latent heat storage [3,4]. Therefore, it is widely used in building energy conservation, power peak shifting and valley filling, industrial waste heat and waste heat recovery, electronic device overheating protection, solar and geothermal energy utilization, and other thermal energy storage and temperature control fields [5–7]. At the core of latent heat energy storage are phase-change materials, and various inorganic, organic, and mixed phase-change materials have been studied in building energy conservation, such as paraffin [8,9], polyols [10,11], inorganic salts [12,13], and fatty acids [14,15]. Fatty acids have garnered significant interest because of their favorable phase transition temperature, high latent phase transition heat, lack of toxicity, corrosion, low undercooling, minimal or no volume change, excellent thermal stability, and wide-ranging raw material availability [16,17]. In addition, two or more fatty acids can form low-eutectic mixtures and obtain different phase-change temperatures to adapt to different temperature requirements in engineering practice. They are frequently utilized in low-temperature thermal energy storage systems, which include those used for building energy conservation and solar energy usage [18,19]. Although the thermal behavior of eutectic mixes of fatty acids is good, the thermal conductivity is low, and the material is more likely to leak out of the container when it undergoes a phase change from a solid to a liquid form. Therefore, it is necessary to mix and shape fatty acid eutectic mixtures, prevent leakage, and improve thermal conductivity. To create composite phase-change materials with good thermal conductivity, high-thermal-conductivity media are typically added to phase-change materials. Carbon materials (such as expanded graphite and carbon nanotubes) and metal particles and oxides are the most prevalent materials employed as high-thermal-conductivity media. Metal materials and metal oxides have excellent thermal conductivity; therefore, Fe, Cu, Al, Ag, Al_2O_3, Fe_2O_3, Fe_3O_4, C_uO, MgO, TiO_2, etc., are often used as thermal-conductivity-enhancing media for phase-change materials [20]. The addition of porous structural elements with high thermal conductivity to fatty acid phase-change materials is a technology that is currently being employed, and it not only improves the poor thermal conductivity of fatty acids but also stops fatty acid leakage during phase change [21–24]. Expanded graphite (EG) is a high-thermal-conductivity material with a porous structure and stable morphology at high temperatures, making it one of the most commonly used materials [25,26].

Zhu, H et al. prepared a lauric acid–stearic acid/EG (LA–SA/EG) CPCM using different mass fractions of LA–SA as PCMs and the addition of 10 wt.% EG to improve the thermal conductivity. The test results showed that, after the addition of EG, the thermal conductivity was nearly 10 times higher than before [27]. Wang, Z et al. also prepared an LA–SA/EG CPCM. Their results demonstrated that the heat storage and release effects of the heat storage device were optimal at an EG ratio of 10% weight percent and that the phase-change temperature and latent heat of the CPCM were stable when the EG content exceeded 5% [28]. Meng, X et al. used a eutectic mixture of CA, LA, and PA as PCMs to make fatty acid/expanded graphite CPCMs; as a result, the composite materials' thermal conductivity was greatly increased [29]. Wen RL et al. prepared CA–LA/EP (capric acid–lauric acid/expanded property) and CA–LA/EVM (capric acid–lauric acid/expanded verticillite), which were predicted to have an expanded property and expanded verticillite, and added 5 wt.% expanded graphite to both PCMs. They found that the thermal conductivity of CA–LA/EP and CA–LA/EVM increased by 89.14% and 87.41%, respectively [30]. Luo, K. et al. used expanded graphite as a heat-conductive material, gas-phase silica (FS) as a carrier material, and LA–SA as a PCM to make lauric acid–stearic acid gas-phase silica expanded graphite (LA–SA–FS–EG). They compared this with an LA–SA–FS composite material without the addition of EG, and it was found that the thermal conductivity of LA–SA–FS–EG–7% containing 7% EG increased by 454%, the thermal storage rate increased by 46.1%, and the thermal release rate increased by 59.4% [31]. Fei, H et al. prepared a series of composite phase-change materials of capric acid–stearic acid octadecanol/expanded graphite (CSOD/EG), and they determined the optimal mass ratio of CSOD to be 81.9:9.1:9. The test results showed that CSOD was evenly breathed into porous EG without causing a chemical reaction and CSOD/EG exhibited good thermal reliability and chemical structural stability after 1000 cycles [32]. Fei, H et al. prepared MA–PA–TD/EG (myristic acid-palmitic acid-tetradecanol/expanded graphic) and MA–SA–LA/EG (myristic acid-stearic acid-lauric acid/expanded graphic CPCM), and the results showed that both CPCMs had good stability and were suitable for use in the field of building energy conservation [33]. Badenhorst, H believes that graphite materials can potentially alleviate the low-thermal-conductivity problem in phase-change materials (PCMs) when used for solar thermal energy storage, and the most economical option appears to be compressed expanded graphite composite materials [34]. Liu S et al. prepared a CA–SA/EG CPCM, and the results showed that the thermal conductivity of the CPCM with a 10% EG mass content was 3.28 times higher than that of the CA–SA PCM [35]. Ao, C et al. prepared SA–EG/EG CPCMs at different ratios, and the results showed that the CPCMs had a heat release rate 3.1 times that of SA and a heat storage rate 2.3 times that of SA [36]. Sari, A prepared PA/EG CPCMs with different mass contents (5%, 10%, 15%, and 20%), and the results showed that, when the mass content of EG was 20%, the thermal conductivity of the CPCM was 60 W/(m·k), which was 3.5 times that of PA [37]. The research on graphite or other materials as supporting materials in previous works and in this study is summarized in Table 1.

Table 1. The research on graphite or other materials as supporting materials in previous works and this study.

PCM	Supporting Material	Mass Content	T_m (°C)	ΔH_m (J/g)	Application	Literature
Stearic acid (SA) + benzamide (BN)	Boron nitride + expanded graphite (BG)	15 wt.% BN + 20 wt.% BG	65.21	132.35	Solar hot-water system	[38]
PA	Polyvinyl butyral + expanded graphite	3 wt.% EG	59.1	125.88	Low-temperature solar energy systems	[39]
		5 wt.% EG	58.5	124.99		
		7 wt.% EG	56.1	122.05		
Paraffin wax	Graphite	0–7 wt.%	/	/	TOS device applications	[40]
	Graphene	0.001–0.07 wt.%	/	/		
Paraffin wax	Graphite	0.007–7 wt.%	Around 52	143.03–131.35	Future networks and electric noise-free remote aerial laser switching applications	[41]
	Graphene	0.001–0.7 wt.%	59.7–52.1	134.4–108.8		
Capric acid + myristic acid	Expanded graphite	0 wt.%	19.4	150.9	Low-temperature LHTES systems and backfill materials in ground source heat pump systems	This study
		1 wt.%	18.4	148.4		
		3 wt.%	18.8	142.5		
		5 wt.%	18.5	142.8		
		8 wt.%	18.5	138.0		
		12 wt.%	18.6	129.9		
		16 wt.%	19.2	128.8		
		20 wt.%	19.2	111.8		

Numerous studies have shown that adding an appropriate amount of EG to fatty acid PCMs can significantly improve the thermal conductivity of composite materials, and composite materials have good thermal reliability and thermal cycling stability. However, there are few reports on the impact of the EG mass content on the thermal performance of CPCMs in systematic research. In this study, a series of fatty acid CPCMs with different EG mass contents (1%, 3%, 5%, 8%, 12%, 16%, and 20%) were prepared by taking capric–myristic acid/expanded graphite as an example. The minimum EG mass content in the CPCMs was determined using the diffusion–exudation circle method, and the structure and thermal properties of the prepared CPCMs were characterized. To verify the thermal conductivity of the composite materials, further tests to determine heat storage and release were conducted. This research represents an interesting background for the application of fatty acid phase-change materials in building energy-saving systems.

2. Materials and Methods

2.1. Materials

Shanghai Shanpu Chemical Co., Ltd., Shanghai, China, supplied the following acids: capric acid (CA, ≥98.5% purity), lauric acid (LA, ≥98.5% purity), myristic acid (MA, ≥98.5% purity), palmitic acid (PA, ≥98.5% purity), and stearic acid (SA, ≥98.5% purity). Table 2 displays the characteristics of the CA, LA, MA, PA, and SA. We bought expandable graphite from Qingdao Hengrunda Graphite Products Co, Ltd. in Qingdao, China. It had 350 meshes, a 100 mL/g expansion rate, a 99% carbon content, and a density of 1.1 g/cm^3.

Table 2. Thermal properties of capric, lauric, myristic, palmitic, and stearic acids.

Fatty Acid	Melting		Freezing	
	T_m (°C)	ΔH_m (J/g)	T_f (°C)	ΔH_f (J/g)
Capric acid	31.40	169.4	31.69	170.3
Lauric acid	43.10	183.6	44.06	183.2
Myristic acid	52.68	188.6	51.63	193.1
Palmitic acid	60.60	198.1	61.10	199.3
Stearic acid	68.90	209.8	67.60	202.2

2.2. Preparation of Fatty Acid Binary Eutectic Mixtures/EG CPCMs

A specific volume of expandable graphite was subjected to a drying process in a vacuum drying oven. After drying, the expanded graphite could be obtained via high-temperature heating in a muffle furnace, with an expansion temperature of 900 °C and an expansion time of 30–50 s. The apparent density of the expanded graphite was estimated to be about 10 kg/m^3, the pore diameter was 2–100 nm, and the specific pore surface area was about 30 m^2/g. Binary low-eutectic mixtures of fatty acids were created by combining melt mixing and ultrasonic oscillation with the raw materials of the CA, LA, MA, PA, and SA [42,43].

The prepared binary low-eutectic mixture of fatty acids and expanded graphite was mixed in a beaker, thoroughly stirred, sealed with a thin film, and placed in a vacuum drying oven at 50 °C. The low-eutectic mixture of the binary fatty acid/EG composite phase-change energy storage material was obtained by heating the combination for 48 h, stirring it every 2 h, and cooling it to room temperature [44]. This process ensured that the mixture uniformly adsorbed EG.

2.3. Methods

The thermal properties of the CPCMs were measured by using a differential scanning calorimeter (DSC, NETSZCH 214Polyma, Selb, Germany) at a heating rate of 5 °C/min from 0 °C to 100 °C under a nitrogen atmosphere. The surface morphological structure of the CPCM samples was observed using a scanning electron microscope (SEM, phenom LE, Thermo Fisher Scientific, Waltham, MA, USA). The thermostability of the CPCM samples was investigated using thermogravimetric analysis (TGA, TA TGA5000IR, New Castle, DE, USA). The samples were heated at a rate of 10 °C/min, from 20 to 400 °C, in a nitrogen atmosphere, with an error of ±0.2%. A Fourier transform infrared spectrometer (FT–IR, Thermo Scientific Nicolet iS10, Waltham, MA, USA), operating in the 400–4000 cm^{-1} range with a resolution of 2 cm^{-1}, was used to examine the material's chemical structure.

The thermal conductivity of the CPCM samples was measured at room temperature using a DRE–III thermal conductivity tester (Xiangtan Xiangyi Instrument & Instrument Co., Ltd., Xiangtan, China) based on the transient plane heat source technique, with a measurement accuracy within ±3%. A dry sample and a probe were placed within a cake-shaped mold with a 3 cm diameter, and this was placed on a test bench after pressing it to the necessary density. After adjusting the settings based on the sample, the device was turned on to heat the probe. The software (Transient planar heat source method thermal conductivity testing system 2019.01) swiftly and precisely analyzed the sample, reported the results, and recorded the voltage and temperature increases over the measurement period.

Heat energy storage/release tests were carried out on the CPCM samples to confirm that the addition of EG improved their thermal conductivity. The experimental setup is shown in Figure 1. Temperature probes were placed into the center of the beaker with the same volume of CPCM. Then, the beakers were put into the same constant-temperature environment for a sufficient amount of time to achieve the set temperature at the center of the materials. The temperature was set to 42 °C for the heat energy release experiment and 12 °C for the heat energy storage experiment. Afterwards, the beakers were placed in a water bath to start the testing process. A steady temperature was maintained in the water bath throughout the testing procedure.

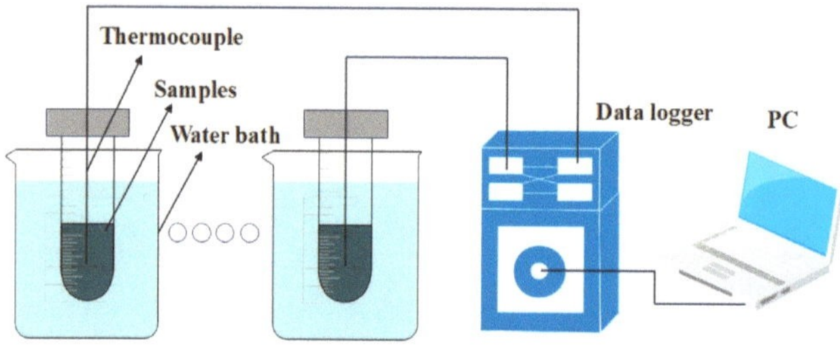

Figure 1. Experimental setup for heat energy storage/release tests.

2.4. Experimental Uncertainty

There was a degree of uncertainty surrounding the experiment because of the measuring device's precision. The uncertainty of the temperature measurement was assessed using the root-sum-square (RSS) approach [45,46]. Temperature variations were measured using a K-type thermocouple, which had an uncertainty of ±1.5 °C. The uncertainty of the temperature data logger was ±0.5 °C. The accuracy of the constant-temperature water bath was ±1.0 °C. The overall uncertainty e_{Temp} of the temperature measurement system was calculated as follows:

$$e_{Temp} = \sqrt{1.5^2 + 0.5^2 + 1^2} = 1.87\ ^\circ\text{C} \tag{1}$$

3. Results and Discussion

3.1. The Adsorption Properties of EG on PCMs

In fatty acid/expanded graphite CPCMs, fatty acids, as heat storage materials, are adsorbed in porous graphite, a support material. The higher the mass content of the heat storage material in the composite, the better the heat storage capacity of the CPCM. We propose a diffusion–exudation circle method to determine the adsorption characteristics of EG in PCMs [47]. Taking CA–MA/EG as an example, six groups of CA–MA/EG CPCMs with different EG mass contents (1%, 3%, 5%, 8%, 12%, 16%, and 20%) were prepared. The sample with a 1% EG mass content was obviously in a liquid form; hence, no heat treatment test was conducted. The results of the other six groups before and after heat treatment are depicted in Figure 2. Figure 2a shows an image before heat treatment, Figure 2b shows an image after heat treatment, and Figure 2c shows an image after boiling off the material. According to the method in Reference [47], a permeability stability evaluation was conducted, and the calculation results are shown in Table 2. In Figure 2, it can be observed that there was significant leakage in the samples with EG mass contents of 3% and 5%, and a large amount of the PCM had still not been adsorbed into the pores. This is because the adsorption capacity of the pores for the liquid PCM reached a saturation state. The samples with an EG mass content of 8% had almost no leakage, while the samples with an EG mass content of 10% or greater remained stable, without leakage. Based on the above analysis and the results in Table 3, it could be determined that the minimum EG mass content is around 8%. To ascertain the precise minimum mass content of EG in the CA–MA/EG CPCM, we prepared four samples with mass contents of 7.2%, 7.4%, 7.6%, and 7.8%, and we determined that the minimum mass content of EG in the CA–MA/EG CPCM was 7.6% by employing the aforementioned method [48].

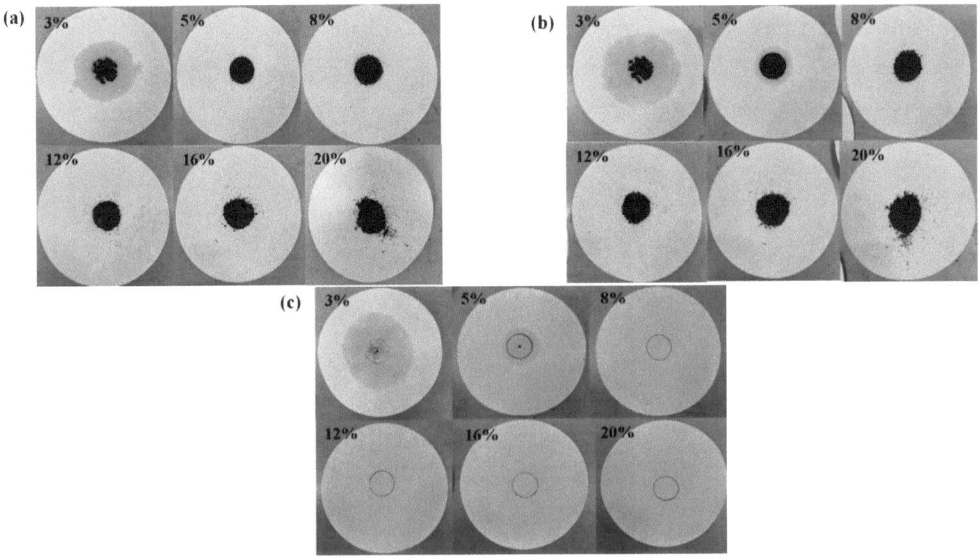

Figure 2. Photographs of CA–MA/EG CPCM samples with different EG mass contents before and after heat treatment. (**a**) Before heat treatment, (**b**) after heat treatment, (**c**) boiling off the material.

Table 3. CA–MA/EG composite phase-change material seepage stability assessment.

Mass Content of EG (%)	Exudation Circle Average Diameter (mm)	Leakage Percentage (%)	Assessment Standard	Assessment Result
1%	/	/	/	/
3%	88.0	193.0	$\Phi > 50$	Extremely unstable
5%	46.5	55.0	$\Phi > 50$	Extremely unstable
8%	0	0	$\Phi \leq 0$	Very stable
12%	0	0	$\Phi \leq 0$	Very stable
16%	0	0	$\Phi \leq 0$	Very stable
20%	0	0	$\Phi \leq 0$	Very stable

3.2. Thermal Properties of the CPCMs

Taking the CA–MA/EG CPCM as an example, the DSC curves of the CA–MA/EG CPCMs with different EG mass contents are shown in Figure 3, and the phase-change temperature and latent heat are shown in Table 4. It can be seen from the table that the phase-change temperature of the CPCMs with the addition of EG changes very little, indicating that the addition of EG has almost no effect on the phase-change temperature of CPCMs and that EG only plays a supporting role in the phase-change material. The latent heat of phase change can also be calculated using Equation (2). Table 4 and Figure 4 present the theoretical latent heat of the phase-change calculation findings for the CA–MA/EG CPCMs with different EG mass contents and compare them with the experimental test data.

$$\Delta H_{CPCM} = (1 - w)\Delta H_{PCM} \tag{2}$$

Here, ΔH_{CPCM} is the calculated latent heat of the CA–MA/EG CPCMs, w is the EG mass content in the CA–MA/EG CPCMs, and ΔH_{PCM} is the mass percentage of MA in the latent heat of CA–MA.

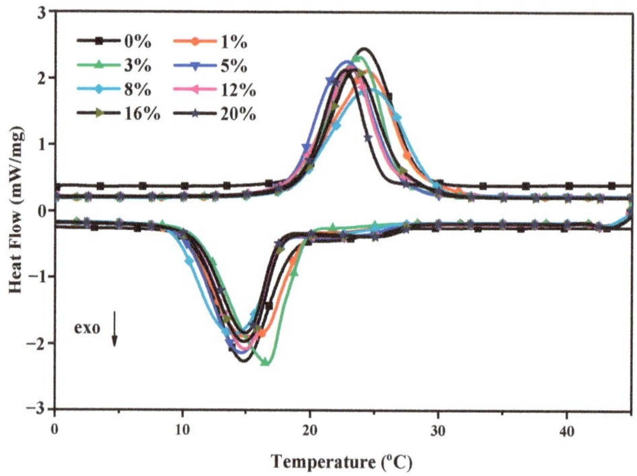

Figure 3. DSC curves of CA–MA and CA–MA/EG CPCMs with different EG mass contents.

Table 4. Thermal properties of CA–MA/EG with different EG mass contents.

W (EG)%	Melting				Freezing			
	T_m (°C)	ΔH_m Experimental Value (J·g^{-1})	ΔH_m Calculated Value (J·g^{-1})	ΔH_m Difference (%)	T_f (°C)	ΔH_f Experimental Value (J·g^{-1})	ΔH_f Calculated Value (J·g^{-1})	ΔH_f Difference (%)
0	19.4	150.9	/	/	18.4	149.2	/	/
1	18.4	148.4	149.4	0.66	19.6	146.8	147.7	0.61
3	18.8	142.5	146.4	2.65	19.4	143.0	144.7	1.19
5	18.5	142.8	143.4	0.39	17.6	141.2	141.7	0.38
8	18.5	138.0	138.8	0.60	18.3	135.4	137.3	1.36
12	18.6	129.9	132.8	2.18	17.7	125.0	131.3	4.80
16	19.2	128.8	126.8	−1.61	18.0	127.1	125.3	−1.41
20	19.2	111.8	120.7	7.39	17.7	110.0	119.4	7.84

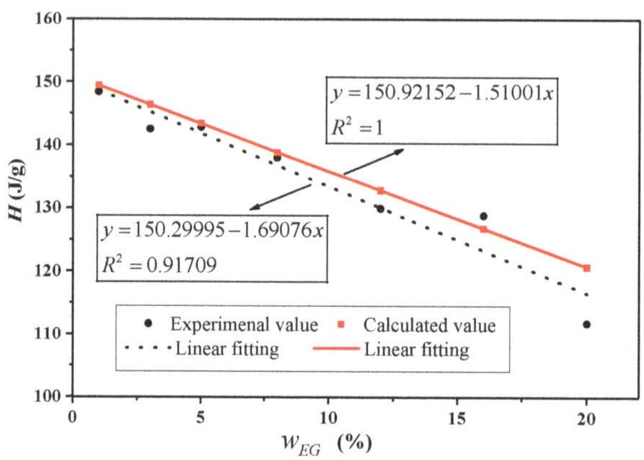

Figure 4. Comparison of the calculated and experimental values of phase transition latent heat for CA–MA/EG CPCMs with different EG mass contents.

It can be seen in Table 3 and Figure 4 that the experimental data are very close to the theoretical values, with a small error, within 10%, indicating that the addition of EG has little effect on the latent heat of the CA–MA/EG CPCM. The latent heat of the CA–MA/EG CPCM decreases with the increase in the EG mass content, and it shows an almost-linear relationship. For example, when the mass content of EG is 20%, the latent heat of the sample is 109.8 J/g, while pure CA–MA has a latent heat of 150.9 J/g, with a ratio of approximately 73%, which is close to the mass fraction of CA–MA in the CPCM of 80%. This suggests that the adsorption of CA–MA in EG is only physical absorption and EG has little effect on the phase-change latent heat.

3.3. Microstructure of the CPCMs

The CA–MA/EG CPCM samples with EG mass contents of 3%, 5%, 8%, 12%, 16%, and 20% were placed under a scanning electron microscope (SEM) to observe their microstructures. The SEM images can be seen in Figure 5. In Figure 5, it can be seen that EG is a loose and porous worm-like grid structure with a large specific surface area, making it easy for molten CA–MA to adsorb into its micro-porous structure. The network pore structure of the EG is composed of graphite flakes and a large number of irregular pores. Under the capillary force of the pores, the molten fatty acid PCM can be easily adsorbed into the microporous structure of the EG. When the EG mass content in the CPCM is low, EG adsorption becomes saturated, and its surface is covered by excess liquid fatty acids. After the fatty acids solidify, block-like fatty acid clusters form on the surface of EG, covering its worm-like structure. When the EG mass content in the CPCM is 8%, the liquid fatty acids basically fill the micro-porous structure of EG, reaching a saturated state. However, as the EG mass content in the CPCM increases to 12%, 16%, and 20%, the fatty acids uniformly and completely adsorb into the pore size of EG. The liquid fatty acids do not completely fill the micro-porous structure of EG, the CPCM still retains the original worm-like form of EG, and the molten liquid fatty acids do not leak out from EG.

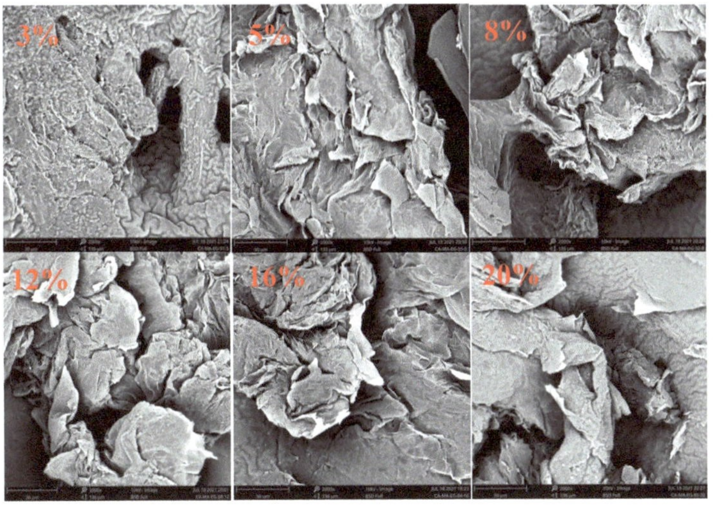

Figure 5. SEM pictures of fatty acid/EG CPCMs with different EG mass contents.

3.4. Thermostability of the CPCMs

Figure 6 displays the TGA curves of CA–MA/EG with an EG mass content of 7.6%. It is evident from this figure that, at a temperature of about 107.4 °C, CA–MA/EG started to lose weight. As the temperature increased, there was a significant loss in mass. At a temperature of roughly 205.2 °C, the PCM nearly entirely dissipated, while the weight loss rate hit its maximum at 184.3 °C. Thus, in low-temperature applications below 100 °C, CA–MA/EG composite materials exhibit good thermal stability.

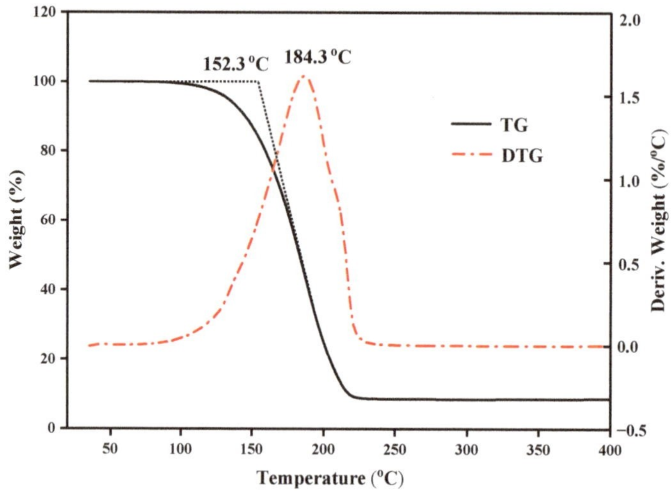

Figure 6. TGA curves of CA–MA/EG CPCMs.

3.5. Infrared Spectral Analysis

CA, MA, CA–MA, and CA–MA/EG were characterized using Fourier transform infrared spectroscopy (FT–IR). Figure 7 displays the FT–IR curves. The spectra show that the C=O absorption peak is around 1708 cm^{-1}, and the bending vibration peak of $-CH_2-$ is about 1463 cm^{-1}. The absorption peak in the wavenumber range of about 2715–3105 cm^{-1} overlaps with the aliphatic group's C–H stretching vibration absorption peak. The peak that is absorbed by the hydroxyl O–H stretching vibration is this C–H one. The spectral curves of CA, MA, CA–MA, and CA–MA/EG are comparable, and the characteristic peaks correspond one on one, according to the data for C–O above. This shows that, following the addition of expanded graphite, the structures of CA, MA, and CA–MA remained constant, and no new compounds were produced.

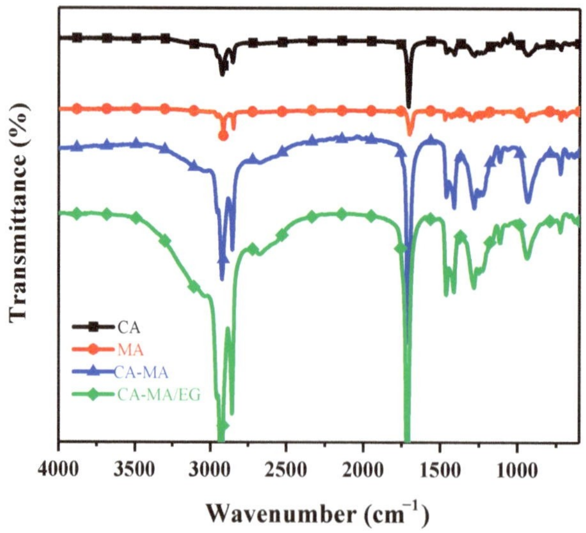

Figure 7. FT–IR curves of CA, MA, CA–MA, and CA–MA/EG.

3.6. Heat Transfer Performance Experiments

The amount of EG added has a significant impact on the thermal conductivity of form-stable phase-change materials. Figure 8 shows the thermal conductivity curves of the CA–MA/EG CPCMs with EG mass contents of 5%, 8%, 12%, 16%, and 20%, and the corresponding data are listed in Table 5. When the mass content of EG in the CPCM was 1%, the CPCM was clearly a flowable liquid, and when the mass content of EG was 3%, the CPCM was a viscous fluid; therefore, the CPCMs with EG mass contents of 1% and 3% were not measured. From Figure 8, we can see that the density of the CPCMs used for measurement was 300 Kg/m^3. The thermal conductivity of the CPCMs with EG mass contents of 5%, 8%, 12%, 16%, and 20% was 0.2683, 0.3827, 0.7656, 1.1227, and 1.2130 W/(m·k), respectively. It can be seen that the thermal conductivity of the composite material gradually increased with the increase in the EG mass content. The thermal conductivity of the CPCM with a mass content of 20% was 418.5% higher than that with a mass content of 5%. Table 4 also shows the thermal conductivity of CPCMs after adding expanded graphite to the PCMs in some of the literature, which also indicates that EG can effectively improve the heat transfer performance of fatty acid PCMs. As mentioned above, when the mass content of EG in a CPCM is low, EG is in a state of adsorption saturation, its surface is covered with excess liquid fatty acids, and heat transfer mainly occurs through convection. As the mass content of EG in a CPCM increases, liquid fatty acids gradually fill the micro-porous structure of EG until adsorption is completed, and heat transfer mainly occurs through conduction.

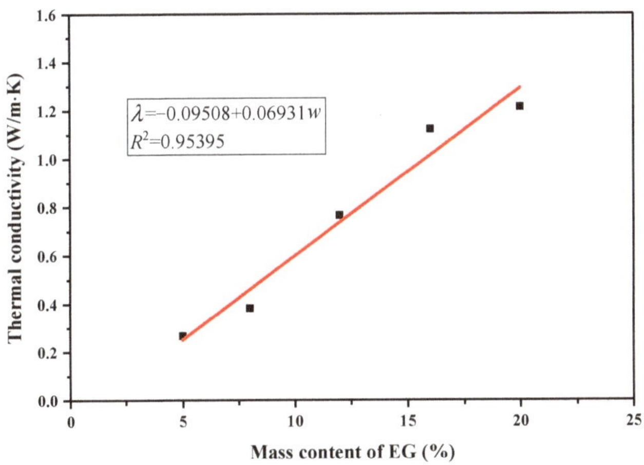

Figure 8. Variations in thermal conductivity of CPCMs with different EG mass contents.

Table 5. Thermal conductivity of CPCMs with different EG mass contents.

CPCMs	Mass Content of EG	Thermal Conductivity (W·m^{-1}·k^{-1})	Literature
LA–SA/EG	5% 15%	0.466 0.585	[26]
CA–LA–PA/EG	5%	0.738	[27]
CA–LA/EP/EG CA–LA/EVM/EG	5% 5%	0.244 0.253	[28]
LA–SA–FA–EA	3% 5% 7%	2.545 3.229 6.045	[29]
CSOD/EG	8.33%	3.136	[30]
CA–MA/EG	5% 8% 12% 16% 20%	0.2683 0.3827 0.7656 1.1227 1.2130	This study

3.7. Heat Energy Storage and Release of the CPCMs

In the processes of storing and releasing heat energy, the temperature settings were 42 °C and 12 °C, respectively. The test results are shown in Figure 9, with Figure 9a showing the heat energy storage curves and Figure 9b showing the heat energy release curves. In Figure 9, it can be seen that, when the EG mass contents in the CPCMs were 3%, 5%, 8%, and 12%, the melting curves had a clear "sensible heat–latent heat–sensible heat" three-stage characteristic similar to that of a pure PCM. However, as the EG mass content in the CPCM increased, the latent heat stage characteristics gradually weakened. When the EG mass contents were 16% and 20%, the latent heat stage times of the curve were extremely short, and the curve changed almost linearly. The heat energy storage times of the six sample groups were 4490 s, 4000 s, 3360 s, 2610 s, 1200 s, and 1140 s, and the heat energy release times were 7430 s, 6490 s, 5320 s, 4420 s, 3310 s, and 3090 s, respectively. From the morphological changes in the curves, it can be seen that the heat transfer law of the CPCM changed with the increase in the EG mass content. The heat transfer during the phase-change stage of the heat storage/release process was dominated by the phase-change heat transfer of the CPCM, and the pore structure of EG played an auxiliary role in enhancing the thermal conductivity. When the phase-change heat transfer characteristics of the curves disappeared, the PCM wrapped in the pores of EG became liquid, and the heat transfer of the material was dominated by the thermal conductivity of EG.

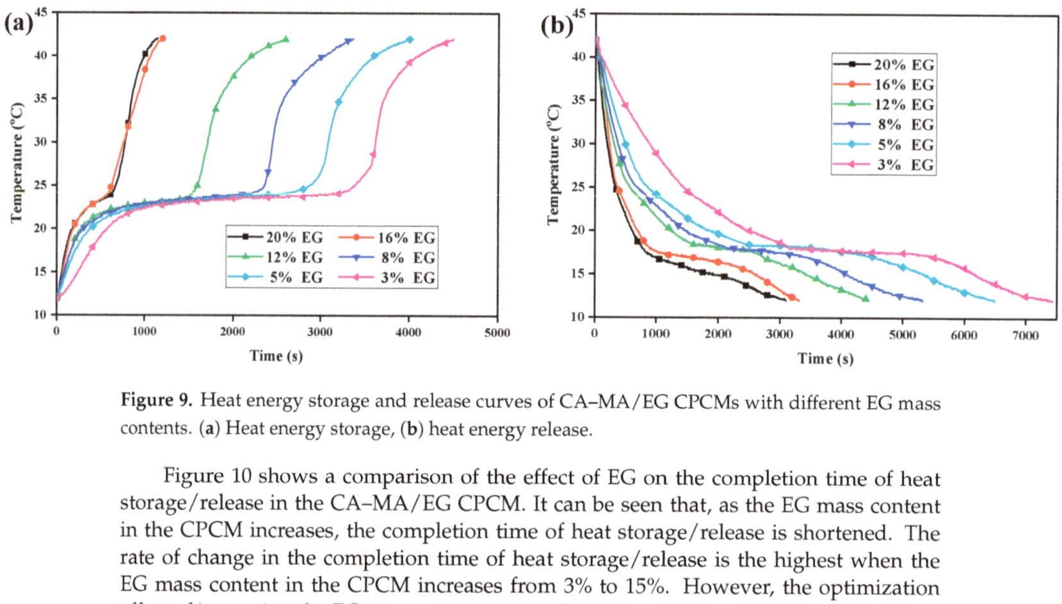

Figure 9. Heat energy storage and release curves of CA–MA/EG CPCMs with different EG mass contents. (**a**) Heat energy storage, (**b**) heat energy release.

Figure 10 shows a comparison of the effect of EG on the completion time of heat storage/release in the CA–MA/EG CPCM. It can be seen that, as the EG mass content in the CPCM increases, the completion time of heat storage/release is shortened. The rate of change in the completion time of heat storage/release is the highest when the EG mass content in the CPCM increases from 3% to 15%. However, the optimization effect of increasing the EG mass content in the CPCM from 15% to 20% is not significant. Considering that leakage occurs when the EG mass content in the CPCM is less than 8%, it can be considered that 8–10% is the optimal mass content range in the CA–MA/EG CPCM for EG.

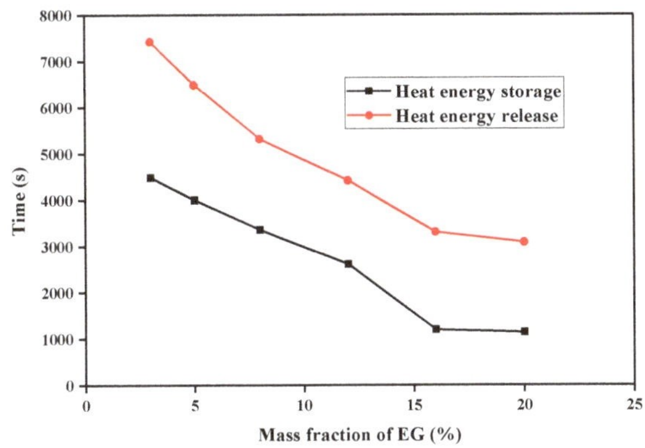

Figure 10. Relationship between EG content in CA–MA/EG CPCMs and heat storage/release time.

4. Conclusions

In this study, a series of fatty acid composite phase-change materials with different expanded graphite mass contents (1%, 3%, 5%, 8%, 12%, 16%, and 20%) were prepared by taking capric–myristic acid/expanded graphite as an example. The main conclusions drawn from the adsorption properties of expanded graphite on phase-change materials, thermal properties, microstructure, thermostability, infrared spectral analysis, heat transfer performance, and heat energy storage and release in composite phase-change materials are as follows:

(1) The minimum expanded graphite mass content in capric–myristic acid/expanded graphite composite phase-change materials is 7.6%. When the mass content of expanded graphite in composite phase-change materials exceeds the minimum content, the liquid fatty acid phase-change materials are completely filled into the porous structure of expanded graphite. Following the addition of expanded graphite, the structures of phase-change materials remain constant, and no new compounds are produced.

(2) The latent heat of composite phase-change materials decreases almost linearly with an increase in the expanded graphite mass content, and expanded graphite has almost no effect on the phase-change temperature. In low-temperature applications below 100 °C, capric–myristic acid/expanded graphite composite phase-change materials exhibit good thermal stability.

(3) The thermal conductivity of composite phase-change materials gradually increases with an increase in the expanded graphite mass content. With an increase in the expanded graphite mass content in composite phase-change materials, the heat transfer mainly transitions from phase-change heat transfer to thermal conductivity, but when the expanded graphite mass content exceeds 15%, an increase in the expanded graphite mass content has no significant effect on the composite phase-change material heat storage/release time.

The research presented above leads us to the conclusion that the prepared composite phase-change materials can be widely used in building energy-saving systems, such as floors in floor radiation heating systems, and as backfill materials for buried pipe wells in ground source heat pump systems. This will be explored by the authors in future work.

Author Contributions: Conceptualization, D.Z. and Y.L.; methodology, D.Z. and S.X.; validation, D.Z. and S.X.; formal analysis, D.Z. and S.X.; data curation, D.Z. and S.X.; writing—original draft preparation, D.Z. and S.X.; writing—review and editing, D.Z. and S.X.; funding acquisition, D.Z. All authors have read and agreed to the published version of the manuscript.

Funding: This research was funded by the Natural Science Foundation of Hunan Provincial (Grant No. 2022JJ50237).

Institutional Review Board Statement: Not applicable.

Informed Consent Statement: Not applicable.

Data Availability Statement: Data are contained within the article.

Conflicts of Interest: The authors declare no conflicts of interest.

References

1. Pielichowska, K.; Pielichowski, K. Phase change materials for thermal energy storage. *Prog. Mater. Sci.* **2014**, *65*, 67–123. [CrossRe
2. Zhou, D.; Zhao, C.; Tian, Y. Review on thermal energy storage with phase change materials (PCMs) in building application *Appl. Energy* **2012**, *92*, 593–605. [CrossRef]
3. Gao, J.; Xu, Z.; Wang, R. Experimental study on a double–stage absorption solar thermal storage system with enhanced energ storage density. *Appl. Energy* **2020**, *262*, 114476. [CrossRef]
4. Mohamed, S.A.; Al–Sulaiman, F.A.; Ibrahim, N.I.; Zahir, M.H.; Al–Ahmed, A.; Saidur, R.; Yilbas, B.S.; Sahin, A.Z. A review o current status and challenges of inorganic phase change materials for thermal energy storage systems. *Renew. Sustain. Energy Re* **2017**, *70*, 1072–1089. [CrossRef]
5. Baskar, I.; Chellapandian, M.; Jaswanth, S. Development of a novel composite phase change material based paints and mortar fo energy storage applications in buildings. *J. Energy Storage* **2022**, *55*, 105829. [CrossRef]
6. Horibe, A.; Jang, H.; Haruki, N.; Sano, Y.; Kanbara, H.; Takahashi, K. Melting and solidification heat transfer characteristics c phase change material in a latent heat storage vessel: Effect of perforated partition plate. *Int. J. Heat Mass Transf.* **2015**, *82*, 259–26 [CrossRef]
7. Chung, O.; Jeong, S.G.; Kim, S. Preparation of energy efficient paraffinic PCMs/expanded vermiculite and perlite composites fo energy saving in buildings. *Sol. Energy Mater. Sol. Cells* **2015**, *137*, 107–112. [CrossRef]
8. Yang, J.; Jia, Y.; Bing, N.; Wang, L.; Xie, H.; Yu, W. Reduced graphene oxide and zirconium carbide co–modified melamin sponge/paraffin wax composites as new form–stable phase change materials for photothermal energy conversion and storag *Appl. Therm. Eng.* **2019**, *163*, 114412. [CrossRef]
9. George, M.; Pandey, A.K.; Abd Rahim, N.; Tyagi, V.V.; Shahabuddin, S.; Saidur, R. A novel polyaniline (PANI)/paraffin wax nan composite phase change material: Superior transition heat storage capacity, thermal conductivity and thermal reliability. *So Energy* **2020**, *204*, 448–458. [CrossRef]
10. Gunasekara, S.N.; Pan, R.J.; Chiu, J.N.; Martin, V. Polyols as phase change materials for surplus thermal energy storage. *App Energy* **2016**, *162*, 1439–1452. [CrossRef]
11. Lv, L.; Huang, S.; Cen, K.; Zhou, H. Experimental study of screening polyols and their binary eutectic phase change materials fo long–term thermal energy storage. *J. Clean. Prod.* **2023**, *399*, 136636. [CrossRef]
12. Tao, Y.; Lin, C.; He, Y. Preparation and thermal properties characterization of carbonate salt/carbon nanomaterial composit phase change material. *Energy Convers. Manag.* **2015**, *97*, 103–110. [CrossRef]
13. Li, M.; Jin, B.; Ma, Z.; Yuan, F. Experimental and numerical study on the performance of a new high–temperature packed–bec thermal energy storage system with macroencapsulation of molten salt phase change material. *Appl. Energy* **2018**, *221*, 1–15 [CrossRef]
14. Yang, T.; King, W.; Miljkovic, N. Phase change material–based thermal energy storage. *Cell Rep. Phys. Sci.* **2021**, *2*, 10054C [CrossRef]
15. Yuan, Y.; Zhang, N.; Tao, W.; Cao, X.; He, Y. Fatty acids as phase change materials: A review. *Renew. Sustain. Energy Rev.* **2014**, *29* 482–498. [CrossRef]
16. Akeiber, H.; Nejat, P.; Abd Majid, M.Z.; Wahid, M.A.; Jomehzadeh, F.; Famileh, I.Z.; Calautit, J.K.; Hughes, B.; Zaki, S.A. A review on phase change material (PCM) for sustainable passive cooling in building envelopes. *Renew. Sustain. Energy Rev.* **2016**, *60* 1470–1497. [CrossRef]
17. Sun, N.; Xiao, Z. Synthesis and performances of phase change materials microcapsules with a polymer/BN/TiO_2 hybrid shell fo thermal energy storage. *Energy Fuels* **2017**, *31*, 10186–10195. [CrossRef]
18. He, H.; Yue, Q.; Gao, B.; Zhang, X.; Li, Q.; Wang, Y. The effects of compounding conditions on the properties of fatty acids eutectic mixtures as phase change materials. *Energy Convers. Manag.* **2013**, *69*, 116–121. [CrossRef]
19. Tang, F.; Su, D.; Tang, Y.; Fang, G. Synthesis and thermal properties of fatty acid eutectics and diatomite composites as shape-stabilized phase change materials with enhanced thermal conductivity. *Sol. Energy Mater. Sol. Cells* **2015**, *141*, 218–224. [CrossRef]
20. Qiu, L.; Ning, Z.; Feng, Y.; Michaelides, E.E.; Żyła, G.; Jing, D.; Zhang, X.; Norris, P.M.; Markides, C.N.; Mahian, O. A review of recent advances in thermophysical properties at the nanoscale: From solid state to colloids. *Phys. Rep.* **2020**, *843*, 1–81. [CrossRef]
21. Gorbacheva, S.N.; Makarova, V.V.; Ilyin, S.O. Hydrophobic nanosilica–stabilized graphite particles for improving thermal conductivity of paraffin wax–based phase–change materials. *J. Energy Storage* **2021**, *36*, 102417. [CrossRef]
22. Bulk, A.; Odukomaiya, A.; Simmons, E.; Woods, J. Processing Compressed Expanded Natural Graphite for Phase Change Material Composites. *J. Therm. Sci.* **2022**, *32*, 1213–1226. [CrossRef]

23. Guo, Y.L.; Yang, W.B.; Jiang, Z.N.; He, F.F.; Zhang, K.; He, R.; Wu, J.Y.; Fan, J.H. Silicone rubber/paraffin@silicon dioxide form–stable phase change materials with thermal energy storage and enhanced mechanical property. *Sol. Energy Mater. Sol. Cells* **2019**, *196*, 16–24. [CrossRef]
24. Ben Khedher, N.; Bantan, R.A.; Kolsi, L.; Omri, M. Performance investigation of a vertically configured LHTES via the combination of nano–enhanced PCM and fins: Experimental and numerical approaches. *Int. Commun. Heat Mass Transf.* **2022**, *137*, 106246. [CrossRef]
25. Hua, J.; Yuan, C.; Zhao, X.; Zhang, J.; Du, J. Structure and thermal properties of expanded graphite/paraffin composite phase change material. *Energy Sources Part A Recovery Util. Environ. Eff.* **2019**, *41*, 86–93. [CrossRef]
26. Jeon, J.; Park, J.H.; Wi, S.; Kim, K.H.; Kim, S. Thermal performance enhancement of a phase change material with expanded graphite via ultrasonication. *J. Ind. Eng. Chem.* **2019**, *79*, 437–442. [CrossRef]
27. Zhu, H.; Zhang, P.; Meng, Z.; Li, M. Thermal Characterization of Lauric–Stearic Acid/Expanded Graphite Eutectic Mixture as Phase Change Materials. *J. Nanosci. Nanotechnol.* **2015**, *15*, 3288–3294. [CrossRef] [PubMed]
28. Wang, Z.; Huang, G.; Jia, Z.; Gao, Q.; Li, Y.; Gu, Z. Eutectic Fatty Acids Phase Change Materials Improved with Expanded Graphite. *Materials* **2022**, *15*, 6856. [CrossRef] [PubMed]
29. Meng, X.; Zhang, H.; Zhao, Z.; Sun, L.; Xu, F.; Zhang, J.; Jiao, Q.; Bao, Y.; Ma, J. Preparation, Encapsulation and Thermal Properties of Fatty Acid/Expanded Graphite Composites as Shape–stabilized Phase Change Materials. *Chem. J. Chin. Univ.* **2012**, *33*, 526–530. [CrossRef]
30. Wen, R.; Zhang, X.; Huang, Y.; Yin, Z.; Huang, Z.; Fang, M.; Liu, Y.; Wu, X. Preparation and properties of fatty acid eutectics/expanded perlite and expanded vermiculite shape–stabilized materials for thermal energy storage in buildings. *Energy Build.* **2017**, *139*, 197–204. [CrossRef]
31. Luo, K.; Wu, D.; Wang, Y.; Fei, H.; Jiang, H.; Ye, Z. Preparation and characterization of lauric acid–stearic acid/fumed silica/expanded graphite thermally conductive enhanced composites. *J. Energy Storage* **2023**, *73*, 109151. [CrossRef]
32. Fei, H.; He, Q.; Du, W.; Li, P.; Zhou, J.; Pan, Y.; Liang, X. Structural characteristics and thermal performances of capric acid–stearic acid–octadecanol adsorbed into porous expanded graphite under vacuum condition. *J. Energy Storage* **2023**, *72*, 108326. [CrossRef]
33. Fei, H.; Zhou, J.; He, Q.; Wang, L.; Liang, X.; Pan, Y. Characteristic and Properties of Ternary Shape–Stabilized Composite Phase Change Materials Based on Expanded Graphite. *ACS Omega* **2021**, *6*, 29215–29222. [CrossRef] [PubMed]
34. Badenhorst, H. A review of the application of carbon materials in solar thermal energy storage. *Sol. Energy* **2019**, *192*, 35–68. [CrossRef]
35. Liu, S.; Zhang, X.; Zhu, X.; Xin, S. A Low–Temperature Phase Change Material Based on Capric–Stearic Acid/Expanded Graphite for Thermal Energy Storage. *ACS Omega* **2021**, *6*, 17988–17998. [CrossRef] [PubMed]
36. Ao, C.; Yan, S.; Zhao, S.; Hu, W.; Zhao, L.; Wu, Y. Stearic acid/expanded graphite composite phase change material with high thermal conductivity for thermal energy storage. *Energy Rep.* **2022**, *8*, 4834–4843. [CrossRef]
37. Sari, A.; Karaipekli, A. Preparation, thermal properties and thermal reliability of palmitic acid/expanded graphite composite as form–stable PCM for thermal energy storage. *Sol. Energy Mater. Sol. Cells* **2009**, *93*, 571–576. [CrossRef]
38. Huang, D.; Ma, G.; Yu, Z.; Lv, P.; Zhou, Q.; Liu, Q.; Peng, C.; Xiong, F.; Huang, Y. Highly thermal conductive shape–stabilized composite phase change materials based on boron nitride and expanded graphite for solar thermal applications. *RSC Adv.* **2023**, *13*, 13252–13262. [CrossRef]
39. Lin, Y.; Zhu, C.; Alva, G.; Fang, G. Palmitic acid/polyvinyl butyral/expanded graphite composites as form–stable phase change materials for solar thermal energy storage. *Appl. Energy* **2019**, *228*, 1801–1809. [CrossRef]
40. Said, A.; Salah, A.; Fattah, G. Thermo–optic switching properties of paraffin–wax hosting carbon fillers. *J. Energy Storage* **2018**, *19*, 260–271. [CrossRef]
41. Said, A.; Salah, A.; Fattah, G. Enhanced Thermo–Optical Switching of Paraffin–Wax Composite Spots under Laser Heating. *Materials* **2017**, *10*, 525. [CrossRef] [PubMed]
42. Atinafu, D.G.; Dong, W.; Huang, X.; Gao, H.; Wangc, G. Introduction of organic–organic eutectic PCM in mesoporous N–doped carbons for enhanced thermal conductivity and energy storage capacity. *Appl. Energy* **2018**, *211*, 1203–1215. [CrossRef]
43. Zhou, D.; Xiao, S.; Xiao, X.; Liu, Y. Preparation, Phase Diagrams and Characterization of Fatty Acids Binary Eutectic Mixtures for Latent Heat Thermal Energy Storage. *Separations* **2023**, *10*, 49. [CrossRef]
44. Zhou, D.; Xiao, S.; Xiao, X. Preparation and thermal performance of fatty acids binary eutectic mixtures/expanded graphite as form–stable phase change materials for thermal energy storage. *ACS Omega* **2023**, *8*, 8596–8604. [CrossRef]
45. Venkitaraj, K.; Suresh, S. Experimental study on thermal and chemical stability of pentaerythritol blended with low melting alloy as possible PCM for latent heat storage. *Exp. Therm. Fluid Sci.* **2017**, *88*, 73–87. [CrossRef]
46. Li, J.; Wang, W.; Deng, Y.; Gao, L.; Bai, J.; Xu, L.; Chen, J.; Yuan, Z. Thermal Performance Analysis of Composite Phase Change Material of Myristic Acid–Expanded Graphite in Spherical Thermal Energy Storage Unit. *Energies* **2023**, *16*, 4527. [CrossRef]

47. Zhou, D.; Yuan, J.; Zhou, Y.; Liu, Y. Preparation and characterization of myristic acid/expanded graphite composite phase change materials for thermal energy storage. *Sci. Rep.* **2020**, *10*, 10889. [CrossRef]
48. Zhou, D.; Yuan, J.; Zhou, Y.; Liu, Y. Preparation and Properties of Capric–Myristic Acid/Expanded Graphite Composite Phase Change Materials for Latent Heat Thermal Energy Storage. *Energies* **2020**, *13*, 2462. [CrossRef]

molecules MDPI

Article

Improvement in Electrochemical Performance of Waste Sugarcane Bagasse-Derived Carbon via Hybridization with SiO_2 Nanospheres

Muhammad Mudassir Ahmad Alwi [1], Jyoti Singh [2], Arup Choudhury [2,*], SK Safdar Hossain [3,*] and Akbar Niaz Butt [1]

[1] Department of Materials Engineering, College of Engineering, King Faisal University, P.O. Box 380, Al-Ahsa 31982, Saudi Arabia; malwi@kfu.edu.sa (M.M.A.A.); abutt@kfu.edu.sa (A.N.B.)
[2] Department of Chemical Engineering, Birla Institute of Technology, Ranchi 835215, India; jyotisingh161198@gmail.com
[3] Department of Chemical Engineering, College of Engineering, King Faisal University, P.O. Box 380, Al-Ahsa 31982, Saudi Arabia
* Correspondence: arup@bitmesra.ac.in (A.C.); snooruddin@kfu.edu.sa (S.S.H.); Tel.: +91-9973277473 (A.C.); Fax: +91-651-2276053 (A.C.)

Abstract: Sugar industries generate substantial quantities of waste biomass after the extraction of sugar water from sugarcane stems, while biomass-derived porous carbon has currently received huge research attention for its sustainable application in energy storage systems. Hence, we have investigated waste sugarcane bagasse (WSB) as a cheap and potential source of porous carbon for supercapacitors. The electrochemical capacitive performance of WSB-derived carbon was further enhanced through hybridization with silicon dioxide (SiO_2) as a cost-effective pseudocapacitance material. Porous WSB-C/SiO_2 nanocomposites were prepared via the in situ pyrolysis of tetraethyl orthosilicate (TEOS)-modified WSB biomass. The morphological analysis confirms the pyrolytic growth of SiO_2 nanospheres on WSB-C. The electrochemical performance of WSB-C/SiO_2 nanocomposites was optimized by varying the SiO_2 content, using two different electrolytes. The capacitance of activated WSB-C was remarkably enhanced upon hybridization with SiO_2, while the nanocomposite electrode demonstrated superior specific capacitance in 6 M KOH electrolyte compared to neutral Na_2SO_4 electrolyte. A maximum specific capacitance of 362.3 F/g at 0.25 A/g was achieved for the WSB-C/SiO_2 105 nanocomposite. The capacitance retention was slightly lower in nanocomposite electrodes (91.7–86.9%) than in pure WSB-C (97.4%) but still satisfactory. A symmetric WSB-C/SiO_2 105//WSB-C/SiO_2 105 supercapacitor was fabricated and achieved an energy density of 50.3 Wh kg^{-1} at a power density of 250 W kg^{-1}, which is substantially higher than the WSB-C//WSB-C supercapacitor (22.1 Wh kg^{-1}).

Keywords: waste sugarcane bagasse; activated carbon; SiO_2 nanospheres; capacitance; supercapacitors

1. Introduction

Globally, fossil fuel consumption has led to an increase in energy demand and environmental sustainability [1–3]. Over the last few decades, a number of factors including population growth, fossil fuel exhaustion, and waste material production have contributed to the increasing demand for limitless energy sources and storage devices [4,5]. There is a growing demand for environmentally friendly, affordable, and renewable energy systems. To make these systems reliable, an efficient and cost-effective energy storage system is necessary for securing power at low cost [6,7]. Research and commercial interest are growing in the use of waste biomass to produce porous carbon-based electrode materials for cost-effective and efficient energy storage systems, since biomass-derived carbon offers low cost, sustainability, porosity, and excellent electrical conductivity [8]. This strategy can boost up the global economy as well as reduce the waste biomass-derived environmental pollution [9]. Furthermore, it will help to reduce the environmental impact of burning waste biomass for energy [10].

For supercapacitors, carbon is the most commonly used electrode material because it offers excellent electrical conductivity, chemical resistance, and high thermal and mechanical stability. Various carbon materials including carbon nanotubes, graphene, and activated carbon have been widely investigated as electrode materials for supercapacitors [11]. For supercapacitor manufacturers, low material costs are the most important issue, in addition to different technical issues. Recently, biomass-derived carbon has become increasingly popular for manufacturing supercapacitors, due to its low cost, abundance, and environmental friendliness [12]. However, the capacitive performance of these carbon materials is very much dependent on their physical properties such as surface area, mesoporosity, electrical conductivity, and electrochemical stability [13–15]. Researchers have explored various chemical and physical activation processes to manipulate the porosity and surface area of biomass-derived carbon materials and thus enhance their electrochemical capacitive performances [16]. However, the activation treatment can somewhat improve the specific capacitance and charge–discharge capacity of biomass-derived carbon, but it requires further improvement in the capacitance and energy density of biomass-derived carbon to meet commercial demand [17,18].

Hybridization of biomass-derived carbon with redox-active materials can enhance capacitance through the combined effects of electrical double-layer capacitance (EDLC) and pseudocapacitance [19–21]. Numerous pseudocapacitance materials including transition metal oxides/sulfides [22–25], conducting polymers [26,27], and transition metal MXenes [28–30] have been extensively investigated to enhance the electrochemical capacitance of carbon materials via hybridization. Nonetheless, there have been limited attempts to enhance the energy density of biomass-derived carbon by incorporating pseudocapacitance materials. In two studies, Wu et al. [16] and Hu et al. [31], bamboo-derived carbon was enhanced by hybridization with nickel hydroxide and copper oxide. Tang et al. [32] achieved a specific capacitance of 234.2 F/g at 1 A/g and a rate capability of 71.4% at 20 A/g for bamboo shoot shell-derived carbon/SiC composites. Recently, the banana peel-derived carbon/MnO_2 composite demonstrated a specific capacitance of 139.6 F/g at 300 mA/g with 92.3% capacitance retention after 1000 cycles [33]. Recently, agro-waste biomass-derived carbon materials have gained increasing interest for supercapacitor applications because of their high surface area and tannable porosity [34–46]. Among various agro-wastes, sugarcane bagasse (*Sachharum officinarum*) is a typical waste biomass generated in huge quantities in the sugar industries. Most of this waste sugarcane bagasse is disposed of on land, causing serious environmental pollution. Hence, it is a great opportunity to investigate these waste bagasse fibers as a possible precursor for producing porous carbon and then to make use of this carbon as a supercapacitor electrode material. Earlier, researchers studied the electrochemical capacitive performance of SB-derived carbon and achieved low capacitance [47–50]. Recently, a few research groups tried to enhance the capacitance of SB-derived carbon by the incorporation of pseudocapacitance materials like metal oxides [51] and conducting polymers [52].

Indian sugar mills produce ~3 tons of waste sugarcane bagasse per annum, and each mill discards this huge waste bagasse as residue or uses it as a source of heat to meet the sugar mill's heat demand. In the present investigation, we explored waste sugarcane bagasse as a cheap carbon source and SiO_2 as an environmentally friendly and cheap pseudocapacitance material to prepare WSB-C/SiO_2 nanocomposites with enhanced capacitive performance. The nanocomposites were prepared via the one-step pyrolytic decomposition of TEOS-modified bagasse fibers. The effects of SiO_2 concentrations on the morphology and porosity of WSB-C/SiO_2 nanocomposites were evaluated. The electrochemical measurements of different formulated WSB-C/SiO_2 nanocomposites were studied on both alkaline and neutral electrolytes to compare their capacitive performance and electrochemical stability.

2. Results and Discussions

The functional groups generated in the different formulated waste SB-derived carbon materials were identified by FTIR spectroscopic analysis. Figure 1a illustrates the FTIR spectra of porous WSB-C and the different formulated WSB-C/SiO_2 nanocomposites. Several IR peaks appear in the FTIR spectrum of porous WSB-C (Figure 1a). In addition, a strong peak at 1191 cm^{-1} and a weak peak at 907 cm^{-1} were observed for the stretching and bending vibration of C-C bond conjugated to C=C bond, respectively, while a medium IR band at 1643 cm^{-1} appeared for the stretching mode of C=C bonds [53]. The peak at 622 cm^{-1} raised in the spectra was perhaps due to the out-of-plane deformation of -OH groups [54]. In contrast to C-SB, the WSB-C/SiO_2 nanocomposites revealed many IR peaks related to SiO_2. The peaks at 1092 and 487 cm^{-1} can be attributed to the intrinsic vibration of Si-O-Si and Si-O bonds [30], indicating the pyrolytic formation SiO_2 nanospheres from TEOS. During the carbonization process of TEOS-modified SB fibers, the -O-Si(OEt)$_3$ groups were detached from cellulose chains by the pyrolytic dissociation of C-O bonds, which are weaker than Si-O bonds, followed by hydrolysis into orthosilicic acid ($Si(OH)_4$) and finally thermal decomposition into SiO_2 [55,56]. Raman spectra of WSB-C and the different WSB-C/SiO_2 nanocomposites are displayed in Figure 1b. Two characteristic Raman active bands at around 1593 and 1356 cm^{-1} appeared for all nanocomposite samples. The low frequency G band is related to graphitic lattice vibration with E_{2g} mode, while the D band represents defects in graphitic structure. The degree of disorder in the different formulated samples was characterized by determining their I_D/I_G ratio [57]. As shown in Figure 1b, the I_D/I_G value of WSB-C is significantly lower than that of the WSB-C/SiO_2 nanocomposites, indicating that there is a greater degree of graphitization in WSB-C as compared to the nanocomposites. The in situ pyrolytic growth of SiO_2 nanospheres during carbonization of the TEOS-modified SB matrix perhaps impedes the graphitization kinetics. However, the higher I_D/I_G values of the nanocomposites suggest the greater extent of an amorphous phase in the nanocomposites compared to WSB-C, which can facilitate their electrochemical capacitive performances. The X-ray diffraction patterns of the formulated WSB-C/SiO_2 nanocomposites are illustrated in Figure 1c. The XRD pattern of WSB-C reveals a strong peak at $2\theta = 26.4^\circ$, assigned to the (002) plane of the graphite lattice [58]. This high intense graphitic (002) peak indicates the formation of highly crystalline carbon as a result of the carbonization of SB-derived crystalline cellulose, which was prepared by alkaline treatment of raw SB prior to the carbonization process. In contrast, the WSB-C/SiO_2 nanocomposites display several diffraction peaks at $2\theta = 20.4^\circ$, 36.5°, 39.3°, 40.2°, 42.3°, and 50° corresponding to the (100), (110), (102), (111), (200), and (112) crystalline planes of SiO_2, respectively [59], along with a (002) peak of crystalline graphite. The intensity of the graphitic (002) peak gradually decreases with increasing SiO_2 concentration in the nanocomposites, which might be related to the degree of graphitization of the SB matrix. It can be noticed that the position of the (002) crystalline peak slightly shifted toward lower 2θ values for the WSB-C/SiO_2 nanocomposites compared to WSB-C. The lower 2θ values of the WSB-C/SiO_2 nanocomposites indicate their higher *d*-spacing compared to WSB-C. The *d*-spacing of WSB-C, WSB-C/SiO_2 1025, WSB-C/SiO_2 105, and WSB-C/SiO_2 11 are calculated to be 0.33, 0.341, 0.358, and 0.347 nm, respectively. The results indicate the disordered layer structure of graphite with multi-layered stacking [60], which is well in agreement with the Raman spectroscopy results. The greater *d*-spacing of the WSB-C/SiO_2 nanocomposites can facilitate the easy diffusion of electrolyte ions into electrodes during electrolysis, making them potential candidates for energy storage applications. Furthermore, there is a weak and broad peak appearing in the 2θ range of 21–25° for the WSB-C/SiO_2 nanocomposites, indicating some degree of amorphous SiO_2 and carbon [61]. The surface characteristics of SB-derived carbon and its nanocomposites were evaluated by BET analysis. The N_2 adsorption–desorption isotherms of WSB-C and the formulated WSB-C/SiO_2 nanocomposites are illustrated in Figure 2a. A mixture of type I and type IV isotherms is observed for both WSB-C and the WSB-C/SiO_2 nanocomposites, indicating a typical combination of micropores and mesopores [62]. The initial part of the isotherm follows type I adsorption, which is attributed to the controlled monolayer adsorption by micropores [63]. However, the middle of the isotherm shows a characteristic hysteresis loop indicating the presence of large mesopores with a wide pore size distribution. The specific surface areas and total pore volumes of the different samples are presented in Figure 1. The specific surface areas and pore volumes were decreased for the WSB-C/SiO_2 nanocomposites compared to WSB-C. The maximum S_{BET} and total pore volume of 342 m^2/g and 0.0426 cm^3/g were recorded for WSB-C, while the lowest values of 115.7 m^2/g and 0.0219 cm^3/g were obtained for the WSB-C/SiO_2 11 nanocomposite containing higher concentrations of SiO_2 nanoparticles. This could be attributed to the blocking of micropores by SiO_2 nanoparticles, as evidenced in the FE-SEM images. In contrast to microporosity, the mesoporosity of the WSB-C/SiO_2 nanocomposites is significantly increased compared to WSB-C, and the C/SiO_2 105 nanocomposite exhibits the highest amount of mesoporosity with 74.2%. The BJH pore size distributions of WSB-C and the different formulated WSB-C/SiO_2 nanocomposites are displayed in Figure 2b. The average diameters of the pores of the different samples are presented in Table 1. The pore size is gradually increased with increasing SiO_2 content, which might be the result of pore enlargement caused by heterogeneous shrinkage during the carbonization process [64].

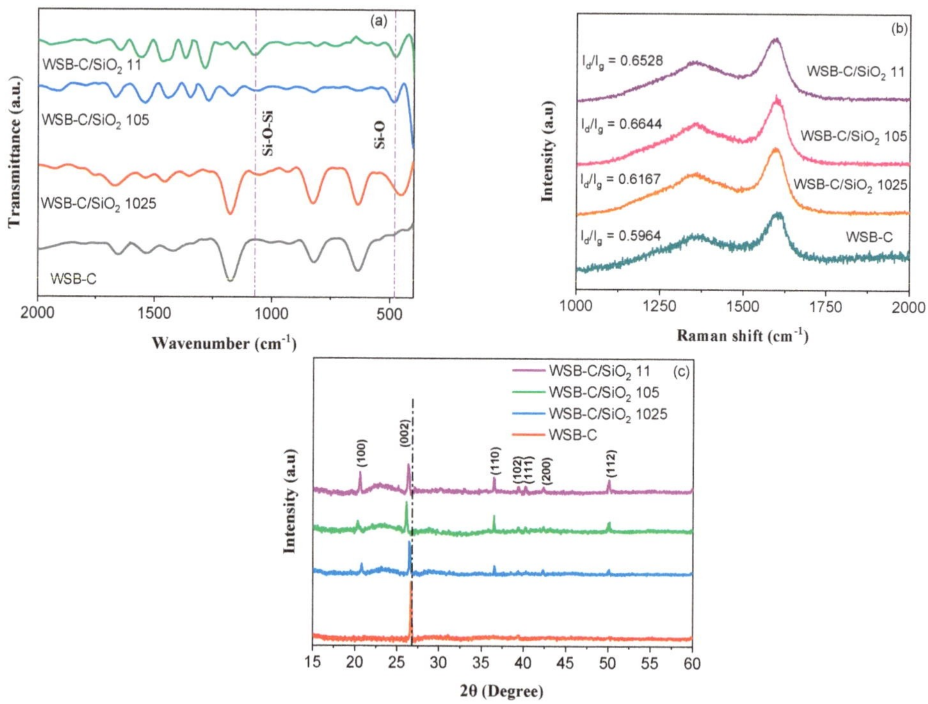

Figure 1. (**a**) FTIR spectra, (**b**) Raman spectra, and (**c**) XRD patterns of pure WSB—C and the different formulated WSB—C/SiO_2 nanocomposites.

Table 1. Summary of BET characteristics of the as-prepared WSB-C and its nanocomposites.

Sample	WSB-C	WSB-C/SiO_2 1024	WSB-C/SiO_2 105	WSB-C/SiO_2 11
S_{BET} (m^2/g)	342.8	279.9	207.7	115.7
V_{total} (cm^3/g)	0.0426	0.389	0.0288	0.0219
Mesopore (%)	44.7	57.4	74.2	68.8
D_{av} (nm)	3.46	3.63	3.92	4.47

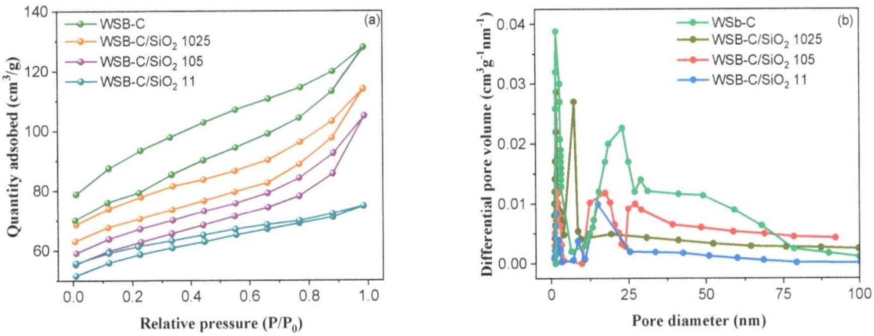

Figure 2. (**a**) Nitrogen adsorption and desorption isotherms and (**b**) Barrett–Joyner–Halenda (BJH) pore size distribution of WSB–C and the WSB–C/SiO_2 nanocomposites.

The surface chemical compositions of the as-synthesized WSB-C/SiO_2 nanocomposites were determined by XPS analysis. Figure 3a demonstrates the XPS survey curves of activated SB-derived carbon and the different WSB-C/SiO_2 nanocomposites. The XPS signals for three major elements such as carbon (C), oxygen (O), and silicon (Si) at ~285 eV, ~531 eV, and ~104 eV are observed in the survey curves of the WSB-C/SiO_2 nanocomposites, while WSB-C contained no elemental silicon. As is obvious, the peak intensity of elemental Si for the nanocomposites gradually increased with increasing TEOS loading in the modified-SB fibers. Figure 3b displays the core level Si 2p spectra of WSB-C/SiO_2 105 consisting of a single Gaussian peak at 104.3 eV, corresponding to the Si-O-Si bond of SiO_2 nanospheres [62]. The C 1s spectra is deconvoluted into four peaks at 284.1 eV, 284.9 eV, and 287.5 eV, shown in Figure 3c, which are associated with C=C, C-C, and C=O bonds, respectively. The atomic percentages of elemental C, Si, and O in the different formulated carbon samples are demonstrated in Figure 3d. The results indicate that the atomic percentage of carbon is reduced for the WSB-C/SiO_2 nanocomposites compared to WSB-C. The lowest yield of 77.21% was recorded for WSB-C/SiO_2 11, which is 17% lower than that achieved for WSB-C. Furthermore, the at% of carbon is gradually decreased with increasing TEOS content in modified-SB. These findings suggest that the degree of carbonization/graphitization of SB was affected by the in situ formation of SiO_2, which might be due to the change in the heat of carbonization during the pyrolysis process.

The microstructures of SB-derived carbon and its hybrids with SiO_2 were observed with SEM analysis and the resulting images are shown in Figure 4. The orderly aligned channels appear in the cross-sectional images of the nanocomposites with multiple sizes of pores on their surfaces. The top surface of the WSB-C displays slightly crumpled features with multiple size pores (Figure 4a). In contrast, the WSB-C/SiO_2 nanocomposites clearly exhibit a uniform distribution of SiO_2 nanospheres on the porous carbon surface. The concentration of SiO_2 nanospheres gradually increases with increasing TEOS loading in the modified-SB precursor. For the WSB-C/SiO_2 1025 and WSB-C/SiO_2 105 nanocomposites, the majority of SiO_2 nanospheres remain discrete and embedded into the carbon matrix (Figure 4d), while the SiO_2 nanospheres are largely agglomerated in the WSB-C/SiO_2 11 nanocomposite (Figure 4e). The extensive agglomeration of SiO_2 nanospheres reduced the specific surface area and limited the pseudocapacitive effects of the WSB-C/SiO_2 11 nanocomposite, which have been observed in the BET and CV results. The cross-sectional view of the WSB-C/SiO_2 nanocomposites in Figure 4f–h exhibits three-dimensional (3D) channels within the carbon matrix. This 3D structure provides plenty of space for growing SiO_2 nanospheres. Like tree trunks, these channels remain aligned along their longitudinal axis, which can facilitate the transport of electrolyte through them. Furthermore, the cross-sectional images reveal a uniform dispersion of SiO_2 nanospheres throughout the channel wall, which can facilitate the access of electrolyte ions to redox-active SiO_2 and thus enhance the capacitance efficiency of the nanocomposites.

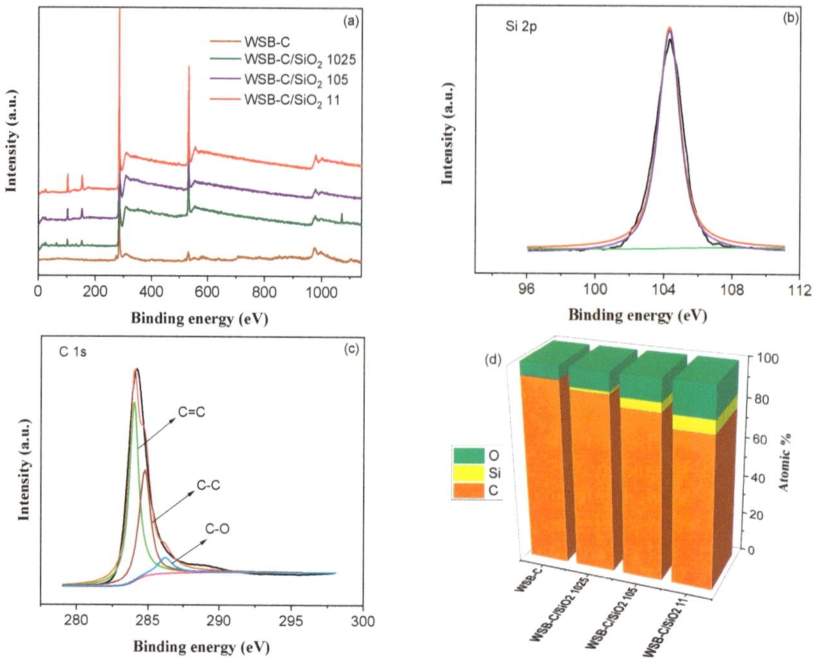

Figure 3. (**a**) XPS survey spectra of WSB-C and the WSB−C/SiO_2 nanocomposites; the high-resolution spectra of (**b**) Si 2p and (**c**) C 1s along with the fitting peaks of the as-prepared WSB−C/SiO_2 105; (**d**) atomic compositions of WSB−C and the WSB−C/SiO_2 nanocomposites.

The effects of SiO_2 nanospheres on the electrochemical properties of SB-derived activated carbon were evaluated using CV and GCD measurements. Figure 5a demonstrates the CV profiles of WSB-C and the different WSB-C/SiO_2 nanocomposites in a potential range of -1.0 to 0 V at 50 mV/s. The WSB-C electrode reveals a nearly rectangular CV curve with no obvious redox peaks, suggesting the behavior of an electrical double-layer capacitor with fast charge and discharge processes. On the other hand, the CV curves of the WSB-C/SiO_2 nanocomposite electrodes reveal a distorted rectangular shape with a pair of faradaic redox peaks, suggesting both EDLC and pseudocapacitance characteristics of the WSB-C/SiO_2 electrode. The oxidation and reduction peaks appear at −0.12 and −0.30 V, respectively. The redox reactions involved during the electrochemical process in the KOH electrolyte can be illustrated as follows [65]:

$$Si^{II}O_2 + K^+ + e^- \leftrightarrow Si^{III}OOK \quad (1)$$

For the WSB-C/SiO_2 electrodes, the anodic peak current gradually increased for the WSB-C/SiO_2 1025 and WSB-C/SiO_2 105 electrodes but it decreased for the WSB-C/SiO_2 11 electrode. The WSB-C/SiO_2 105 electrode exhibited better pseudocapacitive characteristics in terms of current response to voltage. However, the decrease in anodic current of the WSB-C/SiO_2 11 electrode was the result of the agglomeration of SiO_2 nanospheres (as shown in the FE-SEM image) and the subsequent decrease in surface area of active SiO_2. The CV curves of the WSB-C electrode in a wide range of scan rates are displayed in Figure 5b. The CV curves retained a nearly rectangular shape even at a high scan rate of 125 mV/s, indicating the fast and reversible EDLC behavior of the WSB-C electrode. The shape of the CV profiles of the WSB-C/SiO_2 105 electrode remained unchanged with a small shift in redox peak position upon increasing the scan rate (Figure 5c), suggesting its high-rate capability over a wide range of scan rates.

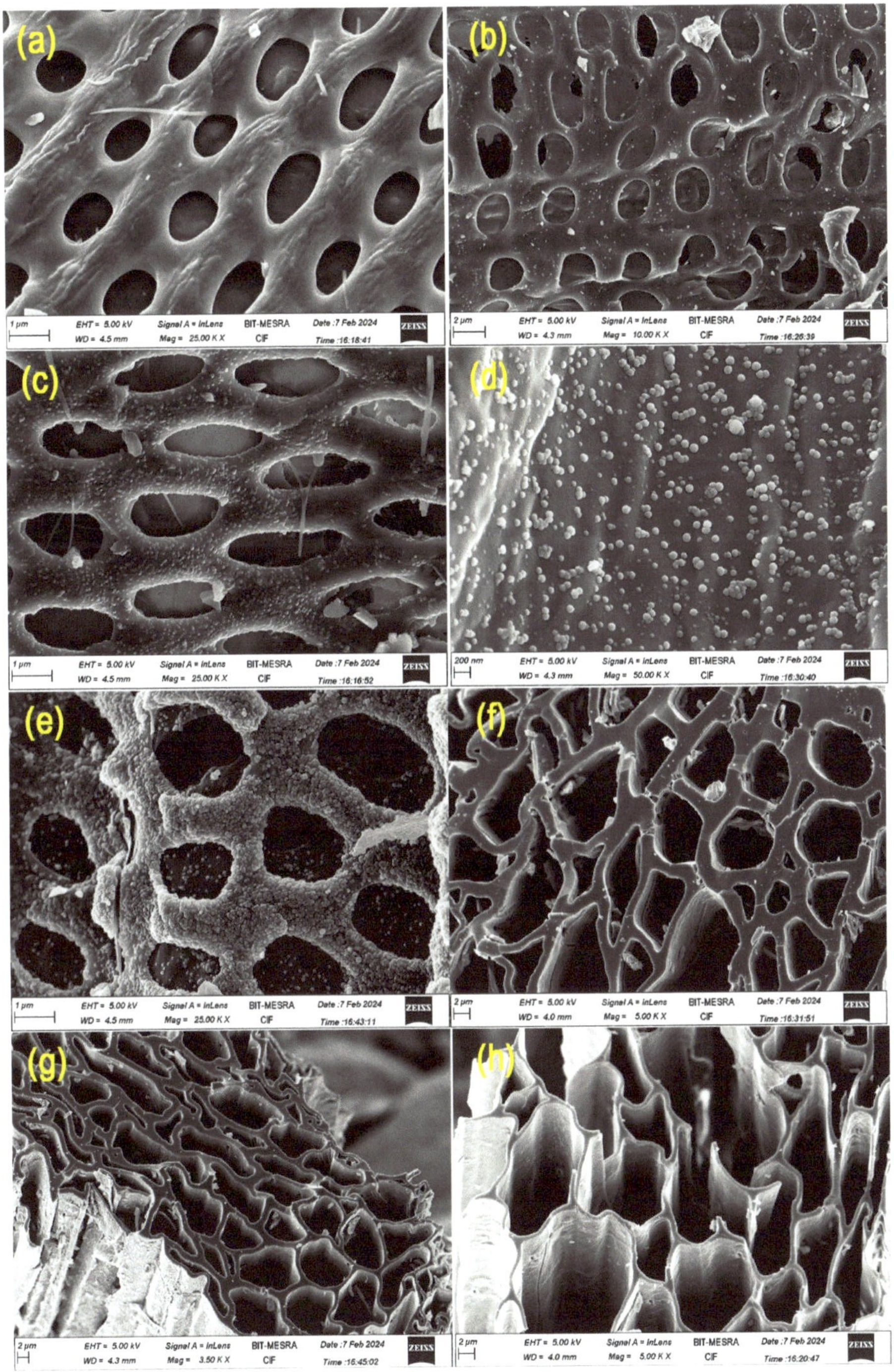

Figure 4. SEM images of surface views of (**a**) WSB−C, (**b**) WSB−C/SiO_2 1025, (**c**,**d**) WSB−C/SiO_2 105, and (**e**) WSB−C/SiO_2 11. SEM images of cross-sections of (**f**) WSB−C/SiO_2 1025, (**g**) WSB−C/SiO_2 105, and (**h**) WSB−C/SiO_2 11.

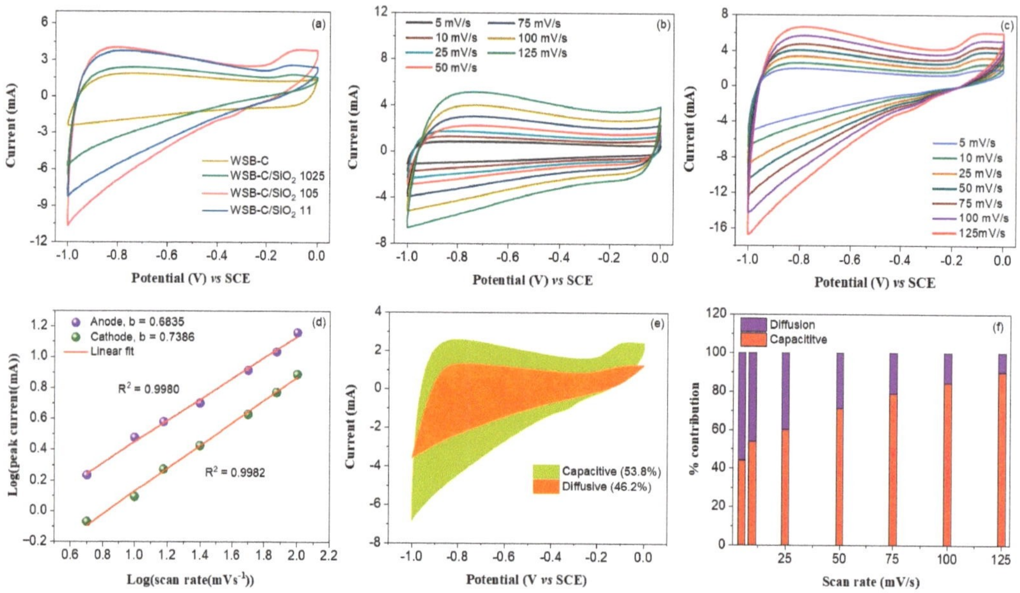

Figure 5. (**a**) CV curves of WSB−C, WSB−C/SiO_2 1025, WSB−C/SiO_2 105, and WSB−C/SiO_2 11 electrodes in 6 M KOH electrolyte at a scan rate of 50 mV/s; (**b**) CV profiles of the WSB−C electrode at different scan rates; (**c**) CV curves of the WSB−C/SiO_2 105 electrode at different scan rates; (**d**) linear fitting curves between ln (i) vs. ln (sweep rate) for the WSB−C/SiO_2 105 electrode; (**e**) Dunn's method analysis of the capacitance contribution of the WSB−C/SiO_2 105 electrode at 10 mV/s; and (**f**) contribution ratio of capacitive and diffusion-controlled processes at various scan rates.

It is necessary to study the kinetics of charge storage in WSB-C/SiO_2 electrodes at different potentials in order to identify the mechanism behind the charge stored in them. In general, there are two different mechanisms through which the total charge is stored: (i) diffusion mechanism (intercalation/deintercalation) and (ii) capacitive mechanism [64]. The Power's law was used to characterize this phenomenon [65]:

$$i\,(V) = a\vartheta^b \tag{2}$$

where ϑ is the scan rate and b is the slope of the linear plot of Log (i) vs. Log (ϑ). Two conditions based on the b values determine the mechanism responsible for charge storage: if $b = 0.5$, the diffusion-controlled intercalation process is dominant over the capacitive process, and when $b = 1.0$, the capacitive contribution is higher than the diffusion contribution. Figure 5d demonstrates the logarithmic linear plots between redox current and scan rate. In the present case, the b values for anodic and cathodic processes are 0.6835 and 0.7386, respectively, which suggests that the charge storage process is governed by both surface-controlled and diffusion-controlled mechanisms. Furthermore, the diffusion-controlled faradaic process made a considerable contribution to enhancing the charge storage capability of the WSB-C/SiO_2 105 electrode. The large fraction of mesopores in the WSB-C/SiO_2 105 nanocomposite facilitated the intercalation/deintercalation of electrolyte ions. The relative currents from the capacitive and diffusion processes at different scan rates can be determined using the given equation [66]:

$$i\,(V) = k_1(\vartheta) + k_2\vartheta^{1/2} \tag{3}$$

where $k_1(\vartheta)$ and $k_2\vartheta^{1/2}$ values correspond to the current contribution from the diffusion and capacitive processes, respectively. Figure 5e reveals the percent contribution of the diffusion and capacitive processes to the CV area at 5 mV/s. The low scan rate favors the diffusion process and thus diffusion is dominant over the capacitive contribution. As scan rates increase, the diffusion contribution gradually decreases, and the capacitive contribution steadily increases (Figure 5f).

The impedance analyses (EIS) were further carried out to investigate the ionic diffusion kinetics for the as-synthesized WSB-C and its nanocomposites with SiO_2 [67]. The Nyquist plots, shown in Figure 6a, allow us to compare the impedance properties of the different formulated electrode materials. WSB-C exhibits a solution resistance (R_s) of 2.86 Ω due to the relatively low hydrophilicity of carbon material, while the low R_s values of the WSB-C/SiO_2 nanocomposites suggest their better hydrophilic characteristics. At high frequencies, Nyquist plots represent charge-transfer resistance (R_{CT}), while at low frequencies a line with a 45° slope represents capacitive behavior. Table 2 presents the fitted values for each component for each material. The charge-transfer resistance of WSB-C was significantly reduced upon the inclusion of the SiO_2 nanospheres. The lower charge-transfer resistance of the WSB-C/SiO_2 electrodes reflects their electrochemical performance. The equivalent circuit model for WSB-C/SiO_2 105 is shown in the inset of Figure 6a.

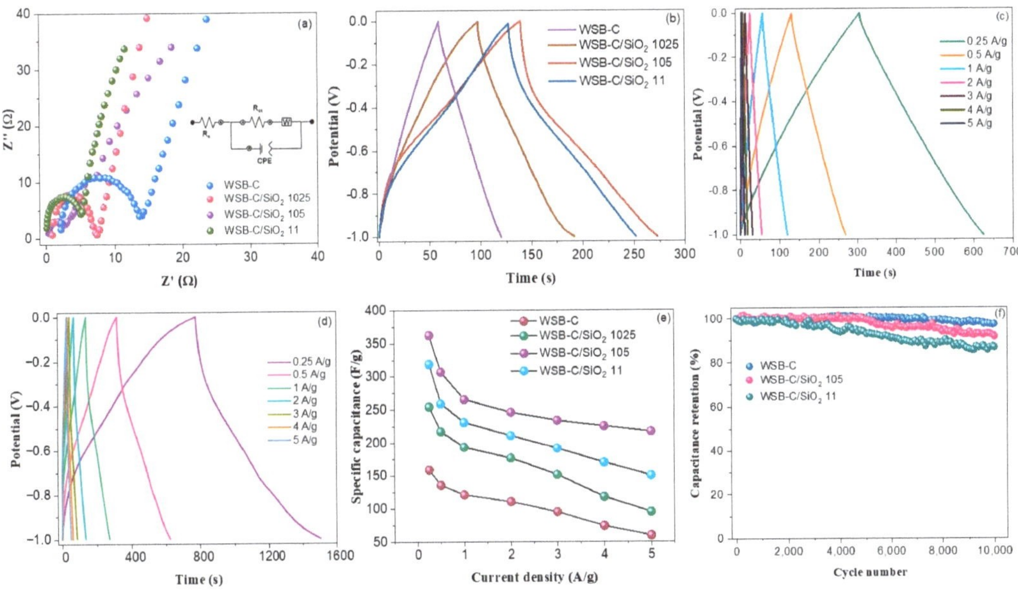

Figure 6. (**a**) Nyquist plots of the WSB−C and the different WSB−C/SiO_2 nanocomposite electrodes; (**b**) GCD curves of the WSB−C and the different WSB-C/SiO_2 nanocomposite electrodes in 6 M KOH electrolyte; (**c**) GCD profiles of the WSB−C electrode at different current densities; (**d**) GCD curves of the WSB−C/SiO_2 105 electrode at different current densities; (**e**) variations in gravimetric capacitances of the WSB−C and the different WSB−C/SiO_2 nanocomposite electrodes as a function of current density; (**f**) cyclic stability up to 10,000 cycles at 1 A/g for the WSB−C and the different WSB−C/SiO_2 nanocomposite electrodes.

Table 2. Calculated values of R_s, R_{CT}, Z_W, and CPE by fitting the experimental data.

Sample	WSB-C	WSB-C/SiO$_2$ 1024	WSB-C/SiO$_2$ 105	WSB-C/SiO$_2$ 11
R_s (Ω)	2.17	0.876	0.114	0.064
R_{CT} (Ω)	11.59	6.39	2.57	4.38
Z_w (Ω/s$^{0.5}$)	48.7	12.67	3.23	8.18
CPE (μF)	11.46	15.63	23.92	19.47

The gravimetric charge–discharge (GCD) profiles of the pure WSB-C and the different WSB-C/SiO_2 nanocomposite electrodes at a current density of 1 A/g are demonstrated in Figure 6b. A nearly symmetric triangular charge–discharge curve appears for the WSB-C electrode, indicating pure non-faradaic EDLC behavior. The WSB-C electrode reveals a Coulombic efficiency of 100.6% at 1 A/g, suggesting its excellent electrochemical reversibility. In contrast, the WSB-C/SiO_2 nanocomposite electrodes display asymmetric distorted triangular charge–discharge curves with an obvious discharge plateau at ~0.4 V, suggesting faradaic contribution to the overall charge storage process. Furthermore, there was a longer discharge time for the WSB-C/SiO_2 electrodes than for the WSB-C electrodes, indicating that the nanocomposites have superior capacity for storing charge than the WSB-C electrodes. Among the nanocomposite electrodes, the WSB-C/SiO_2 105 electrode took the maximum time to discharge, indicating greater charge storage capability. Figure 6c illustrates the GCD curves of the WSB-C electrode at different current densities. The symmetric features of the GCD curves remain unaffected upon increasing the current density, indicating excellent rate capability. The GCD profiles of the WSB-C/SiO_2 105 nanocomposite electrode at various current densities between 0.25 and 5 A/g are demonstrated in Figure 6d. The asymmetric features of the GCD curves appear for all GCD curves due to the pseudocapacitance effects. The gravimetric specific capacitances (C_{gsp}) of the WSB-C and the WSB-C/SiO_2 nanocomposite electrodes were calculated from their respective GCD curves, using Equation (4):

$$C_{gsp} = \frac{I\Delta t}{m\Delta V} \quad (4)$$

Figure 6e demonstrates the variation in specific capacitances as a function of current densities for the different electrodes. The results clearly exhibit a significant improvement in the specific capacitance of SB-derived carbon upon hybridization with SiO_2. The specific capacitances of the WSB-C/SiO_2 1025, WSB-C/SiO_2 105, and WSB-C/SiO_2 11 electrodes are ~160%, ~240%, and ~210% higher than those of the pure WSB-C electrodes, respectively. The highest specific capacitance of 362.3 F/g at 0.25 A/g was achieved for the WSB-C/SiO_2 105 electrode. Table 3 presents that the specific capacitance of the WSB-C/SiO_2 105 electrode is higher than those previously reported for similar types of electrode materials. The capacitance values gradually reduced from 362.3 to 220.6 F/g when the current density increased from 0.25 to 5 A/g, indicating an excellent capacitance retention of 61% even with a 20-fold increase in current density. However, a relative lower capacitance retention was observed for WSB-C/SiO_2 1025 (49% retention) and WSB-C/SiO_2 11 (37% retention). The lower capacitance retention in the WSB-C/SiO_2 1025 and WSB-C/SiO_2 11 electrodes might be ascribed to the presence of a smaller fraction of mesopores and some fraction of unused SiO_2 at high current densities. The cycling stability of the WSB-C/SiO_2 nanocomposite electrodes was compared with that of the WSB-C electrode, shown in Figure 6f. The nanocomposite electrodes revealed somewhat lower cycling stability than the WSB-C electrode, which might be due to their limited pseudocapacitance contribution at higher cycles [68]. The highest cycling stability of 97.4% was achieved for the WSB-C electrode, while the cycling stability went down to 91.7% and 86.9% for the WSB-C/SiO_2 105 and WSB-C/SiO_2 11 electrodes, respectively. This might be due to the relatively low electrochemical stability of metal oxide compared to carbon.

Table 3. Comparison of the electrochemical performance of the present WSB–C/SiO_2 composites and those of similar biomass-derived carbon composite materials reported earlier.

Biomass	Biomass-Derived Carbon ($C_{biomass}$) Composites	Electrolyte	Potential Range (V)	Sp. Capacitance (F/g)	Cycling Stability	Ref.
Bamboo leaves	$C_{biomass}/CuO/Cu_2O$	1 M HCl	−1.0 to +0.3	147@1 A/g	93% after 5000 cycles	[69]
Wasted litchi shell	$C_{biomass}/MnO$	6 M KOH	−1.0 to +0.2	162.7@0.5 A/g	93.5% after 5000 cycles	[70]

Table 3. Cont.

Biomass	Biomass-Derived Carbon ($C_{biomass}$) Composites	Electrolyte	Potential Range (V)	Sp. Capacitance (F/g)	Cycling Stability	Ref.
Vegetable sponge	$C_{biomass}/Mn_3O_4$ Composite	1 M Na_2SO_4	0 to 1.0	162.8@0.5 A/g	89.5% after 4500 cycles	[71]
Watermelon	$C_{biomass}/MnO_2$	6 M KOH	−1.0 to 0	123.5@0.5 A/g	60% after 1000 cycles	[72]
Waste bamboo shoot shells	$C_{biomass}$/PEDOT	1 M H_2SO_4	0 to 1.0	302.5@0.5 A/g	87% after 10,000 cycles	[73]
Loofah	$C_{biomass}/TiO_2$	1 M H_2SO_4	0 to 1.0	250.8@1 A/g	84% after 100 cycles	[74]
Wheat flour	$C_{biomass}/Co_3O_4$	2 M KOH	−0.8 to +0.4	161.4@0.5 A/g	80% after 1000 cycles	[75]
Waste sugarcane bagasse	$C_{biomass}/SiO_2$	6 M KOH	−1.0–0.0	307.1@0.5 A/g	91.7 after 10,000 cycles	Present

The capacitive performance of the WSB-C/SiO_2 105 electrode was further evaluated in neutral electrolyte, i.e., 1 M Na_2SO_4, and the results were compared with those obtained in alkaline electrolyte, i.e., 6 M KOH. The CV profiles of the WSB-C/SiO_2 105 electrode over the potential window of −1.0 to 0 V in 1 M Na_2SO_4 electrolyte at different scan rates are illustrated in Figure 7a. Nearly rectangular-shaped CV curves with a pair of oxidation and reduction peaks centered at −0.49 V and −0.64 V are observed, indicating the involvement of both EDLC and pseudocapacitance mechanisms in the charge storage process. The redox peaks are associated with the reactions given in Equation (5) [76]:

$$Si^{II}O_2 + Na^+ + e^- \leftrightarrow Si^{III}OONa \quad (5)$$

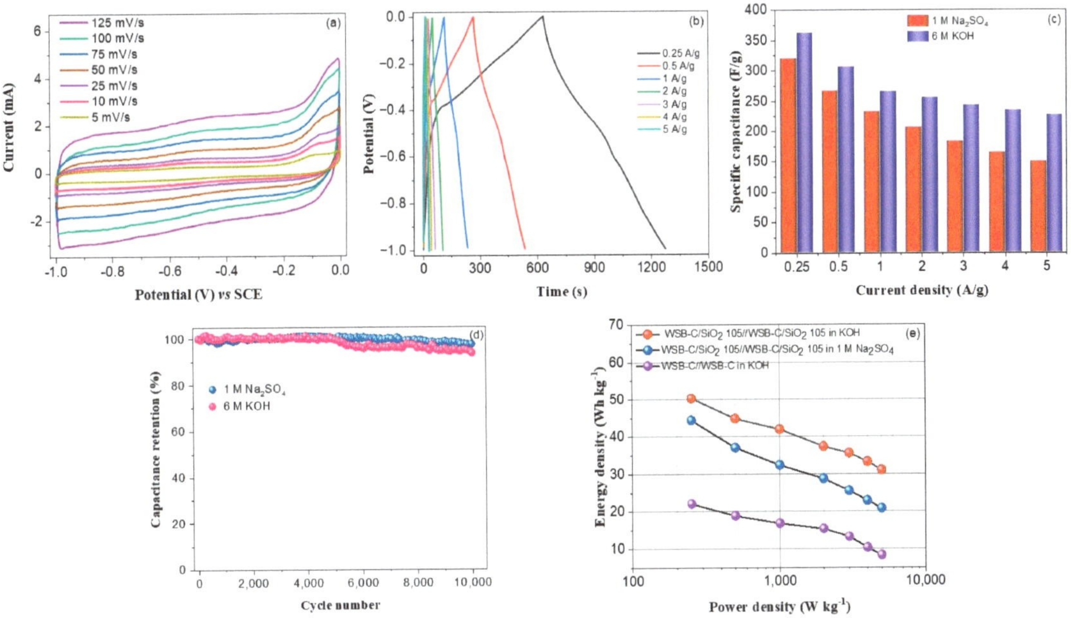

Figure 7. (**a**) CV profiles of the WSB−C/SiO_2 105 nanocomposite electrode in 1 M Na_2SO_4; (**b**) GCD profiles of the WSB−C/SiO_2 105 nanocomposite electrode in 1 M Na_2SO_4; (**c**) variations in gravimetric capacitances of the WSB−C/SiO_2 105 electrode as a function of current density in two different electrolytes; (**d**) cyclic stability up to 10,000 cycles at 1 A/g for WSB−C/SiO_2 105 in two different electrolytes; and (**e**) Ragone plots of the symmetric WSB−C/SiO_2 105// WSB−C/SiO_2 105 supercapacitor in two different electrolytes and the WSB−C//WSB−C supercapacitor in alkaline electrolyte.

There is almost no change in the shape of the CV profiles when the scan rate is increased from 5 to 125 mV/s, indicating the high-rate capability and electrochemical stability of the nanocomposite electrode in Na_2SO_4 electrolyte. The total current gradually increases with the increasing scan rate. The GCD profiles for the two-electrode system at various current densities are displayed in Figure 7b. The distorted triangular shape of the GCD curves with a voltage plateau further suggests the occurrence of faradaic surface-redox reactions during the charge storage process. The Coulombic efficiency was determined to be 100.9%. The specific capacitances at different current densities were determined from the corresponding GCD curves, using Equation (4). Figure 7c reveals the gravimetric specific capacitance of the WSB-C/SiO_2 105 electrode at different current densities in two different electrolytes, i.e., 6 M KOH and 1 M Na_2SO_4. The maximum specific capacitance of 319.8 F/g at 0.25 A/g was achieved in 1 M Na_2SO_4 electrolyte. As can be seen in Figure 7c, the capacitive performance of the nanocomposite electrode is somewhat lower in the Na_2SO_4 electrolyte than in the KOH electrolyte. Compared to K^+ ions, Na^+ ions have a bigger size and higher internal resistance, resulting in a slower transport rate, which results in lower capacitance. With the Na_2SO_4 electrolyte, the WSB-C/SiO_2 105 electrodes show a lower rate capability of 46.8% than with the KOH electrolytes (61%). For hydrated Na^+ ions, their larger size is indeed unfavorable to smooth transportation and diffusion, especially at higher current densities, resulting in lower capacitance. The nanocomposite electrode exhibited superior cycling stability with capacitance retention of 95.6% after 10,000 cycles in 1 M Na_2SO_4 compared to that in 6 M KOH (91.7%), as shown in Figure 7d. The relatively lower cycling stability in the alkaline electrolyte might be due to the harsh and corrosive nature of the strong alkaline electrolyte (KOH) compared to the neutral Na_2SO_4 electrolyte. Figure 7e demonstrates the Ragone plots for the as-assembled symmetric WSB-C//WSB-C and WSB-C/SiO_2 105//WSB-C/SiO_2 105 devices in KOH and Na_2SO_4 electrolytes. The WSB-C/SiO_2 105//WSB-C/SiO_2 105 device in 6 M KOH electrolyte delivered a maximum energy density of 50.3 WH kg^{-1} at a power density of 250 W kg^{-1}, which is significantly greater than that produced by the WSB-C//WSB-C device (22.1 Wh kg^{-1} at 250 W kg^{-1}). The as-achieved energy density is superior to earlier reported biomass-derived carbon-based symmetric supercapacitors such as silkworm cocoon-derived carbon (34.4 Wh kg^{-1}) [77]; wheat bran-derived carbon (32.7 Wh kg^{-1}) [78]; peanut meal-derived carbon (24.9 Wh kg^{-1}) [79]; rice straw-derived carbon (7.8 Wh kg^{-1}) [80]; coconut shell-derived carbon (14.7 Wh kg^{-1}) [81]; sugarcane bagasse-derived carbon (37.5 Wh kg^{-1}) [82]; and cornstalk-derived carbon (10 Wh kg^{-1}) [83]. Furthermore, the SC device achieved higher energy density in KOH electrolyte than in Na_2SO_4 electrolyte (44.4 Wh kg^{-1}). The mechanism of electrochemical process and LED lightening of the as-fabricated symmetric WSB-C/SiO_2 105//WSB-C/SiO_2 105 device is displayed in Scheme 1. The LED bulb (1.5 V) was lit with two symmetric WSB-C/SiO_2 105//WSB-C/SiO_2 105 electrodes after charging to 1.5 V, where the dimension of each electrode used in the device was 1.5×1.5 cm^2. The LED bulb showed stable brightness for 30 min. The results indicate that the present WSB-C/SiO_2//WSB-C/SiO_2 symmetric supercapacitors have great potential for energy storage applications with impressive market value. This impressive energy density of waste sugarcane bagasse-derived carbon/SiO_2 105 makes it a potential cost-effective and environmentally friendly electrode material for next-generation supercapacitors.

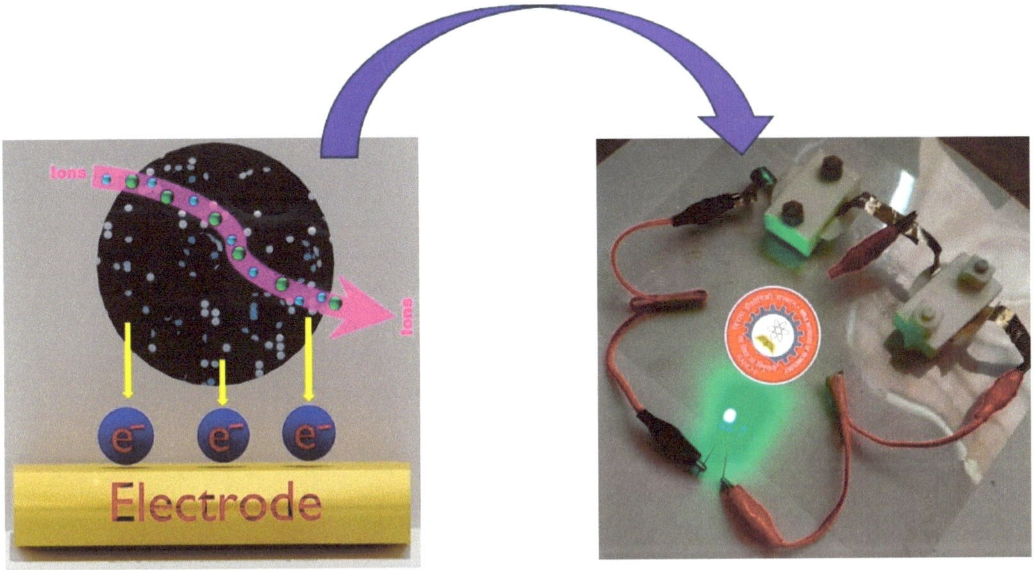

Scheme 1. Mechanism of electrochemical process and the as-assembled prototype symmetric WSB-C/SiO_2 105//WSB-C/SiO_2 device in KOH electrolyte with LED lightening.

3. Experimental

3.1. Materials

Waste sugarcane bagasse fibers were collected from a sugar mill in Uttar Pradesh, India. Potassium hydroxide (KOH), glacial acetic acid, triethyl orthosilicate, ethanol, and liquid ammonia solution were collected from Alfa Asare, India. All chemicals were utilized without undergoing additional purification.

3.2. Activation and Chemical Modification of Sugarcane Bagasse Fibers

The as-collected WSB fibers were initially washed thoroughly with DI water and dried at 60 °C. The dried WSB fibers were milled to size 40 mesh. For the activation process, 100 g WSB powder was dispersed in 500 mL 6 M KOH solution under stirring for 2 h. Afterward, the WSB fibers were treated with glacial acetic acid and washed with DI water to neutralize (pH = 7) them. The activated WSB fibers were then chemically modified with TEOS, where activated WSB fibers were mixed with different concentrations of TEOS ($W_{SB\ fiber}/W_{TEOS}$ = 1/0.25, 1/0.5, and 1/1) in ethanol/water (80/20 *v*/*v*) and stirred for 3 h. Liquid NH3 was drop-wisely added to the solution for neutralization. The resulting TEOS-modified WSB fibers were washed with DI water and thermally cured at 120 °C for 3 h.

3.3. Carbonization of Activated SB and TEOS-Modified SB Fibers

The alkaline-activated WSB and TEOS-modified WSB fibers were carbonized at 600 °C under nitrogen flow (20 mL/min) using a tube furnace. The carbonization of activated WSB and TEOS-modified WSB fibers produced porous carbon (WSB-C) and carbon/SiO_2 nanocomposites, respectively. The nanocomposites obtained from TEOS-modified WSB fibers with fiber/TEOS ratios of 1/0.25, 1/0.5, and 1/1 are designated as WSB-C/SiO_2 1025, WSB-C/SiO_2 105, and WSB-C/SiO_2 11, respectively.

3.4. Structural Characterizations

FTIR and Raman analyses were conducted on Shimadzu IR-Prestige 21 (Shimadzu Corp, Japan) and Renishaw Raman System 3000 spectrophotometers (Renishaw, UK), respectively, using powdered samples. The XRD patterns of the samples were obtained with a Rigaku SmartLab diffractometer (Rigaku Corporation, Japan). The surface characteristics of the as-prepared materials were evaluated using a Micromeritics ASAP 2020 adsorption analyzer (Micromeritics Instrument Corporation, USA). An ESCA MultiLab 2000 spectroscopic analyzer (VG Systems Ltd. UK) was used to characterize the surface chemical compositions of the materials. A field-emission scanning electron microscope (FE-SEM, S-4700, Hitachi, Japan) was used to take microstructural images of the as-prepared samples.

3.5. Electrochemical Measurements

The electrochemical measurements were carried out in aqueous 6 M KOH electrolyte at room temperature (25 °C) using 3-electrode and 2-electrode cells for cycling voltametric (CV) and charge–discharge experiments, respectively. A potentiostat–galvanostat (SQUIDSTAT SOLO, Admiral Instruments, USA) was used for these measurements. For the 3-electrode cell, powdery sample-coated GCE with 2 μL Nafion solution (5 wt%), platinum sheet, and calomel electrode served as working, counter, and reference electrodes, respectively. For the 2-electrode configuration, two symmetric electrodes (1×1 cm^2) consisting of active material-loaded Ni-foam were superimposed with a separator. The electrode was prepared by dispersing the WSB-C or WSB-C/SiO_2 powder (~1.8 mg) on nickel foam and pressing under pressure of 15 tons, using a hydraulic press. EIS measurements were performed in a range of 100,000–0.1 Hz and an AC amplitude of 5 mV.

4. Conclusions

In summary, waste sugarcane bagasse-derived carbon/SiO_2 nanocomposites were prepared through carbonization of TEOS-modified activated bagasse fibers. The capacitive performance of WSB-derived activated carbon was remarkably enhanced upon hybridization with SiO_2, indicating a significant pseudocapacitive contribution from redox-active SiO_2. A maximum capacitance of 362.3 F/g at 0.25 A/g was achieved for the WSB-C/SiO_2 105 nanocomposite, which is around 2.5 times higher than that of the WSB-derived activated carbon. However, higher SiO_2 concentrations decrease the capacitance of the nanocomposite due to the agglomeration of SiO_2 particles, as shown for WSB-C/SiO_2 11. The nanocomposite electrode demonstrated superior capacitance and energy density in KOH electrolyte compared to neutral Na_2SO_4 electrolyte. The WSB-C/SiO_2 105//WSB-C/SiO_2 105 supercapacitor achieved an energy density of 50.3 Wh kg^{-1} at a power density of 250 W kg^{-1} in 6 M KOH electrolyte, which is considerably higher than that achieved in 1 M Na_2SO_4 (44.4 Wh kg^{-1}). However, the cycling stability of the supercapacitor was significantly higher in neutral Na_2SO_4 electrolyte than in alkaline electrolyte. Hence, the WSB-C/SiO_2 105 nanocomposite as an efficient, sustainable, and cheap electrode material could potentially be used in next-generation energy storage devices.

Author Contributions: Conceptualization, M.M.A.A. and J.S.; methodology, M.M.A.A. and J.S.; software, M.M.A.A., J.S. and S.S.H.; validation, A.C. and S.S.H.; formal analysis, A.C., S.S.H. and A.N.B.; investigation, J.S., A.C. and S.S.H.; resources, S.S.H. and A.N.B.; data curation, A.C., S.S.H. and A.N.B.; writing—original draft preparation, J.S., A.C. and S.S.H.; writing—review and editing, A.C., S.S.H. and A.N.B.; visualization, S.S.H. and A.N.B.; supervision, A.C.; project administration, S.S.H. and A.N.B.; funding acquisition, S.S.H. and A.N.B. All authors have read and agreed to the published version of the manuscript.

Funding: The research was funded by the Deputyship for Research & Innovation, Ministry of Education, Saudi Arabia, through project number INST087.

Institutional Review Board Statement: Not applicable.

Informed Consent Statement: Not applicable.

Data Availability Statement: Data are contained within the article.

Acknowledgments: The authors extend their appreciation to the Deputyship for Research & Innovation, Ministry of Education, Saudi Arabia, for funding this research work through project number INST087.

Conflicts of Interest: The authors declare no conflicts of interest.

References

1. Sharma, S.; Chand, P. Supercapacitor and electrochemical techniques: A brief review. *Results Chem.* **2023**, *5*, 100885. [CrossRef]
2. Yang, W.; Wang, J.; Gao, S.; Zhang, H.; Wang, H.; Li, Q. Photo-assisted charging of carbon fiber paper-supported CeO_2/MnO_2 heterojunction and its long-lasting capacitance enhancement in dark. *J. Adv. Ceram.* **2022**, *11*, 1735–1750. [CrossRef]
3. Vijayakumar, M.; Shankar, A.B.; Rohita, D.S.; Rao, T.N.; Karthik, M. Conversion of Biomass Waste into High Performance Supercapacitor Electrodes for Real-Time Supercapacitor Applications. *ACS Sustain. Chem. Eng.* **2019**, *7*, 17175–17185. [CrossRef]
4. Castro-Gutierrez, J.; Celzard, A.; Fierro, V. Energy storage in supercapacitors: Focus on tannin-derived carbon electrodes. *Front. Mater.* **2020**, *7*, 217. [CrossRef]
5. Senthil, C.; Lee, C.W. Biomass-derived biochar materials as sustainable energy storage devices. *Renew. Sustain. Ener. Rev.* **2020**, *137*, 110464. [CrossRef]
6. Hatfield-Dodds, S.; Schandl, H.; Newth, D.; Obersteiner, M.; Cai, Y.; Baynes, T.; West, J.; Havlik, P. Assessing global resource use and greenhouse emissions to 2050, with ambitious resource efficiency and climate mitigation policies. *J. Clean. Prod.* **2017**, *144*, 403–414. [CrossRef]
7. Ostergaard, P.A.; Duic, N.; Noorollahi, Y.; Mikulcic, H.; Kalogirou, S. Sustainable Development using renewable energy technology. *Renew. Energy* **2019**, *146*, 2430–2437. [CrossRef]
8. Sattayarut, V.; Wanchaem, T.; Ukkakimapan, P.; Yordsri, V.; Dulyaseree, P.; Phonyiem, M.; Endo, M. Nitrogen self-doped activated carbons via the direct activation of Samanea saman leaves for high energy density supercapacitors. *RSC Advan.* **2019**, *9*, 21724–21732. [CrossRef] [PubMed]
9. Chen, T.; Luo, L.; Luo, L.; Deng, J.; Wu, X.; Fan, M.; Weigang, Z. High energy density supercapacitors with hierarchical nitrogen-doped porous carbon as active material obtained from bio-waste. *Renew. Energy* **2021**, *175*, 760–769. [CrossRef]
10. Kalak, T. Potential Use of Industrial Biomass Waste as a Sustainable Energy Source in the Future. *Energies* **2023**, *16*, 1783. [CrossRef]
11. Iqbal, S.; Khatoon, H.; Pandit, A.H.; Ahmad, S. Recent development of carbon-based materials for energy storage devices. *Mater. Sci. Energy Technol.* **2019**, *2*, 417–428. [CrossRef]
12. Zhang, Y.; Pan, H.; Zhou, Q.; Liu, K.; Ma, W.; Fan, S. Biomass-derived carbon for supercapacitors electrodes—A review of recent advances. *Inorg. Chem. Comm.* **2023**, *153*, 110768. [CrossRef]
13. Yang, H.; Ye, S.; Zhou, J.; Liang, T. Biomass-derived porous carbon materials for supercapacitor. *Front. Chem.* **2019**, *7*, 1–17. [CrossRef] [PubMed]
14. Awasthi, G.P.; Bhattarai, D.P.; Maharjan, B.; Kim, K.S.; Park, C.H.; Kim, C.S. Synthesis and characterizations of activated carbon from Wisteria sinensis seeds biomass for energy storage applications. *J. Ind. Eng. Chem.* **2019**, *72*, 265–272. [CrossRef]
15. Pandolfo, A.G.; Hollenkamp, A.F. Carbon properties and their role in supercapacitors. *J. Power Sour.* **2006**, *157*, 11–27. [CrossRef]
16. Wang, H.; Wen, J. Biomass porous carbon-based composite for high performance supercapacitor. *Mater. Res. Expr.* **2020**, *7*, 115601. [CrossRef]
17. Sahu, R.K.; Gangil, S.; Bhargav, V.K.; Sahu, P.; Ghritalahre, B. Synthesizing biomass into nano carbon for use in high-performance supercapacitors—A brief critical review. *J. Energy Storage* **2023**, *72*, 108348. [CrossRef]
18. Devi, N.S.; Hariram, M.; Vivekanandan, S. Modification techniques to improve the capacitive performance of biocarbon materials. *J. Energy Storage* **2021**, *33*, 101870. [CrossRef]
19. Iqbal, S.; Rasheed, T.; Bilal, M.; Shah, K.H.; Bilal, M.; Sheraz, T.A. Biomass-derived nitrogen-rich porous carbon composite for supercapacitor application. *J. Mater. Sci. Mater. Electron.* **2022**, *33*, 14793–14804. [CrossRef]
20. Chen, L.; Wang, J.; Zhang, J.; Hou, S.; Hao, C.; Zhang, J. Recent advances in flexible supercapacitors. *J. Solid State Electrochem.* **2022**, *26*, 2627–2658. [CrossRef]
21. Borchardt, L.; Oschatz, M.; Kaskel, S. Tailoring porosity in carbon materials for supercapacitor applications. *Mater. Horiz.* **2014**, *1*, 157–168. [CrossRef]
22. Saraf, M.; Natarajan, K.; Mobin, S.M. Robust nanocomposite of nitrogen-doped reduced graphene oxide and MnO_2 nanorods for high-performance supercapacitors and nonenzymatic peroxide sensors. *ACS Sustain. Chem. Eng.* **2018**, *6*, 10489–10504. [CrossRef]
23. Li, L.; Ji, P.; Geng, C.; Li, Y.; Meng, L.; Zhou, B.; Liang, J.; Peng, J.; Su, X. Facile synthesis of high-entropy $(Co_{0.2}Cr_{0.2}Fe_{0.2}Mn_{0.2}Ni_{0.2})_3O_4$ nanopowders and their electrochemical properties as supercapacitor electrode. *J. Energy Storage* **2023**, *73*, 109182. [CrossRef]
24. Ahmad, M.W.; Anand, S.; Dey, B.; Yang, D.J.; Choudhury, A. Asymmetric supercapacitors based on porous $MnMoS_4$ nanosheets-anchored carbon nanofiber and N, S-doped carbon nanofiber electrodes. *J. Alloys Compd.* **2022**, *906*, 164271. [CrossRef]
25. Guo, M.; Liu, Y.; Zhang, F.; Cheng, F.; Cheng, C.; Miao, Y.; Gao, F.; Yu, J. Inactive Al^{3+}-doped $La(CoCrFeMnNiAl_x)_{1/(5+x)}O_3$ high-entropy perovskite oxides as high performance supercapacitor electrodes. *J. Adv. Ceram.* **2022**, *11*, 742–753. [CrossRef]
26. Anand, S.; Ahmad, M.W.; Saidi, A.K.A.A.; Yang, D.J.; Choudhury, A. Polyaniline nanofiber decorated carbon nanofiber hybrid mat for flexible electrochemical supercapacitor. *Mater. Chem. Phys.* **2020**, *254*, 123480. [CrossRef]

27. Choudhury, A.; Dey, B.; Mohapatra, S.S.; Kim, D.W.; Yang, K.S.; Yang, D.J. Flexible and freestanding supercapacitor based o nanostructured poly (m-aminophenol)/carbon nanofiber hybrid mats with high energy and power densities. *Nanotechnolog* **2018**, *29*, 165401. [CrossRef] [PubMed]
28. Saraf, M.; Chacon, B.; Ippolito, S.; Lord, R.W.; Anayee, M.; Wang, R.; Inman, A.; Shuck, C.E.; Gogotsi, Y. Enhancing charge storag of $Mo_2Ti_2C_3$ MXene by partial oxidation. *Adv. Funct. Mater.* **2024**, *34*, 2306815. [CrossRef]
29. Saraf, M.; Zhang, T.; Averianov, T.; Shuck, C.E.; Lord, R.W.; Pomerantseva, E.; Gogotsi, Y. Vanadium and niobiun MXenes—Bilayered V_2O_5 asymmetric supercapacitors. *Small Methods* **2023**, *7*, 2201551. [CrossRef]
30. Saraf, M.; Shuck, C.E.; Norouzi, N.; Matthews, K.; Inman, A.; Zhang, T.; Pomerantseva, E.; Gogotsi, Y. Free-standin α-MoO_3/Ti_3C_2 MXene hybrid electrode in water-in-salt electrolytes. *Energy Environ. Mater.* **2023**, *6*, e12516. [CrossRef]
31. Hu, W.; Wang, B.; Yu, Y.; Wang, N.; Wu, X. Biomass derived carbon containing in-situ constructed nickel-based hydroxid nanostructures based on MnO_2 template for high performance asymmetric supercapacitors. *J. Alloys Compd.* **2021**, *884*, 16114 [CrossRef]
32. Tang, Q.; Chen, X.; Zhou, D.; Liu, C. Biomass-derived hierarchical porous carbon/silicon carbide composite for electrochemica supercapacitor. *Colloids Surf. A.* **2021**, *620*, 126567. [CrossRef]
33. Yang, G.; Park, S.J. MnO_2 and biomass-derived 3D porous carbon composites electrodes for high performance supercapacito applications. *J. Alloys Compd.* **2018**, *741*, 360–367. [CrossRef]
34. Ali, G.A.M.; Manaf, S.A.A.; Divyashree, D.; Feng, C.K.; Hegde, G. Superior supercapacitive performance in porous nanocarbon *J. Energy Chem.* **2016**, *25*, 734–739. [CrossRef]
35. Ali, G.A.M.; Manaf, S.A.B.A.; Kumar, A.; Chong, K.F.; Hegde, G. High performance supercapacitor using catalysis free porou carbon nanoparticles. *J. Phys. D Appl. Phys.* **2014**, *47*, 495307. [CrossRef]
36. Chmiola, J.; Yushin, G.; Gogotsi, Y.; Portet, C.; Simon, P.; Taberna, P.L. Anomalous increase in carbon capacitance at pore sizes les than 1 nanometer. *Science* **2006**, *313*, 1760–1763. [CrossRef] [PubMed]
37. Kumar, A.; Hegde, G.; Manaf, S.A.B.A.; Ngaini, Z.; Sharma, K.V. Catalyst free silica templated porous carbon nanoparticles fron bio-waste materials. *Chem. Commun.* **2014**, *50*, 12702–12705. [CrossRef] [PubMed]
38. Liu, X.; Li, S.; Mi, R.; Mei, J.; Liu, L.M.; Cao, L.; Lau, W.M.; Liu, H. Porous structure design of carbon xerogels for advance supercapacitor. *Appl. Energy* **2015**, *153*, 32–40. [CrossRef]
39. Rodrigues, A.C.; da Silva, E.L.; Oliveira, A.P.S.; Matsushima, J.T.; Cuña, A.; Marcuzzo, J.S.; Gonçalves, E.S.; Baldan, M.R High-performance supercapacitor electrode based on activated carbon fiber felt/iron oxides. *Mater. Today Commun.* **2019**, *21* 100553. [CrossRef]
40. Li, R.; Huang, J.; Li, J.; Cao, L.; Zhong, X.; Yu, A.; Lu, G. Nitrogen-doped porous hard carbons derived from shaddock peel fo high-capacity lithium-ion battery anodes. *J. Electroanal. Chem.* **2020**, *862*, 114044. [CrossRef]
41. Gong, Y.; Li, D.; Fu, Q.; Zhang, Y.; Pan, C. Nitrogen self-doped porous carbon for high-performance supercapacitors. *ACS Appl Ener. Mater.* **2020**, *3*, 1585–1592. [CrossRef]
42. Rodriguez Correa, C.; Otto, T.; Kruse, A. Influence of the biomass components on the pore formation of activated carbon. *Biomass Bioenergy* **2017**, *97*, 53–64. [CrossRef]
43. Cagnon, B.; Py, X.; Guillot, A.; Stoeckli, F.; Chambat, G. Contributions of hemicellulose, cellulose and lignin to the mass and the porous properties of chars and steam activated carbons from various lignocellulosic precursors. *Bioresour. Technol.* **2009**, *100* 292–298. [CrossRef] [PubMed]
44. Wang, Y.; Qu, Q.; Gao, S.; Tang, G.; Liu, K.; He, S.; Huang, C. Biomass derived carbon as binder-free electrode materials fo supercapacitors. *Carbon* **2019**, *155*, 706–726. [CrossRef]
45. Ahmed, M.H.; Byrne, J.A.; McLaughlin, J.A.D.; Elhissi, A.; Ahmed, W. Comparison between FTIR and XPS characterization of amino acid glycine adsorption onto diamond-like carbon (DLC) and silicon doped DLC. *Appl. Surf. Sci.* **2013**, *273*, 507–514 [CrossRef]
46. Zhang, J.; Gao, J.; Chen, Y.; Hao, X.; Jin, X. Characterization, preparation, and reaction mechanism of hemp stem based activated carbon. *Results Phys.* **2017**, *7*, 1628–1633. [CrossRef]
47. Chen, W.; Wang, H.; Lan, W.; Li, D.; Zhang, A.; Liu, C. Construction of sugarcane bagasse-derived porous and flexible carbon nanofibers by electrospinning for supercapacitors. *Indus. Crops Prod.* **2021**, *170*, 113700. [CrossRef]
48. Wang, B.; Wang, Y.; Peng, Y.; Wang, X.; Wang, J.; Zhao, J. 3-dimensional interconnected framework of N-doped porous carbon based on sugarcane bagasse for application in supercapacitors and lithium-ion batteries. *J. Power Sourc.* **2018**, *390*, 186–196 [CrossRef]
49. Wahid, M.; Puthusseri, D.; Phase, D.; Ogale, S. Enhanced capacitance retention in a supercapacitor made of carbon from sugarcane bagasse by hydrothermal pretreatment. *Energy Fuel* **2014**, *28*, 4233–4240. [CrossRef]
50. Okonkwo, C.A.; Menkiti, M.C.; Obiora-Okafo, I.A.; Ezenwa, O.N. Controlled pyrolysis of sugarcane bagasse enhanced mesoporous carbon for improving capacitance of supercapacitor electrode. *Biomass Bioenergy* **2021**, *146*, 105996. [CrossRef]
51. Pongpanyanatea, K.; Roddecha, S.; Piyanirund, C.; Phraewphiphat, T.; Hasin, P. Dispersed MnO_2 nanoparticles/sugarcane bagasse-derived carbon composite as an anode material for lithium-ion batteries. *RSC Adv.* **2024**, *14*, 2354–2368. [CrossRef]
52. Chen, J.; Qiu, J.; Wang, B.; Feng, H.; Yu, Y.; Sakai, E. Polyaniline/sugarcane bagasse derived biocarbon composites with superior performance in supercapacitors. *J. Electroanal. Chem.* **2017**, *801*, 360–367. [CrossRef]

53. Sun, Z.; Zhou, W.; Luo, J.; Fan, J.; Wu, Z.; Zhu, H.; Huang, J.; Zhang, X. High-efficient and pH-sensitive orange luminescence from silicon-doped carbon dots for information encryption and bio-imaging. *J. Colloid Interf. Sci.* **2022**, *607*, 16–23. [CrossRef] [PubMed]
54. Liang, Y.; Ouyang, J.; Wang, H.; Wang, W.; Chui, P.; Sun, K. Synthesis and characterization of core–shell structured SiO_2@YVO_4:Yb^{3+}, Er^{3+} microspheres. *Appl. Surf. Sci.* **2012**, *258*, 3689–3694. [CrossRef]
55. Ishak, S.; Mandal, S.; Lee, H.; Singh, J.K. pH-controlled synthesis of sustainable lauric acid/SiO_2 phase change material for scalable thermal energy storage. *Sci. Rep.* **2021**, *11*, 15012. [CrossRef] [PubMed]
56. Nassifa, N.; Livage, J. From diatoms to silica-based biohybrids. *Chem. Soc. Rev.* **2011**, *40*, 849–859. [CrossRef]
57. Kim, C.; Park, S.H.; Cho, J.K.; Lee, D.Y.; Park, T.J.; Lee, W.J.; Yang, K.S. Raman spectroscopic evaluation of polyacrylonitrile-based carbon nanofibers prepared by electrospinning. *J. Raman Spectrosc.* **2004**, *35*, 928–933. [CrossRef]
58. Gong, X.; Zheng, S.; Zhao, X.; Vomiero, A. Engineering high-emissive silicon-doped carbon nanodots towards efficient large-area luminescent solar concentrators. *Nano Energy* **2022**, *10*, 107617. [CrossRef]
59. Hariyanto, B.; Wardani, D.A.P.; Kurniawati, N.; Har, P.; Darmawan, N.; Irzaman. X-ray peak profile analysis of silica by Williamson–Hall and size-strain plot methods. *J. Phys. Conf. Ser.* **2021**, 012106. [CrossRef]
60. Yin, L.; Wu, M.; Li, Y.; Wu, G.; Wang, Y. Synthesis of SiO_2@carbon-graphene hybrids as anode materials of lithium-ion batteries. *New Carbon Mater.* **2017**, *32*, 311–318. [CrossRef]
61. Huang, S.; Yang, D.; Zhang, W.; Qiu, X.; Li, Q.; Li, C. Dual-templated synthesis of mesoporous lignin-derived honeycomb-like porous carbon/SiO_2 composites for high-performance Li-ion battery. *Microporous Mesoporous Mater.* **2021**, *317*, 111004. [CrossRef]
62. Yulianti, R.T.; Destyorini, F.; Irmawati, Y.; Priyono, S.; Fauzi, M.H.; Umar, A.A.; Uyama, H.; Fauzia, V.; Yudianti, R. High capacitance performance of hierarchically SiO_2 self-doped porous activated carbon derived from palm empty fruit bunches. *J. Energy Stor.* **2023**, *71*, 108153. [CrossRef]
63. Supriyal, S.; Bhat, V.S.; Jayeoye, T.J.; Rujiralai, T.; Chong, K.F.; Hegde, G. An investigation on temperature-dependant surface properties of porous carbon nanoparticles derived from biomass. *J. Nanostruct. Chem.* **2022**, *12*, 495–511. [CrossRef]
64. Chen, L.; Ji, T.; Mu, L.; Shi, Y.; Brisbin, L.; Guo, Z.; Khan, M.A.; Young, D.P.; Zhu, J. Facile synthesis of mesoporous carbon nanocomposites from natural biomass for efficient dye adsorption and selective heavy metal removal. *RSC Adv.* **2016**, *6*, 2259–2269. [CrossRef]
65. Liu, Y.; Jiang, S.P.; Shao, Z. Intercalation pseudocapacitance in electrochemical energy storage: Recent advances in fundamental understanding and materials development. *Mater. Adv.* **2020**, *7*, 100072. [CrossRef]
66. Joshi, A.; Lalwani, S.; Singh, G.; Sharma, R.K. Highly oxygen deficient, bimodal mesoporous silica-based supercapacitor with enhanced charge storage characteristics. *Electrochim. Acta.* **2019**, *297*, 705–714. [CrossRef]
67. Sankar, K.V.; Surendran, S.; Pandi, K.; Allin, A.M.; Nithya, V.D.; Lee, Y.S.; Selvan, R.K. Studies on the electrochemical intercalation/de-intercalation mechanism of $NiMn_2O_4$ for high stable pseudocapacitor electrodes. *RSC Adv.* **2015**, *5*, 27649–27656. [CrossRef]
68. Sathiya, M.; Prakash, A.S.; Ramesh, K.; Tarascon, J.M.; Shukla, A.K. V_2O_5-Anchored carbon nanotubes for enhanced electrochemical energy storage. *Mesoporous J. Am. Chem. Soc.* **2011**, *133*, 16291–16299. [CrossRef] [PubMed]
69. Wang, Q.; Zhang, Y.; Xiao, J.; Jiang, H.; Hu, T.; Meng, C. Copper oxide/cuprous oxide/hierarchical porous biomass-derived carbon hybrid composites for high-performance supercapacitor electrode. *J. Alloys Compd.* **2019**, *782*, 1103–1113. [CrossRef]
70. Zhao, N.; Deng, L.; Luo, D.; Zhang, P. One-step fabrication of biomass-derived hierarchically porous carbon/MnO nanosheets composites for symmetric hybrid supercapacitor. *Appl. Surf. Sci.* **2020**, *526*, 146696. [CrossRef]
71. Yin, Y.; Yan, S.; Ni, Z.; Jin, C.; Zhao, L. Economical synthesized Mn_3O_4/biomass-derived carbon from vegetable sponge composites and its excellent supercapacitive behavior. *Biomass Convers. Bioref.* **2023**, *13*, 12115–12124. [CrossRef]
72. Ren, Y.; Xu, Q.; Zhang, J.; Yang, H.; Wang, B.; Yang, D.; Hu, J.; Liu, Z. Functionalization of biomass carbonaceous aerogels: Selective preparation of MnO_2@CA composites for supercapacitors. *ACS Appl. Mater. Interf.* **2014**, *6*, 9689–9697. [CrossRef] [PubMed]
73. Han, J.; Li, Q.; Peng, C.; Shu, N.; Pan, F.; Wang, J.; Zhu, Y. Increasing S dopant and specific surface area of N/S-codoped porous carbon by in-situ polymerization of PEDOT into biomass precursor for high performance supercapacitor. *Appl. Surf. Sci.* **2020**, *502*, 144191. [CrossRef]
74. Jiang, L.; Ren, Z.; Chen, S.; Zhang, Q.; Lu, X.; Zhang, H.; Wan, G. Bio-derived three-dimensional hierarchical carbon-graphene-TiO_2 as electrode for supercapacitors. *Sci. Rep.* **2018**, *8*, 4412. [CrossRef] [PubMed]
75. Cui, Z.; Gao, X.; Wang, J.; Yu, J.; Dong, H.; Zhang, Q.; Yu, L.; Dong, L. Synthesis and supercapacitance of Co_3O_4 supported on porous carbon derived from wheat flour. *ECS J. Solid State Sci. Technol.* **2018**, *7*, M161–M165. [CrossRef]
76. Ahmad, M.W.; Anand, S.; Dey, B.; Fatima, A.; Yang, D.J.; Choudhury, A. N/P/O/S heteroatom-doped porous carbon nanofiber mats derived from a polyacrylonitrile/l-cysteine/P_2O_5 precursor for flexible electrochemical supercapacitors. *ACS Appl. Energy Mater.* **2021**, *4*, 12177–12190. [CrossRef]
77. Sun, J.; Niu, J.; Liu, M.; Ji, J.; Dou, M.; Wang, F. Biomass-derived nitrogen-doped porous carbons with tailored hierarchical porosity and high specific surface area for high energy and power density supercapacitors. *Appl. Surf. Sci.* **2018**, *427*, 807–813. [CrossRef]
78. Wang, D.; Min, Y.; Yu, Y. Facile synthesis of wheat bran-derived honeycomb-like hierarchical carbon for advanced symmetric supercapacitor applications. *J. Solid State Electrochem.* **2014**, *19*, 577–584. [CrossRef]

79. Zhao, G.; Li, Y.; Zhu, G.; Shi, J.; Lu, T.; Pan, L. Biomass-based N, P, and S self-doped porous carbon for high performance supercapacitors. *ACS Sustain. Chem. Eng.* **2019**, *7*, 12052–12060. [CrossRef]
80. Sudhan, N.; Subramani, K.; Karnan, M.; Ilayaraja, N.; Satish, M. Biomass-derived activated porous carbon from rice straw for a high-energy symmetric supercapacitor in aqueous and non-aqueous electrolytes. *Energy Fuels* **2017**, *31*, 977–985. [CrossRef]
81. Wang, J.; Zhang, Q.; Deng, M. Eco-friendly preparation of biomass-derived porous carbon and its electrochemical properties. *ACS Omega* **2022**, *7*, 22689–22697. [CrossRef] [PubMed]
82. Wang, B.; Wang, Y.; Peng, Y.; Wang, X.; Wang, N.; Wang, J.; Zhao, J. Nitrogen-doped biomass-based hierarchical porous carbon with large mesoporous volume for application in energy storage. *Chem. Eng. J.* **2018**, *348*, 850–859. [CrossRef]
83. Yue, X.; Yang, H.; Cao, Y.; Jiang, L.; Li, H.; Shi, F.; Liu, J. Nitrogen-doped cornstalk-based biomass porous carbon with uniform hierarchical pores for high-performance symmetric supercapacitors. *J. Mater. Sci.* **2022**, *57*, 3645–3661. [CrossRef]

molecules

MDPI

Article

All-Nitrogen Energetic Material Cubic Gauche Polynitrogen: Plasma Synthesis and Thermal Performance

Chenxi Qu [1,2], Jiale Li [1], Kewei Ding [1,*], Songsong Guo [1] and Yating Jia [1]

[1] Xi'an Modern Chemistry Research Institute, Xi'an 710065, China; qcx204@163.com (C.Q.)
[2] School of Chemistry and Chemical Engineering, Beijing Institute of Technology, Beijing 100081, China
* Correspondence: dkw204@163.com

Abstract: Numerous theoretical calculations have demonstrated that polynitrogen with an extending polymeric network is an ultrahigh-energy all-nitrogen material. Typical samples, such as cubic gauche polynitrogen (cg-N), have been synthesized, but the thermal performance of polynitrogen has not been unambiguously determined. Herein, macroscopic samples of polynitrogen were synthesized utilizing a coated substrate, and their thermal decomposition behavior was investigated. Polynitrogen with carbon nanotubes was produced using a plasma-enhanced chemical vapor deposition method and characterized using infrared, Raman, X-ray diffraction X-ray photoelectron spectroscopy and transmission electron microscope. The results showed that the structure of the deposited polynitrogen was consistent with that of cg-N and the amount of deposition product obtained with coated substrates increased significantly. Differential scanning calorimetry (DSC) at various heating rates and TG-DSC-FTIR-MS analyses were performed. The thermal decomposition temperature of cg-N was determined to be 429 °C. The apparent activation energy (E_a) of cg-N calculated by the Kissinger and Ozawa equations was 84.7 kJ/mol and 91.9 kJ/mol, respectively, with a pre-exponential constant ($\ln A_k$) of 12.8 min^{-1}. In this study, cg-N was demonstrated to be an all-nitrogen material with good thermal stability and application potential to high-energy-density materials.

Keywords: all-nitrogen material; polynitrogen; PECVD; thermal decomposition performance; decomposition kinetics

1. Introduction

All-nitrogen materials have emerged as an important avenue for the development of high-energy-density materials (HEDMs). There have been significant experimental and theoretical advances in all-nitrogen materials, which are considered to be a strategic field for energetic materials [1–3]. All-nitrogen materials release enormous quantities of energy during decomposition into nitrogen because of the significant energy differences among the N≡N triple bond (954 kJ/mol), N=N double bond (418 kJ/mol) and N–N single bond (160 kJ/mol) [4]. Theoretical calculations have demonstrated the favorable performance of all-nitrogen materials, such as cg-N polynitrogen with an extending polymeric network, Td-N_4 nitrogen clusters and *cyclo*-N_5^- pentazolate anions. First-principles simulations on cg-N polynitrogen have indicated a density of 3.9 g/cm^3 and an energy 10.6 times that of octahydro-1,3,5,7-tetranitro-1,3,5,7-tetrazocine (HMX) [5]. The theoretical heat of formation of a Td-N_4 nitrogen cluster was determined to be 761 kJ/mol [6]. The detonation velocity and pressure of LiN_5 were 11,362 m/s and 40.9 GPa, respectively [7]. All-nitrogen materials offer the advantages of a high density, high enthalpy of formation, ultrahigh energy, and clean, nonpolluting blast and are therefore considered a new generation of HEDMs for use in explosives and propellants.

Different preparation methods typically produce all-nitrogen materials with different structures. Significant initial progress has been made in the chemical synthesis of all-nitrogen compounds since the end of the 20th century. Christe et al. successfully synthesized a marginally stable all-nitrogen ion, $N_5^+[AsF_6]^-$ in 1999, which was followed by successive synthesis of more pentazeniums, such as $N_5^+[SbF_6]^-$, and $N_5^+[Sb_2F_{11}]$ [8,9]. In 2017, Hu et al. and Lu et al. reported the synthesis and characterization of an atmospherically stable *cyclo*-N_5^- salt, $(N_5)_6(H_3O)_3(NH_4)_4Cl$ [10,11]. Researchers have since synthesized a series of metallic or nonmetallic salts of N_5^- via ion exchange reactions [12,13]. Compounds containing all-nitrogen structural fragments, such as N8 (1,1′-azobis-1,2,3-triazole), N10 (1,1′-azobis(5-methyltetrazole)) and N11 (1,1′-(triaz-1-ene-1,3-diyl)bis(1H-tetrazol-5-amine)), have been synthesized successively [14,15], providing references for the preparation of all-nitrogen materials.

Polynitrogen materials with extended polymeric network structures are typically synthesized under extreme conditions of pressure and temperature. Under high pressure, nitrogen molecules are converted into solid polynitrogen with various crystalline structures, such as cg (cubic gauche), Cmcm (Cmcm chain), and ch (cis-transchain) [16,17]. The development of the diamond anvil cell (DAC) experimental technique and laser heating technology enabled cubic gauche nitrogen (cg-N) samples to be obtained for the first time under extreme experimental conditions of more than 110 GPa and 2000 K by Eremets et al. [18] in 2004. The major interplanar distances of cg-N at 115.4 GPa were found to be 2.4434 Å, 1.7279 Å, 1.4103 Å, 1.2206 Å and 1.0918 Å, corresponding to Miller-indices (hkl) of (110), (200), (211), (220) and (310), respectively. The Raman peak of cg-N at high pressure was reported at approximately 830 cm^{-1} that shifted with decreasing pressure at a rate of 1.53~1.57 cm^{-1}/GPa [19,20]. This method was used in subsequent studies to synthesize various polynitrogen samples. Additional polynitrogen structures, such as hexagonal layered polynitrogen (HLP-N, 244 GPa and 3000 K) [21] and black phosphorus polynitrogen (BP-N, 140 GPa and 2300 K) [22], were discovered with the experimental pressure and temperature variation. In 2014, Yoo et al. [23] discovered a layered polymeric structure polynitrogen (LP-N) at 150 GPa and 2000 K that had two colossal Raman bands at 1000 cm^{-1} and 1300 cm^{-1}. Subsequently, Li et al. [24] observed the post-layered-polymeric nitrogen (PLP-N) by further heating the LP-N to above 2300 K at 161 GPa.

Some all-nitrogen phases, such as cyclic N5 rings and pseudobenzene N6 hexagonal rings, have also been discovered in studies on phase transitions at high pressures [25,26]. To decrease the preparation pressure of polynitrogens, several researchers have used nitrogen-rich compounds as alternative raw materials to nitrogen. Azide compounds, including NaN_3 [27], CsN_3 [28], and KN_3 [29], were initially tested as substrates for high-pressure polynitrogen preparation. However, only high-pressure phases of azides, such as Phase I and Phase II of NaN_3 in the 0 GPa to 50 GPa range, have been found [27]. Conditions of more than 100 GPa and 2000 K are still required to transform azides into an amorphous polynitrogen structure.

Various novel methods for polynitrogen preparation under atmospheric conditions have been recently reported. A series of nitrogen clusters, such as TiN_{12}^+ [30], ZrN_{12}^+ [31], VN_x^+(x = 8, 9, 10) [32], and LiN_x^+ (x = 2, 4, 6, 8) [33], have been obtained by laser ablation of nitrogen-rich compounds. Cyclic voltammetry (CV) has been used to synthesize polynitrogen N_8^- ions on pristine and boron-doped graphene, as well as on multiwalled carbon nanotubes (CNTs) [34,35]. The use of plasma-enhanced chemical vapor deposition (PECVD) to produce cg-N has opened an avenue for the synthesis of cg-N under ambient conditions [36,37]. However, in-depth studies on the thermal behavior of polynitrogens have not yet been conducted.

In conclusion, theoretical calculations prove that the cg-N polynitrogen is a high-energy-density all-nitrogen material. Recent advances have also been made in preparing cg-N polynitrogen at atmospheric pressure. However, the thermal stability and thermal decomposition performance of cg-N polynitrogen, which are crucial for practical applications in the field of energy materials, have not been determined. In this study, atmospherically stable cg-N polynitrogen with multiwall carbon nanotubes (cg-N/CNT) was synthesized by plasma-enhanced chemical vapor deposition. The deposition duration was optimized, and the product structures were characterized. The thermal properties of the polynitrogen samples were characterized, and the decomposition kinetics of the samples were calculated to investigate the application potential of polynitrogens as HEDMs.

2. Results and Discussion

2.1. X-ray Photoelectron Spectroscopy

The N 1s XPS spectrum of the NaN_3/CNT substrate before and after the deposition reaction is shown in Figure 1. Figure 1a shows that the spectrum of the NaN_3/CNT substrate exhibited two peaks at binding energies of 404.2 eV and 400.1 eV. Lorentz-Gaussian fitting yielded a peak ratio of approximately 1:2. These two peaks were thus determined to be generated by the azide ion of the raw material [36]. Figure 1b shows the N 1s XPS spectrum of the product of the deposition reaction. The two peaks with binding energies of 404.6 eV and 400.8 eV correspond to unreacted sodium azide. The slightly higher binding energies of the peaks compared to those in the spectrum obtained before deposition indicates that the sodium azide substrate may have undergone a solid-state phase change during deposition, but the chemical environment of the azide ion did not change significantly. Of particular interest is the peak with a binding energy of 397.9 eV, which does not lie either within the oxynitride binding energy interval or the interval for carbonitrides, such as pyridine or pyrrole [38]. Consequently, the 397.9 eV binding energy was determined to correspond to a single N–N bond and a similar binding energy to that determined by Benchafia et al. [36,39], confirming the presence of a nitrogenous product with a new valence and the same chemical environment for each nitrogen atom.

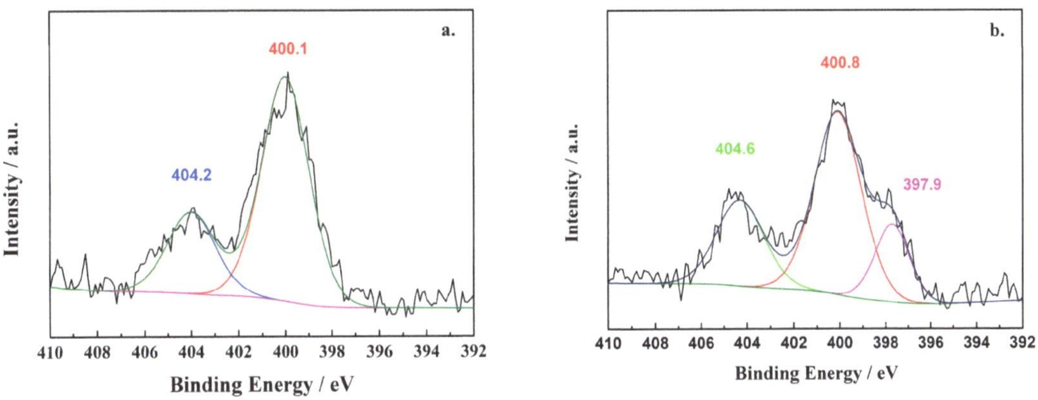

Figure 1. N 1s X-ray photoelectron spectrum of samples ((**a**) Before the deposition reaction, (**b**) after the deposition reaction).

2.2. ATR-FTIR and Raman Spectroscopy

Both IR and Raman are valuable for identifying changes in the azide groups during the deposition process because azide ions have active bending and asymmetric stretch modes. Figure 2 shows the IR spectra of the samples of the substrate and products of deposition with various reaction durations. The IR peaks at 2103 cm^{-1} and 638 cm^{-1} were attributed to the ν2 mode and anti-symmetric ν3 mode of the azide anion. The peaks at approximately 3300 cm^{-1} and 3390 cm^{-1} were assigned to the combination frequency vibration of the 2ν2+ν3 mode. The spectrum of the sample obtained after plasma deposition exhibited two prominent new peaks at approximately 1428 cm^{-1} and 879.8 cm^{-1}. The infrared peak at 879.8 cm^{-1} was determined to be the absorption peak of cg-N at close to ambient pressure because of being identical to the T(TO) mode infrared absorption peak at near zero pressure produced in theoretical simulations carried out by Caracas et al. [40]. The prominent absorption peak near 1428 cm^{-1} was assigned to NaN_3 Phase I at atmospheric pressure [27]. The infrared spectrum demonstrated that absorption of cg-N increased for two hours before decreasing. This result may have been caused by an increase in the atmospheric temperature during the deposition process that made the products disintegrate. Therefore, we selected a deposition duration of 2 h to ensure a sufficient quantity of product was obtained.

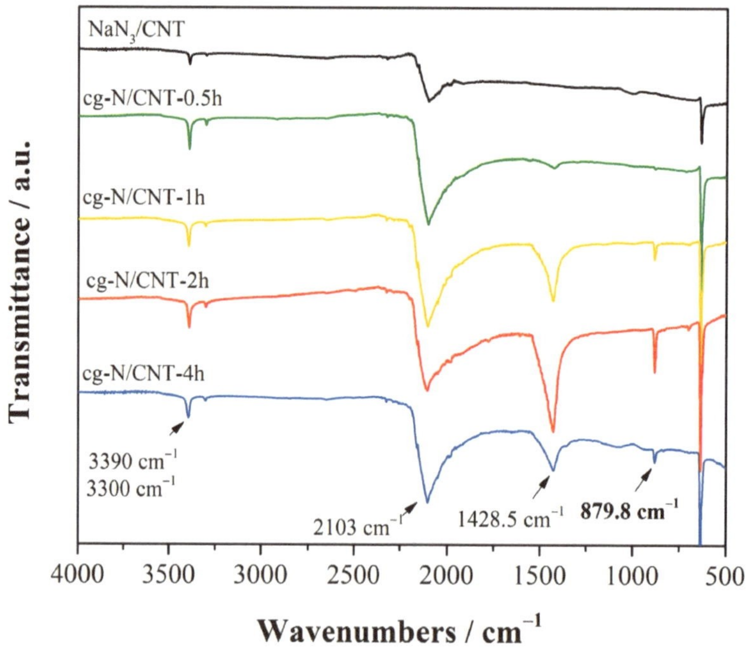

Figure 2. ATR-FTIR spectra of the samples before and after deposition at various reaction times.

The Raman spectra of the plasma samples before and after the two-hour deposition reaction are shown in Figure 3. The Raman peaks of NaN_3/CNT are shown by the black line. The sharp peak near 1340 cm^{-1} and the peaks near 1585 cm^{-1} were assigned to the D- and G-modes of the carbon nanotubes, respectively. The librational lattice mode, the IR active bending ν2 mode, and the symmetric stretching ν1 mode of the unreacted azide anion were characterized by three distinct, intense peaks at wavelengths of 117 cm^{-1}, 1270 cm^{-1}, and 1361 cm^{-1}, respectively [25]. The red line indicates the spectrum obtained after the deposition reaction: the new broad peak at 637 cm^{-1} and the weaker peak at 719 cm^{-1} correspond to the A mode and tilted vibration (T-TO) mode Raman absorption estimated by Caracas [40]. The A-mode Raman peak of cg-N has also been experimentally observed at high pressure, where the peak is shifted considerably toward lower wavenumbers as the pressure decreases [18–20]. Therefore, the broad peak at 637 cm^{-1} and the sharp peak at 719 cm^{-1} in the spectrum of the deposition sample were determined to be the A-mode and T-TO mode Raman peaks under the zero-pressure condition of cg-N, respectively. The broad peaks at 210 cm^{-1} and 625 cm^{-1} were assigned to the nitrogen network in Phase I of NaN_3 at high pressure and temperature [27]. By comparing the Raman peak fitting profiles of cg-N, the peak intensity ratio of cg-N to Phase I of NaN_3 in the product obtained from the coated substrate was 0.75, while that of the product obtained from the powder substrate was 0.25. Preparation of the deposition substrate using the coating method increased the content of the polymerized nitrogen product nearly threefold.

The three NaN_3 peaks are also considerably broader and weaker in intensity in the spectrum of the samples obtained after deposition than those obtained before deposition, indicating the dissociation of the azide anion and the formation of a nonmolecular nitrogen network during the reaction. The simultaneous increase in the I_d/I_g ratio for the CNT peaks from 0.9 to 1.6 implies a significant increase in the defect density of CNTs. The stabilizing effect of the carbon tubes on the internal cg-N is reduced, indicating that the growth of cg-N during deposition results in an increase in carbon tube defects. This result could also be a contributing factor to the rapid depletion of cg-N over time.

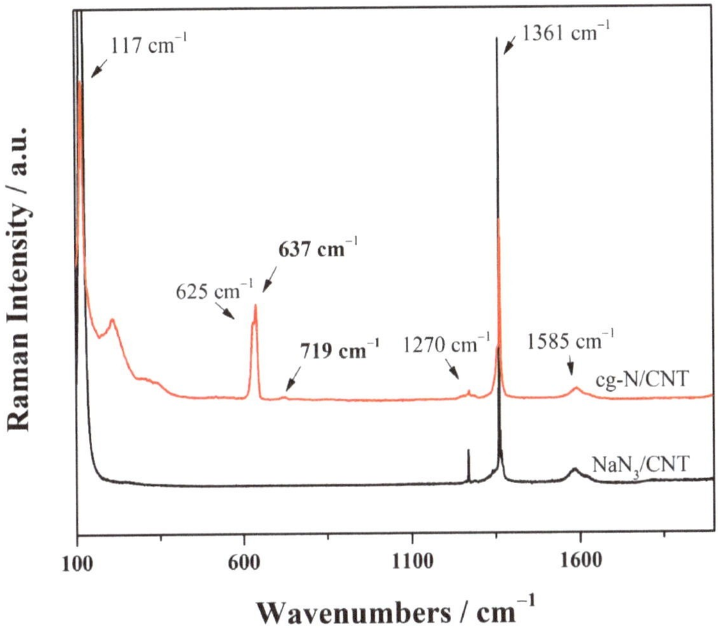

Figure 3. Raman spectra of the samples before and after deposition.

2.3. *X-ray Powder Diffraction*

Figure 4 presents the XRD patterns obtained for the samples before (black line) and after (red line) the plasma deposition reaction. In Figure 4a, the broad peaks (indicated by #) at approximately 26° and 45° are the diffraction peaks of the (002) and (110) crystal planes of the carbon nanotubes [41]. The series of strong peaks near 28°, 31°, and 36° correspond to unreacted sodium azide. Three prominent new reflection peaks (indicated by *) appear near 37°, 66°, and 77° in the spectrum of the sample obtained after plasma deposition. These peaks are considerably weaker than the diffraction peaks of NaN_3 because of the lower crystallinity of the deposition products. Figure 4b shows the diffraction pattern of the sample converted to a wavelength λ = 0.41686 Å, which is the parameter in the diffraction cubic gauche polynitrogen reflection data measured by Eremets et al. [18] The bulk modulus of cg-N can reach 261 GPa at atmospheric pressure [42], which by far exceeds that of conventional ultrahard materials, such as silicon carbide. Therefore, the lattice spacings of cg-N do not change significantly at atmospheric pressure. The three diffraction peaks mentioned above were transformed into 2ndeta values of 9.8, 17.0, and 19.7. Comparison with the diffraction data of cg-N at 115 GPa shows that these new peaks are reflections by the (110), (211), and (220) crystal planes of cg-N. However, the relative intensities of these diffraction peaks are slightly different from those of cg-N, probably because of the crystallographic orientation induced by the carbon nanotubes.

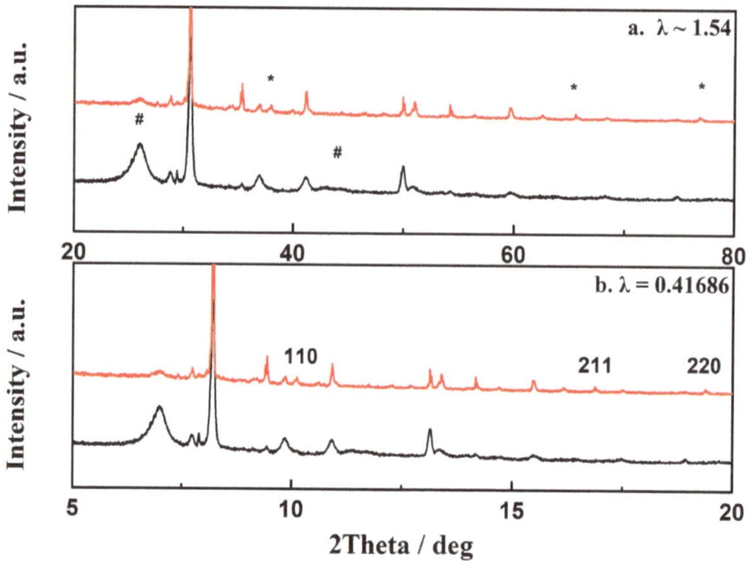

Figure 4. X-ray powder diffraction patterns of plasma deposition samples. (**a**) Diffraction patterns taken with λ ~ 1.54 Å, (**b**) diffraction pattern converted to λ = 0.41686 Å, (*) peaks of cg-N, (#) peaks of CNT.

2.4. Transmission Electron Microscopy

TEM images and selected area electron diffraction (FFT) images of the plasma deposited samples are shown in Figure 5. In the high-resolution TEM images, the deposited polynitrogen with a diameter of approximately 20–30 nm is distributed inside the multi-walled carbon nanotubes. Figure 5b shows the fast Fourier transform (FFT) image of the polynitrogen samples, from which the crystal plane spacing is measured as 1.75 Å. This crystal plane spacing is very close to the (200) plane spacing of 1.73 Å for cg-N obtained by Eremets et al. [18]. However, polynitrogen samples are considerably amorphous because plasma-deposited samples have low crystallinity, which is consistent with the XRD data.

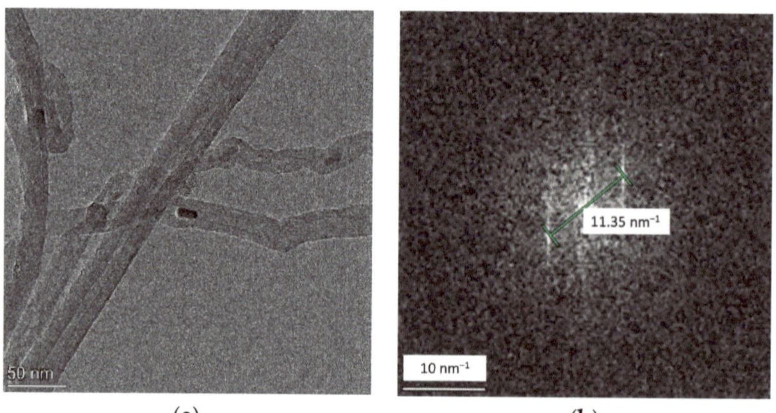

(**a**) (**b**)

Figure 5. (**a**) High-resolution TEM images and (**b**) selected area electron diffraction (FFT) images of samples after deposition.

2.5. Thermal Decomposition Performance of cg-N/CNT

To characterize the thermal stability of the samples, we measured the thermal decomposition temperature of the cg-N/CNT sample and its substrates using differential scanning calorimetry (DSC). Figure 6 presents the DSC curves obtained at a heating rate of 10 °C/min for the pure NaN_3, NaN_3/CNT and cg-N/CNT samples. The decomposition peak of pure sodium azide occurs at 412 °C, which is in reasonable agreement with that reported in the literature [43,44]. A broad exothermic peak between 275 °C and 350 °C appears for the NaN_3/CNT sample obtained by compounding sodium azide with carbon nanotubes. The interaction of the carbon nanotubes with NaN_3 in the interior decreases the thermal stability and therefore the thermal decomposition temperature of NaN_3. There are two prominent exothermic peaks in the DSC curves of cg-N/CNT. The thermal decomposition peak of cg-N obtained by deposition occurs at approximately 429 °C. This result indicates that cg-N/CNT has superior thermal stability to the other two samples and a similar decomposition temperature to that determined using temperature programmed desorption (TPD) by Zhuang et al. [37]. Theoretical calculations suggest that carbon nanotubes stabilize polynitrogens, such as N_8 oligomers [35], and may also contribute to the high thermal decomposition temperature of cg-N.

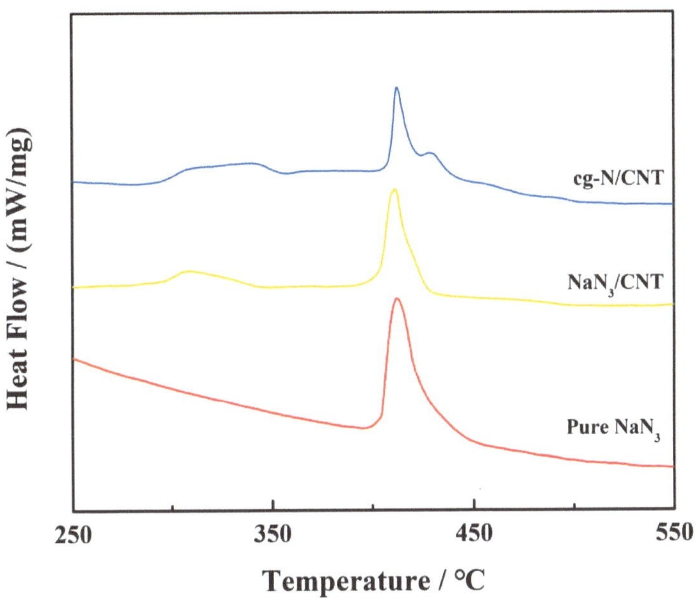

Figure 6. DSC curves of pure NaN_3, NaN_3/CNT and cg-N/CNT obtained at a heating rate of 10 °C/min.

2.6. Non-Isothermal Decomposition Kinetics

To further analyze the behavior of the different components of the sample during thermal decomposition, the nonisothermal kinetic parameters of the exothermic decomposition reaction were investigated. The apparent activation energy (E_a) was calculated using the Kissinger [45] and Ozawa [46] equations shown below:

$$\ln\left(\frac{\beta}{T_p^2}\right) = \ln\left(\frac{A_k R}{E_k}\right) - \frac{E_k}{RT_p} \quad (1)$$

$$\lg\beta + \frac{0.4567 E_o}{RT_p} = C \quad (2)$$

where β is the heating rate for the sample, R is the gas constant (commonly taken as 8.3145 J/mol·K), and T_p is the exothermic peak determined using the heating rate mentioned above. The apparent activation energy (E_a) of the reaction can be calculated from the slope of the linear fit between β and $1/T_p$ for the different equations. The preexponential constant (A_k) can be obtained from the intercepts of the plots.

The DSC measurements obtained at different heating rates are shown in Figure 7, and Table 1 shows the kinetic parameters of cg-N/CNT determined for heating rates of 5, 10, 20 and 25 °C/min. Substituting these parameters into the Kissinger and Ozawa equations yields essentially equivalent E_a values. The apparent activation energy of cg-N/CNT obtained by the Kissinger method is 84.7 kJ/mol, similar to the 91.9 kJ/mol obtained by the Ozawa method. The pre-exponential factor of cg-N/CNT is 12.8 min^{-1}. Furthermore, the apparent activation energy of cg-N/CNT are lower than that of NaN_3, which implies that cg-N has a higher thermal stability and decomposition rate above the decomposition temperature than NaN_3.

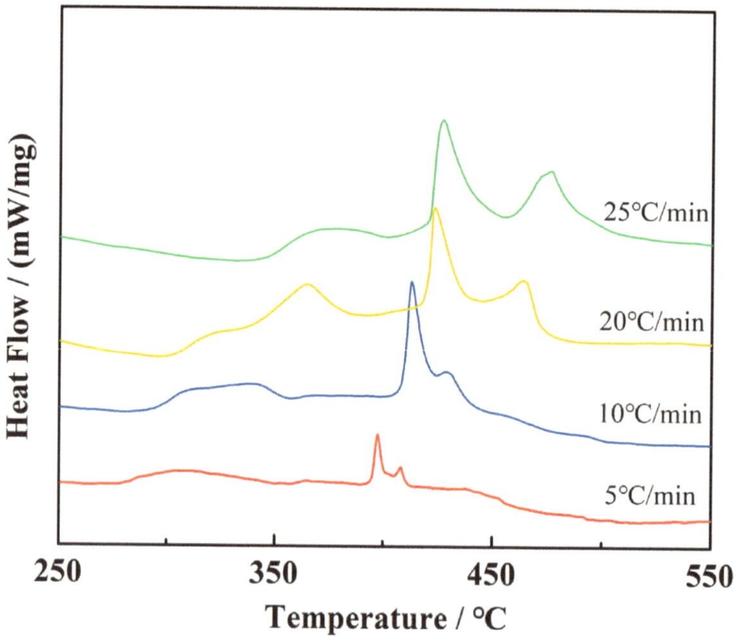

Figure 7. DSC curves of cg-N/CNT obtained at different heating rates.

Table 1. The characteristic parameters for thermal decomposition processes of cg-N.

Sample Component	Heating Rate β (°C/min)	Exothermic Peak T_p (°C)	Apparent Activation Energy E_a (kJ/mol)		Pre-Exponential Constant lnAk
			Kissinger	Ozawa	
cg-N	5	408.1	84.7	91.9	12.8
	10	429.0			
	20	464.8			
	25	476.9			
NaN_3	5	397.6	184.9	191.2	31.8
	10	412.3			
	20	422.9			
	25	426.8			

2.7. TG-DSC-FTIR-MS Analysis

DSC-TG-FTIR-MS analysis was performed to characterize the heat released and the composition of the volatile gas released during the thermal decomposition of cg-N/CNT. Figure 8 shows the TG-DSC curve of cg-N/CNT obtained at a heating rate of 10 °C/min. Three significant thermal weight loss processes can be observed during the decomposition of the sample from room temperature to complete decomposition. The decomposition of NaN_3 in the carbon nanotubes occurs between 275 and 350 °C, corresponding to the initial stage with an approximately 7.98 percent weight loss. The decomposition of the individual sodium azide crystals begins around 400 °C, which corresponds to the second stage with a weight loss of 12.01 percent. The disintegration of cg-N in the carbon nanotubes corresponds to the third stage of thermal weight loss, which starts at approximately 430 °C and results in a mass decrease of approximately 1.91 percent. Whereas XPS can only be used to characterize the surface of a sample, the entire sample is characterized by thermal analysis; thus, a significantly smaller mass percentage of cg-N is determined by thermal analysis than by XPS. The original exothermic peaks were divided into two peaks based on a Bigaussian function [47] to determine the ratios of the quantity of heat released by the various components, where the curves with the fitted peaks are presented in Figure 8b. Comparing the integration areas of the different curves results in an almost 3:2 ratio for the heat released by cg-N to that of NaN_3 with the CNT substrate in an argon gas atmosphere. Considering that the mass ratio of cg-N to NaN_3 is 6.3:1, the heat released per unit mass by cg-N is more than four times that of NaN_3. Although the proportion may not be precise due to the small quantity of products tested, the heat release of cg-N polynitrogen is considerably higher than that of NaN_3.

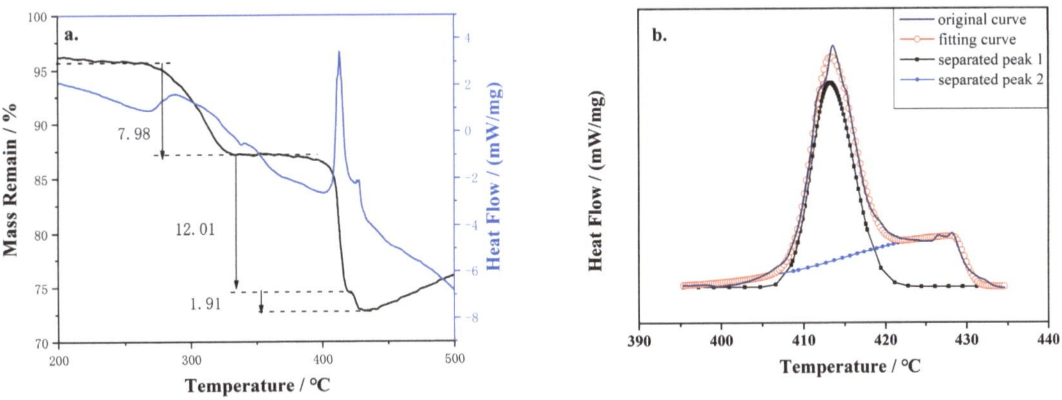

Figure 8. (**a**) TG–DSC curves and (**b**) fitted peaks for the cg-N/CNT samples obtained at a heating rate of 10 °C/min.

Figure 9 shows the MS of the volatile gas released during the thermal decomposition of cg-N/CNT at a heating rate of 10 °C/min. Charge ratios (m/z) of 28, 14, 44, 30, 29, and 27 were detected. Volatile gases with charge ratios of 28 and 14 were the main products of the thermal decomposition of cg-N/CNT at 275–375 °C and 400–440 °C, respectively. These thermal decomposition products were assumed to be gaseous and atomic nitrogen because no other signals appeared in the FT-IR spectra. That is, the major gas product of cg-N/CNT is nitrogen, which is considered a nonpolluting gas.

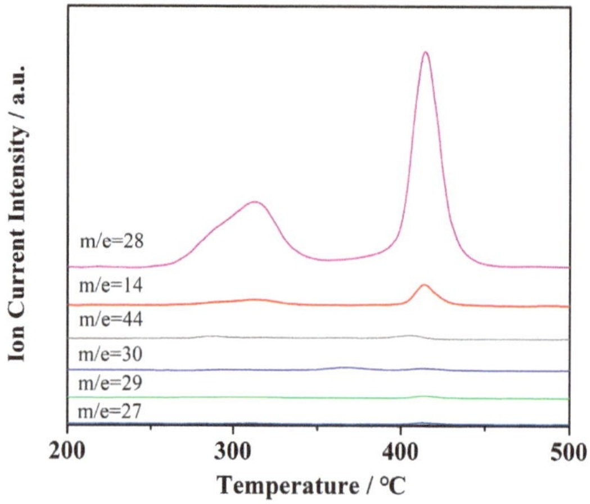

Figure 9. MS of gaseous products of cg-N/CNT samples obtained at a heating rate of 10 °C/min.

3. Materials and Methods

3.1. Materials

Jiangsu Xianfeng Nanomaterials Technology Co., Ltd (Jiangsu, China). supplied multiwalled carbon nanotubes (CNTs) with an inner diameter of 10–20 nm and a length of 0.5–200 μm. Sodium azide (99.5%) was obtained from Sigma-Aldrich Co., LLC (Shanghai, China). High-purity nitrogen and argon gases were provided by Xi'an Weiguang Technology Co. (Xi'an, China), which supplied multiwalled carbon nanotubes (CNTs) with an inner diameter of 10–20 nm and a length of 0.5–200 μm.

3.2. Experimental Methods

Sodium azide (10 g) was dissolved in deionized water (15 g) at room temperature to prepare a saturated aqueous solution of sodium azide. The sodium azide solution (10 g) was added to a Pyrex tube containing short multiwall carbon nanotubes (200 mg). The mixture was sonicated three times for 60 min each time. The resulting suspension was coated on a quartz slide. and completely dried to yield the NaN_3/CNT substrate.

The NaN_3/CNT substrate (50 mg) was placed in a plasma vapor deposition chamber. A gas mixture of 50% nitrogen (N_2) and 50% argon (Ar) was introduced into the deposition chamber at a rate of 30 mL/min for at least 30 min. The pressure in the deposition chamber was adjusted to 40 ± 2 Pa, and a power of 70 W was applied. The reaction time was at least 0.5 h. The system was gradually restored to atmospheric pressure, and plasma-synthesized samples were collected.

3.3. Characterization Methods

Fourier transform infrared spectroscopy (FT-IR) was performed using an infrared spectrometer (Thermo Scientific (Waltham, MA, USA), ATR50) equipped with an attenuated total reflectance (ATR) accessory. Raman spectra were obtained (HORIBA Labram (Tokyo, Japan), HR Evolution) at a 532 nm laser excitation and a spectral resolution of 2 cm^{-1}. Transmission electron microscopy (TEM) images and selected area electron diffraction images were obtained with a transmission electron microscope (Talos F200i). X-ray diffraction (XRD) was conducted on an X-ray powder diffractometer (PANalytical (Almelo, The Netherlands), EMPYREAN) in the 5° to 90° range using Cu K_α radiation (λ = 1.54 Å) and a step size of 0.02°. XPS data were collected using an X-ray photoelectron spectrometer (Thermo Scientific (Waltham, MA, USA), K-Alpha) with Al Kα radiation (1486.6 eV) operating at 225 W.

The thermal properties of the samples were investigated by differential scanning calorimetry (METTLER TOLEDO (Giessen, Germany), DSC 2). The temperature ranged from ambient temperature to 600 °C, and argon gas at a flow rate of 50 mL/min was used. Samples weighing 1.5 ± 0.2 mg were placed in an Al_2O_3 crucible. Pure sodium azide and the NaN_3/CNT samples were heated at a rate of 10 °C/min. The deposited samples were heated at rates of 5, 10, 20, and 25 °C/min to determine the kinetic parameters. DSC-TG-MS-FTIR analysis was performed on a simultaneous thermal analyzer (NETZSCH (Selb, Germany) STA449F3 STA). The deposited samples weighing approximately 1 mg were heated from 50 °C to 600 °C at a heating rate of 10 °C/min. The volatile products were analyzed by mass spectrometry (NETZSCH (Selb, Germany), QMS403) and in situ IR (Thermo Scientific (Waltham, MA, USA), NEXUS870).

4. Conclusions

Polynitrogen samples were synthesized under atmospheric pressure utilizing coated substrates as precursor by plasma-enhanced chemical vapor deposition on carbon nanotubes. The characterization results of the deposition product were all consistent with those of cg-N. The polnitrogen content increased nearly three times compared to the samples obtained from the powder substrate. Thermal analysis at a heating rate of 10 °C/min showed a decomposition temperature of 429 °C for the cg-N sample. The apparent activation energy of cg-N calculated by the Kissinger and Ozawa equations was 84.7 kJ/mol and 91.9 kJ/mol, respectively, with a pre-exponential constant of 12.8 min^{-1}. Therefore, cg-N is an all-nitrogen material with a high energy density and ideal thermal stability, for which applications of high energy density materials are worth exploring.

Author Contributions: Conceptualization, K.D.; methodology, K.D. and C.Q.; validation, C.Q., J.L. and Y.J.; formal analysis, Y.J., J.L. and C.Q.; investigation, J.L.; resources, K.D.; data curation, C.Q. and S.G.; writing—original draft preparation, C.Q.; writing—review and editing, S.G.; visualization, K.D.; supervision, K.D.; project administration, K.D. All authors have read and agreed to the published version of the manuscript.

Funding: This research received no external funding.

Institutional Review Board Statement: Not applicable.

Informed Consent Statement: Not applicable.

Data Availability Statement: Data are contained within the article.

Conflicts of Interest: The authors declare no conflicts of interest.

References

1. Yao, Y.; Adeniyi, A.O. Solid Nitrogen and Nitrogen-rich Compounds as High Energy Density Materials. *Phys. Status Solidi B* **2021**, *258*, 2000588. [CrossRef]
2. Li, Y.; Pang, S.P. Progress of All-nitrogen Ultrahigh-energetic Materials. *Chin. J. Energ. Mater.* **2012**, *1*, 9–16. [CrossRef]
3. Pagoria, P.F.; Lee, G.S.; Mitchell, A.R.; Schmidt, R.D. A review of energetic materials synthesis. *Thermochim. Acta* **2002**, *384*, 187–204. [CrossRef]
4. Samartzis, P.C.; Wodtke, A.M. All-nitrogen Chemistry: How Far are we from N60? *Int. Rev. Phys. Chem.* **2006**, *4*, 527–552. [CrossRef]
5. Christe, K. Recent Advances in the Chemistry of N_5+, N_5 and High-Oxygen Compounds. *Propellants Explos. Pyrotech.* **2007**, *32*, 194–204. [CrossRef]
6. Östmark, H.; Wallin, S.; Goede, P. High Energy Density Materials (HEDM): Overview, Theory and Synthetic Efforts at FOI. *Cent. Eur. J. Energetic Mater.* **2007**, *4*, 83–108.
7. Xu, Y.; Ding, L.; Yang, F.; Li, D.; Lu, M. LiN_5: A novel pentazolate salt with high nitrogen content. *Chem. Eng. J.* **2021**, *429*, 132399. [CrossRef]
8. Christe, K.O.; Wilson, W.W.; Sheehy, J.A.; Boatz, A. N_5+: A Novel Homoleptic Polynitrogen Ion as a High Energy Density Material. *Angew. Chem. Int. Ed.* **1999**, *38*, 2004–2009. [CrossRef]
9. Haiges, R.; Schneider, S.; Schroer, T.; Christe, K.O. High-Energy-Density Materials: Synthesis and Characterization of $N_5^+[P(N_3)_6]^-$, N_5^+ $[B(N_3)_4]^-$, N_5^+ $[HF_2]^-$·n HF, N_5^+ $[BF_4]^-$, N_5^+ $[PF_6]^-$, and N_5^+ $[SO_3F]^-$. *Angew. Chem. Int. Ed.* **2004**, *37*, 4919–4924. [CrossRef]
10. Xu, Y.; Wang, Q.; Shen, C.; Lin, Q.; Wang, P.; Lu, M. A series of energetic metal pentazolate hydrates. *Nature* **2017**, *549*, 78–81. [CrossRef]

11. Zhang, C.; Sun, C.; Hu, B.C.; Yu, C.; Lu, M. Synthesis and characterization of the pentazolate anion cyclo-N_5 in $(N_5)_6(H_3O)_3(NH_4)_4Cl$. *Science* **2017**, *355*, 374–376. [CrossRef]
12. Sun, C.; Zhang, C.; Jiang, C.; Yang, C.; Du, Y.; Zhao, Y.; Hu, B.C.; Zhang, Z.; Christe, K.O. Synthesis of AgN_5 and its extended 3D energetic framework. *Nat. Commun.* **2018**, *9*, 1269. [CrossRef] [PubMed]
13. Zhang, C.; Yang, C.; Hu, B.; Yu, C.; Sun, C. A Symmetric $Co(N_5)_2(H_2O)_4$ $4H_2O$ High-Nitrogen Compound Formed by Cobalt(II) Cation Trapping of a Cyclo-N5 Anion. *Angew. Chem. Int. Ed.* **2017**, *56*, 4512–4514. [CrossRef] [PubMed]
14. Tang, Y.; Yang, H.; Shen, J.; Wu, B.; Ju, X.; Lu, C.; Cheng, G. 4-(1-Amino-5-aminotetrazolyl)methyleneimino-3-methylfuroxan and Its Derivatives: Synthesis, Characterization, and Energetic Properties. *Eur. J. Inorg. Chem.* **2014**, *7*, 1231–1238. [CrossRef]
15. Tang, Y.; Yang, H.; Wu, B.; Ju, X.; Lu, C.; Cheng, G. Synthesis and Characterization of a Stable, Catenated N 11 Energetic Salt. *Angew. Chem. Int. Ed.* **2013**, *52*, 4875–4877. [CrossRef] [PubMed]
16. McMahan, A.K.; LeSar, R. Pressure dissociation of solid nitrogen under 1 mbar. *Phys. Rev. Lett.* **1985**, *17*, 1929–1932. [CrossRef]
17. Mailhiot, C.; Yang, L.; McMahan, A.K. Polymeric nitrogen. *Phys. Rev. B* **1992**, *22*, 14419–14435. [CrossRef] [PubMed]
18. Eremets, M.I.; Gavriliuk, A.G.; Trojan, I.A.; Dzivenko, D.A.; Boehler, R. Single-bonded cubic form of nitrogen. *Nat. Mater.* **2004**, *3*, 558–563. [CrossRef]
19. Li, L.; Pu, M.; Feng, L.; Qi, L.; Zhang, L. Synthesis of Cubic Gauche Nitrogen (cg-N) under High Pressure and High Temperature. *Chin. J. High Press. Phys.* **2018**, *32*, 020102. [CrossRef]
20. Lipp, M.J.; Klepeis, J.P.; Baer, B.J.; Cynn, H.; Yoo, C.S. Transformation of molecular nitrogen to nonmolecular phases at megabar pressures by direct laser heating. *Phys. Rev. B* **2007**, *76*, 014113. [CrossRef]
21. Laniel, D.; Geneste, G.; Weck, G.; Mezouar, M.; Loubeyre, P. Hexagonal Layered Polymeric Nitrogen Phase Synthesized near 250 GPa. *Phys. Rev. Lett.* **2019**, *122*, 066001. [CrossRef] [PubMed]
22. Laniel, D.; Winkler, B.; Fedotenko, T.; Pakhomova, A.S.; Dubrovinskaia, N. High-pressure polymeric nitrogen allotrope with the black phosphorus structure. *Phys. Rev. Lett.* **2020**, *124*, 216001. [CrossRef]
23. Tomasino, D.; Kim, M.; Smith, J.; Yoo, C.S. Pressure-induced symmetry-lowering transition in dense nitrogen to layered polymeric nitrogen (lp-n) with colossal raman intensity. *Phys. Rev. Lett.* **2014**, *113*, 205502. [CrossRef]
24. Li, L.; Tang, Q.Q.; Zhang, F.; Liu, S.; Wu, B.-B.; Zhou, C.-Y. Evidence for a New Extended Solid of Nitrogen. *Chin. Phys. Lett.* **2020**, *37*, 068101. [CrossRef]
25. Zhou, M.; Liu, S.; Du, M.; Shi, X.; Zhao, Z.; Guo, L.; Liu, B.; Liu, R.; Wang, P.; Liu, B. High-pressure-induced structural and chemical transformations in NaN_3. *J. Phys. Chem. C* **2020**, *124*, 19904–19910. [CrossRef]
26. Zhang, M.; Yan, H.; Wei, Q.; Wang, H.; Wu, Z. Novel high-pressure phase with pseudo-benzene "N6" molecule of LiN_3. *EPL-Europhys. Lett.* **2013**, *101*, 26004. [CrossRef]
27. Eremets, M.I.; Popov, M.Y.; Trojan, I.A.; Denisov, V.N.; Boehler, R.; Hemley, R.J. Polymerization of nitrogen in sodium azide. *J. Chem. Phys.* **2004**, *22*, 10618–10623. [CrossRef]
28. Li, D.; Zhu, P.; Jiang, J.; Li, M.; Liu, B.; Wang, X.; Cui, Q.; Zhu, H. High-Pressure Raman and Infrared Spectroscopic Studies of Cesium Azide. *J. Phys. Chem. C* **2016**, *120*, 27013–27018. [CrossRef]
29. Ji, C.; Zheng, R.; Hou, D.; Zhu, H.; Wu, J.; Chyu, M.; Ma, Y. Pressure-induced phase transition in potassium azide up to 55 GPa. *J. Appl. Phys.* **2012**, *111*, 112613. [CrossRef]
30. Ding, K.W. Experimental observation of TiN_{12}^+ cluster and theoretical investigation of its stable and metastable isomers. *Chem. Sci.* **2015**, *6*, 4723–4729. [CrossRef]
31. Ding, K.W.; Chen, H.J.; Xu, H.; Yang, B.; Ge, Z.X.; Lu, C.; Zheng, W.J. Identification of octahedral coordinated ZrN_{12}^+ cationic clusters by mass spectrometry and structure searches. *Dalton Trans.* **2021**, *50*, 10187–10192. [CrossRef]
32. Ding, K.W.; Xu, H.G.; Yang, Y.; Li, T.Q.; Chen, Z.; Ge, Z.X.; Zhu, W.; Zheng, W.J. Mass Spectrometry and Theoretical Investigation of VN_n^+(n = 8, 9, and 10) Clusters. *J. Phys. Chem. A* **2018**, *122*, 4687–4695. [CrossRef] [PubMed]
33. Ding, K.W.; Li, T.Q.; Xu, H.G.; Su, H.P.; Liu, Y.; Zheng, W.J.; Ge, Z.X. Formation and Detection of Lithium-nitrogen Clusters with High Nitrogen Content. *Chin. J. Energet. Mater.* **2018**, *41*, 447–450. [CrossRef]
34. Wu, Z.; Benchafia, E.M.; Iqbal, Z.; Wang, X.Q. N8 Polynitrogen Stabilized on Multi-Wall Carbon Nanotubes for Oxygen-Reduction Reactions at Ambient Conditions. *Angew. Chem. Int. Ed.* **2014**, *53*, 12555–12559. [CrossRef] [PubMed]
35. Yao, Z.; Wu, Z.; Hu, M.; Alzaim, S.; Young, J.; Dong, J.; Chang, J.; Zhuang, H.; Benchafia, E.M.; Yang, Y. Rational synthesis of polymeric nitrogen N_8^- with ultraviolet irradiation and its oxygen reduction reaction mechanism study with in situ shell-isolated nanoparticle-enhanced raman spectroscopy. *ACS Catal.* **2021**, *21*, 13034–13040. [CrossRef]
36. Benchafia, E.M.; Yao, Z.; Yuan, G.; Chou, T.; Piao, H.; Wang, X.; Iqbal, Z. Cubic gauche polymeric nitrogen under ambient conditions. *Nat. Commun.* **2017**, *8*, 930. [CrossRef] [PubMed]
37. Zhuang, H.; Alzaim, S.; Li, S.; Benchafia, E.L.; Young, J.; Ravindra, N.M.; Iqbal, Z.; Wang, X. Synthesis and Stabilization of Cubic Gauche Polynitrogen under Radio-Frequency Plasma. *Chem. Mater.* **2022**, *34*, 4712–4720. [CrossRef]
38. Zhang, C.; Wang, X.; Liang, Q.; Liu, X.; Weng, Q.; Liu, J.; Yang, Y.; Dai, Z.; Ding, K.; Bando, Y. Amorphous Phosphorus/Nitrogen-Doped Graphene Paper for Ultrastable Sodium-Ion Batteries. *Nano Lett.* **2016**, *16*, 2054–2060. [CrossRef]
39. Moulder, J.F.; Stickle, W.F.; Sobol, P.E.; Bomben, K.D.; King, R.C. *Handbook of X-ray Photoelectron Spectroscop*; Perkin-Elmer Waltham, MA, USA, 1992; p. 227.
40. Razvan, C. Raman spectra and lattice dynamics of cubic gauche nitrogen. *J. Chem. Phys.* **2007**, *127*, 144510. [CrossRef]

41. Onoe, T.; Iwamoto, S.; Inoue, M. Synthesis and activity of the Pt catalyst supported on CNT. *Catal. Commu.* **2007**, *8*, 701–706. [CrossRef]
42. Trojan, I.A.; Eremets, M.I.; Medvedev, S.A.; Gavriliuk, A.G.; Prakapenka, V.B. Transformation from molecular to polymeric nitrogen at high pressures and temperatures: In situ X-ray diffraction study. *Appl. Phys. Lett.* **2008**, *93*, 091907. [CrossRef]
43. Aravind, S.L.; Paramashivan, S.S.; Mahadevan, S. Thermo-kinetic studies of NaN_3/KNO_3 air bag gas generant mixture. *J. Therm. Anal. Calorim.* **2019**, *136*, 2183–2193. [CrossRef]
44. Eslami, A.; Hosseini, S.G.; Asadi, V. The effect of microencapsulation with nitrocellulose on thermal properties of sodium azide particles. *Prog. Org. Coat.* **2009**, *65*, 269–274. [CrossRef]
45. Kissinger, H.E. Reaction Kinetics in Differential Thermal Analysis. *Anal. Chem.* **1957**, *29*, 1702–1706. [CrossRef]
46. Yi, Q.H.; Liang, D.; Ma, Q.; Huang, M.; Tan, B.S.; Liu, Y.; Chi, Y. Synthesis, Crystal Structure, and Thermal Behavior of 3-(4-Aminofurazan-3-yl)-4-(4-nitrofurazan-3-yl)furazan(ANTF). *Propellants Explos. Pyrotech.* **2016**, *41*, 906–911. [CrossRef]
47. Gao, Y.C.; Li, M.; Yang, W.T.; Hu, R.; Zhang, Y.C. Thermal Decomposition Performance of CL-20-Based Ultraviolet Curing Propellants. *Propellants Explos. Pyrotech.* **2022**, *47*, e202100335. [CrossRef]

molecules

MDPI

Article

Effect of Pore Size Distribution on Energy Storage of Nanoporous Carbon Materials in Neat and Dilute Ionic Liquid Electrolytes

Maike Käärik [1,*], Mati Arulepp [2], Anti Perkson [2] and Jaan Leis [1,2]

[1] Institute of Chemistry, University of Tartu, Ravila 14a, 50411 Tartu, Estonia
[2] Skeleton Technologies, Sepise 7, 11415 Tallinn, Estonia
* Correspondence: maike.kaarik@ut.ee; Tel.: +372-737-5279

Abstract: This study investigates three carbide-derived carbon (CDC) materials (TiC, NbC, and Mo_2C) characterized by uni-, bi-, and tri-modal pore sizes, respectively, for energy storage in both neat and acetonitrile-diluted 1-ethyl-3-methylimidazolium bis(trifluoromethylsulfonyl)imide. A distribution of micro- and mesopores was studied through low-temperature N_2 and CO_2 adsorption. To elucidate the relationships between porosity and the electrochemical properties of carbon materials, cyclic voltammetry, galvanostatic cycling, and electrochemical impedance spectroscopy measurements were conducted using three-electrode test cells. The ultramicroporous TiC-derived carbon is characterized by a high packing density of 0.85 g cm^{-3}, resulting in superior cathodic and anodic capacitances for both neat ionic liquid (IL) and a 1.9 M IL/acetonitrile electrolyte (93.6 and 75.8 F cm^{-3}, respectively, in the dilute IL). However, the bi-modal pore-sized microporous NbC-derived carbon, with slightly lower cathodic and anodic capacitances (i.e., 85.0 and 73.7 F cm^{-3} in the dilute IL, respectively), has a lower pore resistance, making it more suitable for real-world applications. A symmetric two-electrode capacitor incorporating microporous CDC-NbC electrodes revealed an acceptable cycle life. After 10,000 cycles, the cell retained approximately 75% of its original capacitance, while the equivalent series resistance (ESR) only increased by 13%.

Keywords: nanoporous carbon; carbide-derived carbon; pore size distribution; electrical double-layer capacitor; ionic liquid electrolytes

1. Introduction

The increasing global interest in high-performance energy storage devices has spurred the exploration of various electrochemical storage solutions, including electrostatic, redox, and hybrid storage systems. Among these, electrostatic energy storage devices, commonly known as ultracapacitors (or supercapacitors), stand out as the most environmentally friendly option with the longest cycle life. These devices comprise large-surface carbon electrodes that can endure thousands of charge and discharge cycles. Their exceptionally low internal resistance contributes to their outstanding power characteristics, giving them a significant advantage over redox storage devices like secondary batteries. However, a major challenge faced by ultracapacitor developers is the relatively low energy density, which remains over 20 times lower than that of advanced lithium batteries.

The energy storage mechanism in ultracapacitors involves ion electrosorption on the surface of porous carbon electrodes. In general, a fundamental principle dictates that the larger the surface area accessible to the electrolyte, the greater the device's capacity [1]. Nanoporous carbon stands out as the most appealing electrode material, primarily due to its high surface-area-to-volume ratio. The impact of pore sizes on the capacitance of carbon materials has garnered significant attention. Research has established that a higher capacitance results from optimal matching between electrolyte ions and carbon pore sizes [1–4]. Carbide-derived carbon (CDC), a well-recognized representative of nanoporous carbons, has consistently demonstrated its suitability for ultracapacitor applications [5,6]. The structure and porosity of CDC can be meticulously controlled by adjusting the structure and composition of the precursor carbide and the synthesis conditions [2]. Moreover, the textural properties of CDC can be finely adjusted during post-treatment using either chemical methods [7–9] or physical activation techniques [10–13].

In accordance with the general definition of capacitance, a larger specific surface area corresponds to a higher capacitance. However, it is crucial to consider that this relationship is typically non-linear for materials with larger specific surface areas [14,15]. Therefore, it is essential to account for the pore sizes of the material as well [2,3,16,17]. In recent years, several studies [2,3,18,19] have introduced combined models for micro- and mesoporous carbon materials to illustrate the relationship between capacitance and pore size. These studies have demonstrated that both gravimetric [2,18] and volumetric [2,3] capacitance can be effectively simulated using data obtained from CO_2 and N_2 adsorption experiments.

The primary requirements for ultracapacitor electrolytes include high ionic conductivity, chemical and electrochemical stability, a wide operating temperature range, low resistance, and environmental compatibility [20]. Liquid electrolytes can be broadly categorized into non-aqueous, aqueous, and ionic liquid electrolytes [20,21].

Finding an electrolyte that fulfills all requirements can be quite challenging. Aqueous electrolytes offer high conductivity and capacitance but suffer from low energy density due to a narrow applicable voltage window (approximately 1 V). In contrast, organic (non-aqueous) electrolytes and ionic liquids (ILs) can operate at higher voltages but generally exhibit lower ionic conductivity [20,22]. One drawback of organic electrolytes is the larger size of their ions, which increases the specific resistance compared to water-based electrolytes and necessitates electrodes with larger pore sizes, thus reducing the electrode density [22]. Nevertheless, owing to their broader applicable voltage window (2.6–2.9 V), organic electrolytes, primarily based on quaternary ammonium salts in propylene carbonate (PC) or acetonitrile (ACN), currently dominate the field of commercial storage devices [1,22,23].

Ionic liquids, which consist of asymmetric organic cations and anions, offer several advantages over organic electrolytes. They exhibit high chemical, thermal, and electrochemical stability, are non-flammable, and permit the use of a higher electrochemical stability window (ESW), typically exceeding 3 V. In accordance with the characteristics commonly observed in organic electrolytes, it is noteworthy that the reduction in ionic size within ionic liquids (ILs) corresponds to decreased electrical resistance and an increase in specific capacitance. This principle has been explored in previous studies, such as those referenced in [24,25], in which the influence of cation size in IL electrolytes was investigated, and in [26], which delves into the electrochemical performance of various anions. Furthermore, an investigation was conducted employing three distinct ionic liquids (EMIm-TFSI, EMIm-BF_4, and BMIm-PF_6) in conjunction with porous carbon electrodes characterized by an average pore size of approximately 1.2 nm. This study revealed that smaller ions are more involved in the charging mechanism near the potential of zero charge (pzc), whereas at higher potential values, the mechanism shifts toward preferred counterion adsorption [27].

A notable drawback of IL-based electrolytes is their high viscosity, especially at lower temperatures [1,20,23,28,29]. In order to address the viscosity challenge, ionic liquids can be blended with low-viscosity organic solvents like PC or ACN [30]. Using a mixture of 1-ethyl-3-methylimidazolium bis(trifluoromethylsulfonyl)imide (EMIm-TFSI) and ACN, mesopores were shown to facilitate fast ion transport in carbon electrodes, achieving a specific capacitance of up to 130 F g^{-1} [31].

Largeot et al. investigated the relationship between carbon pore size and ion size while employing EMIm-TFSI as an electrolyte. In this context, $EMIm^+$ and $TFSI^-$ ions are roughly comparable in size (0.7 and 0.79 nm in their longest dimensions). Their study revealed that the maximum capacitance of 160 F g^{-1} (calculated from a galvanostatic experiment conducted at 5 mA cm^{-2} in a neat EMIm-TFSI electrolyte at 60 °C using a voltage window of 0–3 V) was attained with carbon possessing an average pore width of

0.72 nm. This dimension closely matches the size of the $EMIm^+$ and $TFSI^-$ ions, resulting in optimal ion adsorption efficiency [4]. A separate study [32], employing TiC-derived CDC as the electrode and a 2 M EMIm-TFSI/ACN mixture as the electrolyte in a three-electrode configuration, also concluded that the maximum capacitance is achieved when the pore size of the carbon closely aligns with the ion size. In smaller nanopores, ion adsorption, especially of larger TFSI anions, was diffusion-controlled due to hindered accessibility caused by size effects. A similar observation was reported regarding ion sizes and sub-nanopore matching for EMIm-TFSI and EMIm-BF_4, where secondary carbon nanopores of 0.50 nm were exclusively accessible to small BF_4^- anions (0.48 nm) but not to larger $TFSI^-$ ions (0.79 nm) [33].

The interaction between ionic liquid structures and various carbon materials and their consequential effects have been the subject of prior investigations [34]. These studies emphasized that the selection of a carbon electrode should not solely be based on pore size and surface area; it should also consider the interaction between the surface chemistry of the electrode and the electrolyte. While materials with high specific surface areas, such as mesoporous and activated carbon, exhibited respectable specific capacitance and energy density, materials resembling graphene and carbon nanotubes with open surfaces demonstrated superior power density.

Härmas et al. have reported that ultramicroporous materials are unsuitable for high-power and high-energy-density supercapacitors due to the slow establishment of adsorption equilibrium in viscous ionic liquids [35]. However, when the viscosity of the ionic liquid is reduced by introducing an additional solvent like ACN, ultramicroporous materials become exceedingly attractive due to the increased density at the corresponding electrodes, resulting in enhanced volumetric energy and power density [36].

Previously, the interconnectedness of porosity parameters in carbon materials and their utility in quantitatively predicting EDLC across various organic electrolytes were demonstrated [2,3]. In the present study, a preliminary investigation of three CDC materials, originating from TiC, NbC, and Mo_2C, characterized by unimodal, bimodal, and trimodal pore size distributions, respectively, is undertaken. This investigation encompasses energy storage within both undiluted and diluted ILs. The primary objective is to assess whether the energy storage properties of ultramicroporous carbon materials, akin to the majority of CDCs, are enhanced in IL environments due to the presence of larger transport pores within the carbon material structure.

2. Results and Discussion

The three carbon materials (CDCs) with variable pore size distributions used in this study were made by the chlorination of different metal carbides: TiC at a temperature of 500 °C [10], Mo_2C at a temperature of 800 °C [37], and NbC at a temperature of 1000 °C [2]. The CDCs are distinguished as CDC-TiC, CDC-NbC, and CDC-Mo_2C, respectively.

As previously shown [2,38], the structural order and porosity of CDCs are related to the structure of the precursor carbide and the temperature of chlorination; the higher the temperature, the higher the degree of graphitization and the larger the average pore size. The microstructure has also been studied in CDCs derived from TiC [10,39], NbC [39], and Mo_2C [37,40]. In Figure 1a, one can observe that the N_2 adsorption isotherm for CDC-TiC corresponds to Type I (a) by IUPAC [41], characteristic of microporous materials with a narrow pore size distribution. The CDC-NbC exhibits a Type I (b) isotherm, which shows a broader pore size distribution in the micropore region. The CDC synthesized from Mo_2C (i.e., CDC-Mo_2C) corresponds to Type IV (a), with a noticeable hysteresis loop indicating the presence of mesopores.

The pore size distributions (PSDs) were calculated from the N_2 isotherms and are presented in Figure 1b. The PSD of all three CDCs contains a peak with pore sizes around 0.8 nm. CDC-NbC contains additional pores at 1.6 nm. CDC-Mo_2C represents micromesoporous carbon, whose PSD contains, in addition to micropores at 0.8 and 1.9 nm, a considerable amount of mesopores at 3 nm.

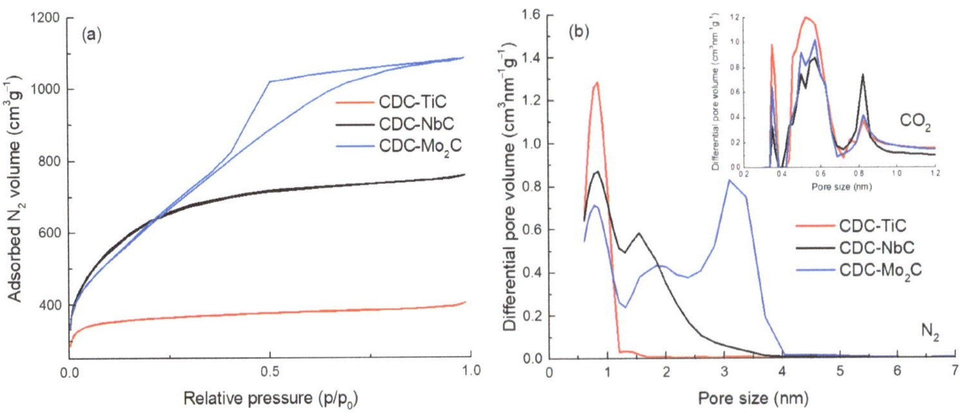

Figure 1. N_2 adsorption isotherms (**a**) and the corresponding pore size distribution of CDC materials calculated from N_2 and CO_2 isotherms (**b**).

The texture characteristics of the CDCs are collected in Table 1. Mo_2C-CDC reveals the largest specific surface area as well as total porosity (2204 m^2g^{-1} and 1.67 cm^3g^{-1}, respectively), whereas CDC-NbC has the most significant amount of micropores (0.90 cm^3g^{-1}). Among all three CDCs, CDC-TiC has a majority of pores smaller than 1 nm. Also, the PSD calculated from CO_2 adsorption (inset on Figure 1b) confirms that the carbon made of TiC has the highest proportion of sub-nanometer-sized pores.

Table 1. Specific surface area (S_{BET} and S_{dft}) including micro- ($S_{dft,\mu}$) and mesopore ($S_{dft,meso}$) ratios, total pore volume (V_t), and volume of different pore size fractions ($V_{<x\,nm}$) calculated from N_2 adsorption isotherms.

ID	S_{BET} ($m^2\,g^{-1}$)	S_{dft} ($m^2\,g^{-1}$)	$S_{dft,\mu}$ ($m^2\,g^{-1}$)	$S_{dft,meso}$ ($m^2\,g^{-1}$)	V_{tot} ($cm^3\,g^{-1}$)	V_{1nm} ($cm^3\,g^{-1}$)	$V_{1.5nm}$ ($cm^3\,g^{-1}$)	V_{2nm} ($cm^3\,g^{-1}$)
CDC-TiC	1257	1318	1306	12	0.61	0.47	0.52	0.53
CDC-NbC	2247	1893	1765	128	1.17	0.40	0.68	0.9
CDC-Mo_2C	2204	1951	1356	595	1.67	0.32	0.48	0.69

An electrochemical evaluation of the CDC electrodes was performed in a three-electrode test cell using the neat ionic liquid EMIm-TFSI as the electrolyte at 60 °C and a 1.9 M EMIm-TFSI/ACN mixture at room temperature. A cyclic voltammetry (CV) study was performed over a potential range of −1.4 V to +1.4 V vs. a carbon electrode.

Figure 2 presents the cyclic voltammetry curves of the CDC materials measured at a voltage scan rate of 1 mV s^{-1} in a potential window of −1.4 V to +1.4 V. A visual examination of these graphs (Figure 2a,b) reveals a characteristic flat minimum in the central region of the curve, near 0 V, corresponding to the zero-charge potential of carbon. This is caused by the fact that, as the electrode potential increases, the number of electrosorbed ions on the surface increases too, which leads to an increase in the density of the electric charge on the electrode surface, resulting in an increase in the capacitance of the electric double layer. This minimum is located somewhat differently on various CDC materials, with the flatness of the minimum increasing as the carbon material becomes more microporous. For example, in the case of CDC-TiC in the EMIm-TFSI/ACN electrolyte, no zero-charge minimum is observed at all. The flatness of this minimum is influenced by the electrolyte viscosity, ion dimensions, and the micromesoporous nature of the carbon (cf. Table 1).

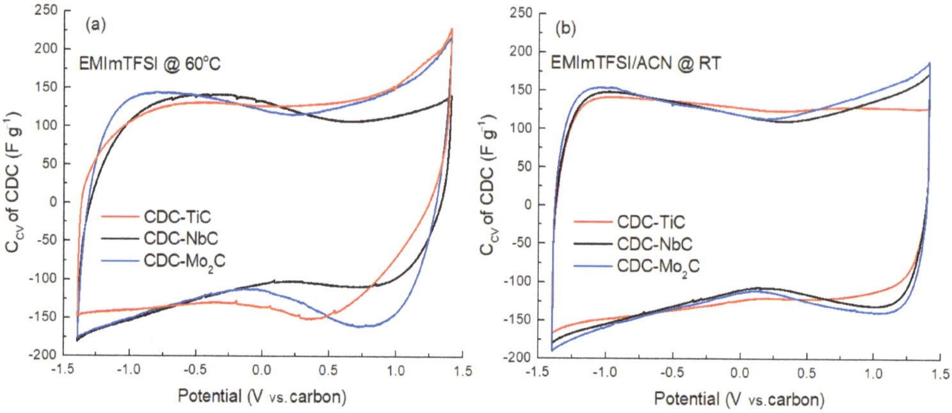

Figure 2. Cycling voltammetry curves of CDC materials in neat IL (**a**) and IL/ACN mixture (**b**), expressed as capacitance (Ccv) vs. potential at a scan rate of 1 mV s^{-1}.

In addition to the minimum, the curves can be characterized by the current response upon potential reversal, specifically the "backswing." For example, it is evident that after the potential reversal (after +1.4 V or −1.4 V) in the neat EMIm-TFSI (Figure 2a), the curves exhibit a flatter curvature compared to dilute electrolyte solutions (Figure 2b). This is again due to the significantly higher viscosity of the neat ionic liquid. Additionally, there is an observable dependence of the slope on the carbon material, where a smaller average pore size (1.23 nm) of CDC-NbC compared to 1.74 nm of CDC-Mo_2C leads to a shallower slope. For instance, a distinctive change in the slope (after potential reversal at +1.4 V) is observed in cells with the neat IL electrolyte, where the slope increases in the sequence CDC-TiC, CDC-NbC, and CDC-Mo_2C. However, in the cathodic (negative potential) region, CDC-TiC stands out due to its very high impedance at negative potentials, which is "amplified" by the high viscosity of the IL and its low conductivity. Therefore, a change in potential causes a sharp current drop to 0 and distorts the CV curve shape. In dilute electrolytes, the slope after potential reversal is steeper, primarily due to the lower electrolyte viscosity and higher conductivity mentioned earlier.

In the mixed IL/ACN electrolyte, when comparing the CV curves of CDC-NbC and CDC-Mo_2C (Figure 2b), they are quite similar, with CDC-TiC, the most microporous, showing some differences. This may be attributed to the ion solvation effects and slightly larger ion dimensions in the EMIm-TFSI/ACN electrolyte compared to the neat IL. Furthermore, the lower anodic capacitance of the CDC-TiC material supports this hypothesis, influenced by the larger dimensions of $TFSI^-$ anions compared to $EMIm^+$ cations (0.6 × 0.9 nm vs. 0.5 × 0.8 nm [42]).

To evaluate the EDL capacitance of CDCs at different current densities (0.5 to 5 mA cm^{-2}), constant current (CC) charge–discharge measurements were conducted in the anodic and cathodic potential ranges. A 5 min holding at the operating potential was applied in order to achieve electrosorption equilibrium (i.e., maximum charging) at the given potential, consequently leading to better reproducibility of the capacitance values. Characteristic CC curves are shown in Figure 3, at a current density of 0.5 mA cm^{-2} and potential ranges of 0 to +1.4 V and 0 to −1.4 V. In Figure 3, CDC-TiC, the most microporous material, stands out somewhat at anodic potentials, manifesting as non-linear charge–discharge curves in the neat IL electrolyte. This behavior is due to the high viscosity of the neat IL (14.2 cP [43]) and the interaction between somewhat larger electrolyte anions and the sub-nanometer-sized pores. After diluting the IL with ACN, the viscosity of the electrolyte significantly decreases (by ~10-fold [43]), and the non-linearity of the charge–discharge curves of CDC-TiC disappears.

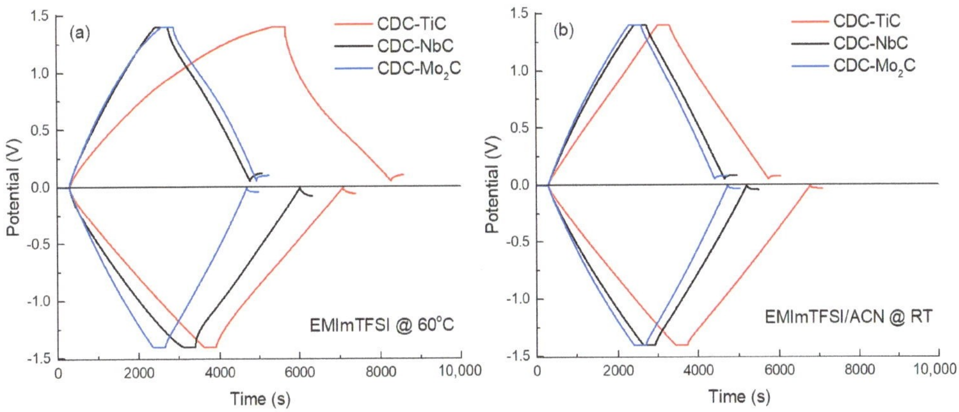

Figure 3. CC charge–discharge curves (I = 0.5 mA cm^{-2}) of CDC materials in neat IL (**a**) and IL/ACN mixture (**b**).

Table 2 provides the specific capacitance values of all three CDCs in this study, calculated using the constant current method and measured in a neat ionic liquid electrolyte at 60 °C. Comparing these values to the gravimetric capacitances, it can be observed that CDC-Mo_2C, with the largest pore dimensions, has nearly equal cathodic and anodic capacitances of 140 vs. 146 F g^{-1}. A similar level of 145 F g^{-1} is achieved by CDC-NbC in the cathodic direction. This uniform level is a result of these materials' pore size and ion dimensions having an almost "ideal" fit in the IL. Moreover, even the high viscosity of the electrolyte does not impose limitations on achieving this capacitance level.

Table 2. Specific capacitance of CDCs at positively and negatively charged electrode potentials by CC (0.5 mA cm^{-2}) in EMIm-TFSI electrolyte.

Carbon Sample	C_{cc}^{-} (F g^{-1})	C_{cc}^{+} (F g^{-1})	C_{cc}^{-} (F cm^{-3})	C_{cc}^{+} (F cm^{-3})
CDC-TiC	127.3	113.3	99.9	88.9
CDC-NbC	145.1	110.0	81.6	61.8
CDC-Mo_2C	140.1	146.0	63.2	65.9

The lower capacitance of the most microporous material (CDC-TiC) is due to its lower specific surface area and significantly lower porosity. The anodic capacitance of CDC-TiC is lower because the pore dimensions of its carbon are very close to the size of the electrolyte anions, and not all micropores are accessible to the electrolyte ions. However, the limited access to the micropores is compensated by the very high carbon packing density in the case of CDC-TiC, which therefore reveals the highest volumetric capacitance values, ranging from 89 to 100 F cm^{-3} for the anodic and cathodic capacitance, respectively. In contrast, CDC-NbC, which has significantly higher overall porosity and surface area than CDC-TiC, exhibits lower density and, therefore, lower volumetric capacitance compared to CDC-TiC. As a rule, the larger the pore volume, the lower the material density (as more pores result in a decrease in material density).

When comparing the capacitance values in the neat IL (Table 2) and the diluted electrolyte (Table 3), it is evident that ion solvation by ACN shifts the capacitance values in both directions, sometimes increasing and sometimes decreasing. For example, the cathodic capacitance is increased in CDC-TiC (in the dilute electrolyte compared to the neat IL), most probably due to a decrease in electrolyte viscosity, which improves ion diffusion in the micropores, while the dilution of IL has a negligible effect on the anodic capacitance, which is practically the same in both electrolytes (113 F g^{-1}).

Table 3. Specific capacitance of CDCs at positively and negatively charged electrode potentials by CC (0.5 mA cm^{-2}) in EMIm-TFSI/ACN electrolyte.

Carbon Sample	C_{cc}^{-} (F g^{-1})	C_{cc}^{+} (F g^{-1})	C_{cc}^{-} (F cm^{-3})	C_{cc}^{+} (F cm^{-3})
CDC-TiC	137.6	113.0	93.6	75.8
CDC-NbC	138.3	119.8	85.0	73.7
CDC-Mo_2C	140.0	127.9	64.8	59.2

CDC-NbC behaves differently from TiC in terms of capacitance. In the dilute IL, CDC-NbC cathodic capacitance is lower (138 vs. 145 F g^{-1}) compared to the solvent-free environment. However, in the case of anodic capacitance, the situation is reversed, with ~120 F g^{-1} vs. 110 F g^{-1} for the IL/ACN and neat IL, respectively. This shows that carbon materials with sufficiently large micropores are suitable for achieving high capacitance in ionic liquids, but in a solvent-diluted environment, the solvation of ions begins to limit the capacitance, which is particularly pronounced in NbC at the anodic potential.

When comparing the capacitances of CDC-Mo_2C in two different electrolytes (see Tables 2 and 3), it appears that its large pores are not limiting and are suitable for cathodic potential, but a capacitance decrease occurs at the anodic potential in the solvent-diluted environment (i.e., capacitance decreases from 146 F g^{-1} to 128 F g^{-1}). However, CDC-Mo_2C provides valuable information due to its very large pore dimensions when fitting with the solvent effects and pore dimensions in the electrolyte. Nevertheless, CDC-Mo_2C has significantly lower volumetric capacitance (60–66 F cm^{-3}) in comparison to the other materials studied, making it less practical for EDL devices due to its low energy density.

The analysis of the CC capacitance vs. current density in the range 0.5–5.0 mA cm^{-2} was performed using ACN-diluted IL (cf. Figure 4). The findings from this study confirm that the cathodic capacitance, attributed to the electrosorption of $EMIm^{+}$ ions, surpasses the anodic capacitance by approximately 20 F g^{-1}. This phenomenon is particularly notable due to the larger dimensions of the $TFSI^{-}$ anions. Furthermore, the influence of the ion size becomes evident in the substantial discrepancy in capacitance observed between the CDCs at the anode. This discrepancy suggests restricted accessibility to micropores when the dimensions of the ions closely approach those of the micropores themselves.

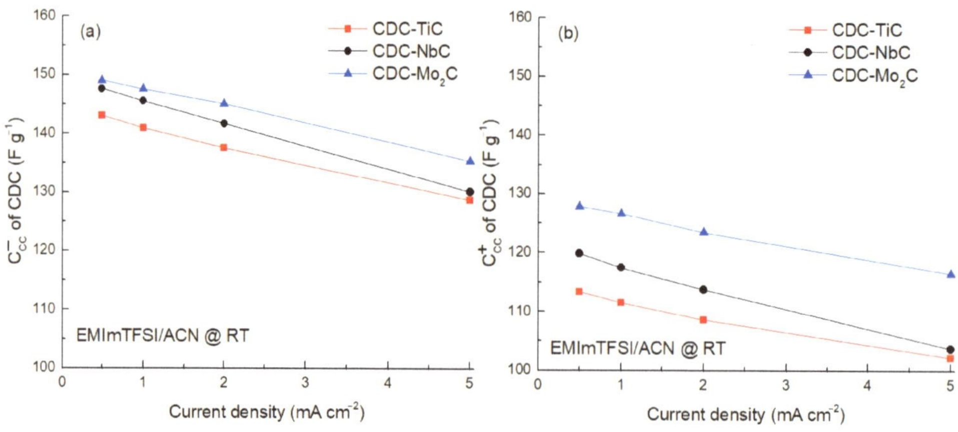

Figure 4. Specific capacitance (C_{cc}) of carbon materials calculated at negative −1.6 V (**a**) and positive 1.4 V (**b**) potentials at different current densities.

Cyclic voltammetry was utilized to investigate the behavior of electrode materials at positive and negative electrode potentials. To ascertain electrochemical stability, measurements were conducted over a wide electrochemical window, i.e., +1.8 V to 0 V and

−1.8 V to 0 V vs. a carbon reference electrode. The CV diagrams for the materials studied are presented in Figure 5.

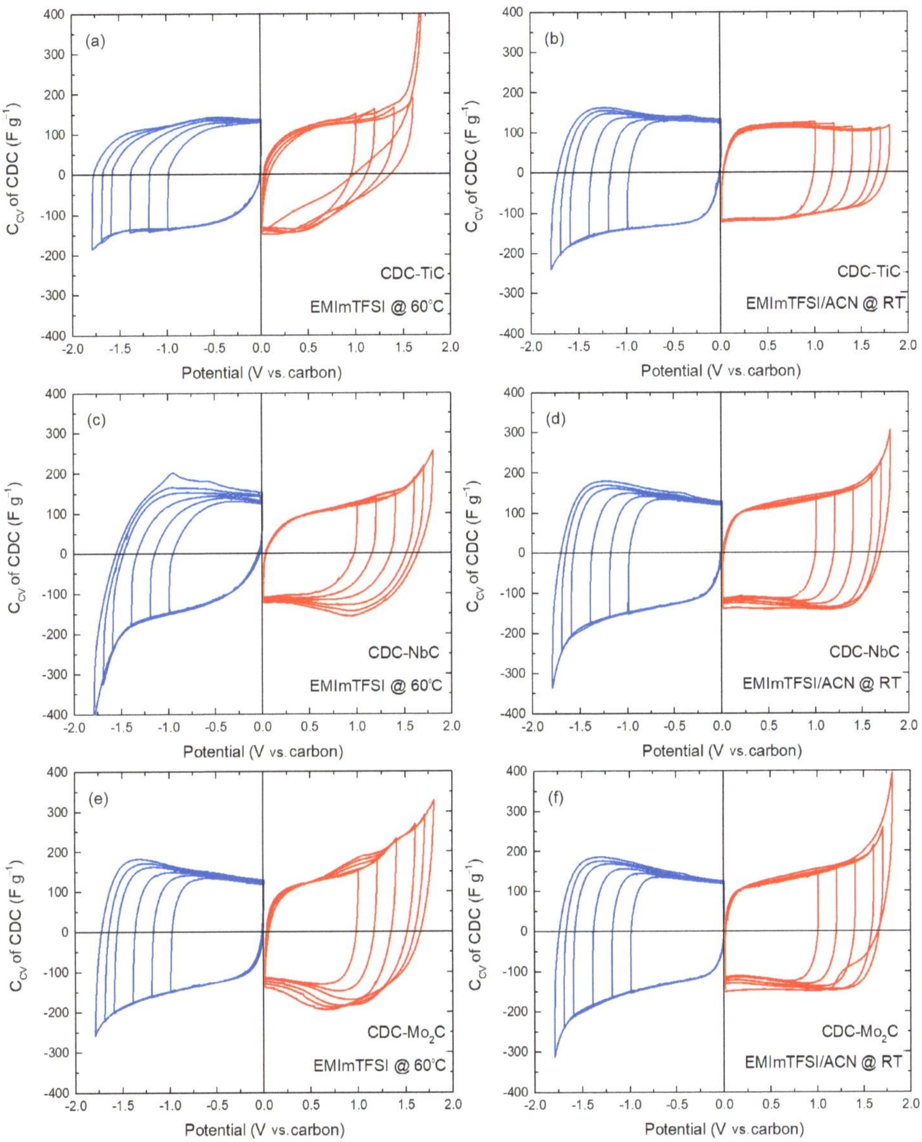

Figure 5. Cycling voltammetry curves (v = 1 mV s^{-1}) for positively (red) and negatively (blue) charged electrodes expressed as gravimetric capacitance (C_{CV}) for CDC-TiC (**a**,**b**), CDC-NbC (**c**,**d**), and CDC-Mo_2C-CDC (**e**,**f**). Electrolyte and measurement temperature are shown on each graph.

Evaluating the electrochemical stability of materials is essential to determining the voltage limits. As these limits are not the same for different carbons, electrochemical window studies were conducted separately at positive and negative electrode potentials. In this process, the electrode potential was incrementally increased, starting from 1.0 V with a 200 mV step and gradually reaching 1.6 V, followed by smaller 100 mV increments up to 1.8 V.

For the CDC-TiC material, its cathodic curves in the neat ionic liquid are relatively similar, and starting from −1.6 V, there is a gradual increase in current during charging (i.e., an increase in capacitance). Such an increase in current results in a flat peak in the reverse scan of the CV cycle in the −0.8 V to −0.6 V range, which slightly grows with an expanding CV potential window, indicating a partially reversible process.

In the positive potential region for the CDC-TiC material, the CV curves up to +1.4 V are relatively similar. With further potential increases beyond +1.6 V, the polarization current increases rapidly, indicating electrolyte decomposition and potential partial surface blocking by system degradation products. This process becomes especially evident in the curve of +1.7 V. Therefore, the shape of the discharge at +1.7 V polarizations is significantly different from lower potentials.

In the case of the EMIm-TSI/ACN electrolyte solution, the anodic end of the CV curve of CDC-TiC is very stable and almost rectangular in shape. The complete absence of peaks in the presence of ACN (Figure 5b) suggests that adding a solvent significantly improves the stability of the carbon/electrolyte interface. In the cathodic CV polarization curves in the dilute IL electrolyte, there is a similar trend as seen in the full-scan curves; achieving capacitance during the reverse scan is faster (see Figure 2a,b), and the numerical value of the specific capacitance is also higher in the dilute electrolyte compared to the neat IL.

The CDC-NbC curves appear relatively similar regardless of the neat or dilute electrolyte used (Figure 5c,d). In both electrolytes, the increase in cathodic current at critical potentials is lower compared to CDC-TiC and CDC-Mo_2C. However, in the neat IL, there is an inexplicable peak (−1.0 V) during the reverse scan of the cathodic region (also observed in the −1.7 V curve), which is not present in the dilute IL electrolyte. Additionally, in both electrolytes, there is a lower cathodic current peak during the reverse scan, around −0.5 V, which is characteristic of all studied carbons. The anodic curves for CDC-NbC show differences compared to the neat and dilute ILs. In the first one, after +1.7 and +1.8 V, there is also a flat peak around +0.8 V, which could be attributed to the desorption of $TFSI^-$ anions after passing higher potentials due to their specific adsorption in CDCs with multi-modal pores (i.e., CDC-NbC and CDC-Mo_2C), though this is not confirmed. In the diluted electrolyte, with a lower ion concentration, such a distinct peak does not seem to form with either material.

In the case of CDC-Mo_2C (Figure 5e,f), the cathodic curve in the neat IL resembles the shape of the CDC-TiC dependence on the CV curves obtained in the EMIm-TFSI/ACN electrolyte. Despite the differences in porosities, it appears that solvated cations in CDC-TiC and CDC-Mo_2C exhibit similar electrosorption properties. The anodic curves of CDC-Mo_2C are clearly different depending on the electrolyte. For instance, in the neat IL, a flat peak forms around +1.0 V during the anodic charging of the material after passing a wider potential window > −1.6 V. However, a similar, albeit very small, peak also forms for CDC-NbC after −1.8 V. However, the cathodic CV curves in CDC-Mo_2C are very similar in both electrolytes, except for a slight difference in the sharper increase in current at −1.8 V during cathodic charging in the neat IL compared to the dilute IL. For CDC-Mo_2C, the anodic CV curves are similar to the anodic curves of CDC-NbC in the diluted electrolyte. However, there is a significant difference after +1.8 V potential, similar to the cathodic potential, indicating that this material has a somewhat smaller electrochemical window in the ACN-containing electrolyte.

The Nyquist plot in different electrolytes (Figure 6) shows a clear differentiation of the curves into two groups (in the neat IL electrolyte characterized by larger semicircles and in the diluted solvent with smaller semicircles). The larger radius of the semicircle is clearly due to the IL electrolyte having nearly 3× higher viscosity and about 2× lower electrical conductivity compared to the ACN electrolyte. In the neat IL, the plot also reveals material-related differences; for instance, the size of the semicircle (related to the charge transfer resistance) is clearly correlated with the material's porosity and the size of the material's particles, which, in the present case, both increase in the order CDC-TiC (Ø < 4 μm), CDC-NbC (Ø < 10 μm), and CDC-Mo_2C (Ø < 20 μm). In addition to the semicircle, the most microporous material also has the largest area of so-called pore resistance, with an angle measuring less than 45°. This lower angle characterizes the simultaneous occurrence of mixed kinetic processes (partially impeded ion diffusion processes and adsorption processes). The other two materials exhibit a noticeable 45° rise. In both electrolytes, the vertical section of the CDC-TiC material does not reach a height of 90° due to steric hindrance to the electrosorption of electrolyte ions in the nanoporous material.

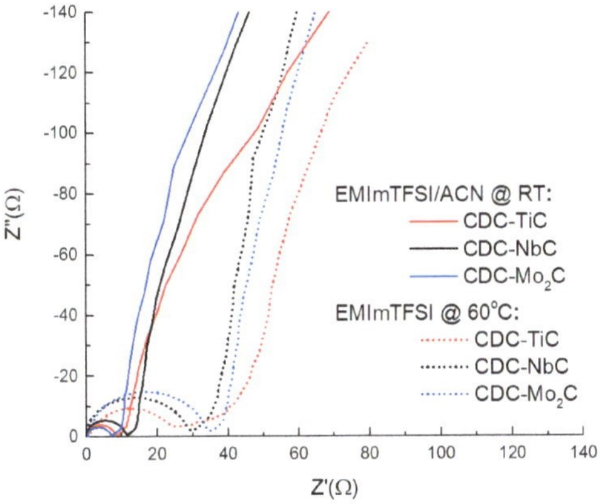

Figure 6. Nyquist plots of CDC materials in neat EMIm-TFSI at 60 °C and 1.9 M EMIm-TFSI/ACN at room temperature (RT).

The R_s values calculated for the uncharged material at a frequency of 1 kHz reveal that the ultramicroporous CDC-TiC has very high electrical resistance (11.4 ohms) in the neat IL electrolyte. This is due to hindered diffusion as observed in the section of the "45-degree slope" (so-called pore resistance), on the Z',Z'' curve in Figure 6, which also causes heat to be released during electrochemical cycling. However, in general, the capacitance of all three CDCs is similar, manifesting in the range of 3.1–5.6 ohms, and it does not matter if they are in solution or not.

For cycle life testing, a symmetric two-electrode cell assembled from CDC-NbC electrode discs was subjected to cycling between 0 and 2.85 V using CV. The evolution of capacitance over the cycles is depicted in Figure 7a. After completing 10,000 cycles, we reversed the cell's terminal polarities, followed by 10 cyclic voltammetry (CV) cycles from 0 to 2 V, referred to as 'repolarization.' Subsequently, we restored the original polarities, and cycling continued within the voltage range of 0 to 2.85 V. During the 10,000 cycles, the capacitance exhibited a reduction of approximately 22% from the initial value of 122 F g^{-1}. This gradual reduction in capacitance over time is a hallmark of most electric double-layer (EDL) capacitors.

Immediately after repolarization, the capacitance recorded in the first cycle exhibited a remarkable surge, almost returning to the initial capacitance level (98%). However, this increase was transient, as subsequent cycling resulted in a rapid decline. This "overshoot" phenomenon can be attributed to the "cleaning" of the electrode surfaces from adsorbed compounds following the polarity reversal [44]. The freshly exposed surface momentarily regains its capacity for readsorbing electrolyte ions. Nevertheless, the redeposition of additives occurs quite swiftly, and the attained capacitance level proves to be temporally unstable.

In addition to the capacitance data obtained via the CV method, this test cell was subjected to electrochemical impedance spectroscopy after varying numbers of cycles (Figure 7b).

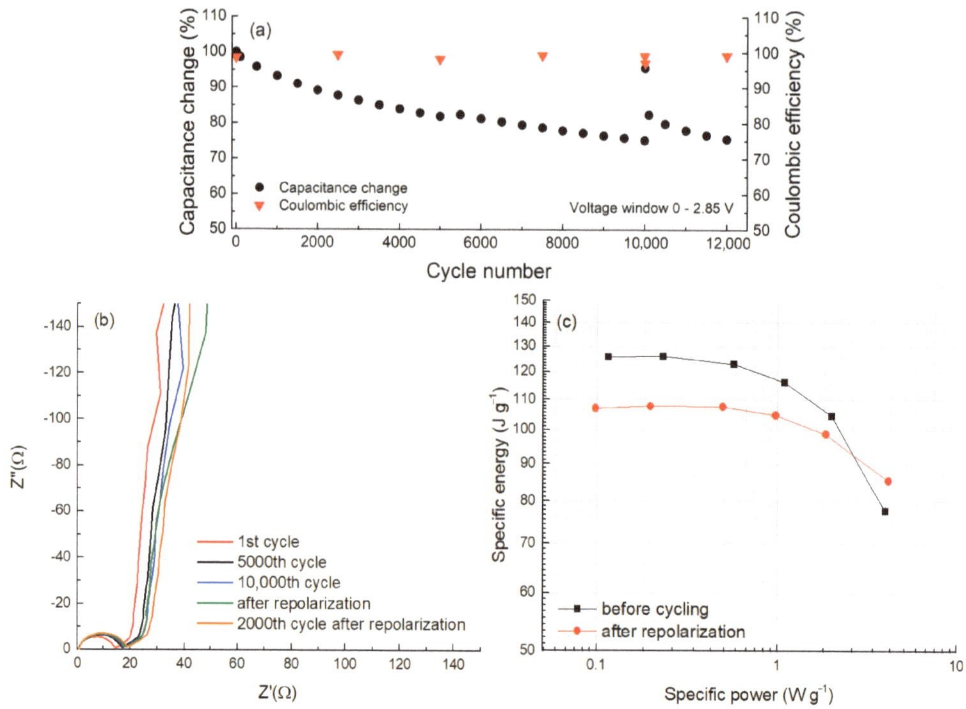

Figure 7. Capacitance and coulombic efficiency evolution (**a**), EIS spectra (**b**), and a Ragone plot (**c**) in cycle life test with a 2-electrode CDC-NbC cell at a voltage scan rate of 50 mV s^{-1} (**a**), including short-term repolarization after 10,000 cycles.

In Figure 7b, a Nyquist plot is presented, obtained before and after a certain number of full charge–discharge cycles. In this plot, the semicircles start from a single point, indicating a very stable system and a constant resistance at the metal/carbon electrode interface. Thus, changes in the carbon/aluminum resistance during the cycle life test can be ruled out. The semicircle slightly grows after around 5000 cycles, suggesting some alterations occurring on the carbon surface. However, the pore resistance (45° rise and its length) remains relatively stable, indicating that any potential changes over time are likely taking place on the external surface of the carbon particles. Furthermore, when observing the vertical "tail", it becomes less vertical just after repolarization (Figure 6b, green curve) compared to the others, suggesting that this process might induce some parallel reaction in the system. Nevertheless, these processes (reactions) ceased to manifest after 2000 cycles of post-repolarization. It is possible that during cycling, surface oxides may attach to cations through anodic hydrogen-bonding-like interactions, hindering anion electrosorption. However, during the repolarization process, the breaking of these hydrogen bonds occurs, resulting in the observed changes.

The values of the series resistance (R_s) measured at 1 kHz are 12.8, 14.2, and 14.5 ohms after the 1st, 5000th, and 10,000th cycles, respectively. Immediately after repolarization, an R_s of 14.9 ohms was measured, with 15.2 ohms after the next 2000 cycles. Therefore, the overall change in cell resistance after completing 12,000 cycles is approximately 19%; this represents only one-fifth of the allowable change in the R_s value, as stipulated by conventional lifetime-testing agreements. In contrast, the decrease in capacitance over the same number of cycles is approximately 20%.

Figure 7c represents the dependence of energy density on the applied power density, also known as the Ragone plot, at the beginning of the cycle life test and immediately after repolarization. The data on the Ragone plot were derived from the dependencies on different cycling rates using the CV method. It was shown that the maximum usable energy of CDC-NbC in a symmetric two-electrode cell at low power is 125 J g^{-1}, which decreases to nearly 108 J g^{-1} after 10,000 cycles (includingrepolarization) at an operating voltage of 2.85 V. The observed increase in power density after 10,000 cycles, particularly at high currents, might be associated with repolarization and the rapid surge in capacitance.

3. Materials and Methods

3.1. Synthesis and Characterization of CDC Materials

The CDC samples of variable pore size distribution were made by a high-temperature chlorination of TiC (H.C. Starck, Ø < 4 μm), NbC (Alfa Aesar, Ø < 10 μm), and Mo_2C (Reahim, Ø < 20 μm) at 500 °C, 1000 °C, and 800 °C, respectively. Carbide powder, placed in the horizontal quartz tube, was reacted with chlorine (Linde Gas, 2.8, 1.5 L min^{-1}) flow at a fixed temperature. During heating and cooling, the reactor was purged with argon (Linde Gas, 4.0, 2 L min^{-1}) flow. After chlorination, the CDC product was additionally treated with hydrogen (Linde Gas, 4.0, 1 L min^{-1}) at 800 °C to deeply dechlorinate the sample. The detailed synthesis conditions of CDC materials are described elsewhere [2].

Textural properties of CDC materials were determined by low-temperature N_2 adsorption at 77 K and CO_2 adsorption at 273 K using a NOVAtouch LX2 (Quantachrome Instruments, Boyton Beach, FL, USA). The BET surface area (S_{BET}) was evaluated based on the BET theory in the P/P_0 interval of 0.02–0.2, and the total pore volume (V_{tot}) was calculated at P/P_0 of 0.97. The specific surface area (S_{dft}), pore size distribution (PSD), and volume of micropores (V_{μ}) were calculated from N_2 adsorption isotherms using a quenched solid density functional theory (QSDFT) equilibrium model for slit-type pores. The PSD from CO_2 adsorption was determined using a non-local density functional theory (NLDFT) model for slit-type pores. In all calculations, the software TouchWin ver. 1.1 was used.

3.2. Electrochemical Characterization

Electrochemical characterization of carbon materials was performed in three-electrode test cells. Working electrodes with a visible surface area of 0.38 cm^2 were made by pressing a slurry of 90% CDC and 10% PTFE (Aldrich, Taufkirchen, Germany, 60% dispersion in water) into a carbon sheet. The thickness of the working electrodes was ~100 μm. The reference and counter electrodes were made from high-surface-area carbon sheets, which, for better electrical contact, were coated on one side with a 2 μm layer of pure Al using PVD. The counter and working electrodes were separated with two layers of SWP glass mat-type separator (Anpei, Tainan, Taiwan, thickness ~1 mm). All electrodes and separators were preliminary dried in vacuum (1 mbar) at 110 °C for a week. After assembling, the cell was vacuumed overnight at 110 °C and filled with the electrolyte 1-ethyl-3-methylimidazolium bis(trifluoromethylsulfonyl)imide (EMIm-TFSI, Sigma-Aldrich, Taufkirchen, Germany, ≥98%) and a 1.9 M solution of EMIm-TFSI in acetonitrile (Sigma-Aldrich, anhydrous 99.5%). The density of the electrolyte solution was 1.18 g cm^{-3}. The neat ionic liquid was used as is, and the IL/ACN mixture was kept on molecular sieves (3 Å) for one week before use. Measurements with neat ionic liquid were performed at 60 °C and with EMIm-TFSI/ACN at room temperature; in both cases, the test cell was soaked for 24 h before measurement.

The electrochemical characteristics of carbon materials were measured using cyclic voltammetry (CV), galvanostatic (CC), and electrochemical impedance spectroscopy (EIS) methods. The potentiostat–galvanostat 1286 with FRA 1255B (Solartron, Hampshire, UK) was used for all electrochemical measurements. The integral capacitance (C_i) was calculated from CC experiments according to the equation $C_i = I\Delta t/\Delta U$, where I is the discharge current and Δt is the discharge time in the potential range of ΔU. The specific capacitance values of CDC ((F g^{-1}), (F cm^{-3})) were calculated as $C_{CC} = C_i/m$ and $C_{CC} = C_i/V$, where m is the weight of the CDC in a working electrode and V is the volume of the working electrode.

The cycling stability of CDC-NbC was evaluated in a two-electrode cell assembled from two electrodes of 7 mm in diameter separated with two layers of cellulose separator TF44-25 (Nippon Kodoshi Corp., Nankoku-City, Kochi, Japan). The CV applying a voltage scan rate of 50 mV s^{-1} was used to evaluate the cycling life in the voltage ranges of 0–2.85 V. Prior to the test, the cell was preconditioned with CV cycling (v = 50 mV s^{-1}), expanding the potential window from 0–1.2 V to 0–2.85 V with an interval of 0.2 V. The capacitance was calculated from CV plots as differential capacitance $C_D = j/v = j(\mathrm{d}E/\mathrm{d}t)$, where j is the current at a fixed cell voltage (1.5 V) and v is the voltage scan rate. The CV data were used to construct the experimental Ragone plots (energy density vs. power density). The coulombic efficiency was calculated as $\eta = Q$ (discharge)/Q (charge), where Q is the charge. The changes in resistance were measured by EIS at 1 kHz using the ZView ver. 3.5i software (Scribner Associates Inc., Southern Pines, NC, USA).

4. Conclusions

In this study, three carbide-derived carbon materials of TiC, NbC, and Mo_2C origin, characterized by uni-, bi-, and tri-modal pore sizes, respectively, were investigated for energy storage in both neat EMIm-TFSI and 1.9 M ACN-diluted EMIm-TFSI electrolytes. To elucidate the relationships between porosity and the electrochemical properties of carbon materials, cyclic voltammetry, galvanostatic cycling, and electrochemical impedance spectroscopy measurements were performed using three-electrode test cells. It was demonstrated that micromesoporous CDC-Mo_2C, with a large fraction of 3–4 nm sized pores, has good access to micropores and, therefore, excellent gravimetric capacitance in both neat and dilute ILs. However, the low packing density of CDC-Mo_2C results in the low energy density of the corresponding storage device, which makes it impractical for large-scale production. Ultramicroporous CDC-TiC exhibits a narrow pore size distribution at approximately 0.8 nm. This carbon material is distinguished by its high packing density of 0.85 g cm^{-3} and demonstrates a notable volumetric capacitance of 93.6 and 75.8 F cm^{-3} for the negatively and positively charged electrodes in the dilute IL, respectively. Regrettably, the significant pore resistance arising from the absence of transport pores results in elevated overall impedance for the associated energy storage device. This elevated impedance is undesirable in numerous applications, particularly in sectors such as transportation. The bimodal pore-sized microporous CDC-NbC has a main fraction of micropores with the same size as CDC-TiC. And although it has a slightly lower capacitance of 85.0 and 73.7 F cm^{-3} for the negatively and positively charged electrodes in the dilute IL, respectively, the lower pore resistance due to the additional fraction of 1–2 nm micropores makes it more suitable for real-world applications. A symmetric two-electrode capacitor incorporating microporous CDC-NbC electrodes revealed an acceptable cycle life. After 10,000 cycles, the cell retained ~75% of its original capacitance, while the ESR increased only by 13% at a coulombic efficiency of 99%.

Author Contributions: Conceptualization, J.L. and M.K.; validation, M.A. and M.K.; investigation, M.A. and M.K.; data curation, M.K.; writing—original draft preparation, M.K.; writing—review and editing, J.L., M.A. and A.P.; supervision, J.L. All authors have read and agreed to the published version of the manuscript.

Funding: This work was financially supported by the Ministry of Education and Research, Republic of Estonia, through the Estonian Research Council (grant number PRG1509).

Institutional Review Board Statement: Not applicable.

Informed Consent Statement: Not applicable.

Data Availability Statement: All data relevant to this publication are included.

Acknowledgments: The authors are thankful for the financial support from the Ministry of Education and Research, Republic of Estonia, through the Estonian Research Council (grant number PRG1509).

Conflicts of Interest: The authors declare no conflict of interest.

References

1. Béguin, F.; Presser, V.; Balducci, A.; Frackowiak, E. Carbons and Electrolytes for Advanced Supercapacitors. *Adv. Mater.* **2014**, *26*, 2219–2251. [CrossRef] [PubMed]
2. Käärik, M.; Arulepp, M.; Käärik, M.; Maran, U.; Leis, J. Characterization and prediction of double-layer capacitance of nanoporous carbon materials using the Quantitative nano-Structure-Property Relationship approach based on experimentally determined porosity descriptors. *Carbon* **2020**, *158*, 494–504. [CrossRef]
3. Käärik, M.; Maran, U.; Arulepp, M.; Perkson, A.; Leis, J. Quantitative Nano-Structure–Property Relationships for the Nanoporous Carbon: Predicting the Performance of Energy Storage Materials. *ACS Appl. Energy Mater.* **2018**, *1*, 4016–4024. [CrossRef]
4. Largeot, C.; Portet, C.; Chmiola, J.; Taberna, P.-L.; Gogotsi, Y.; Simon, P. Relation between the Ion Size and Pore Size for an Electric Double-Layer Capacitor. *J. Am. Chem. Soc.* **2008**, *130*, 2730–2731. [CrossRef] [PubMed]
5. Perez, C.R.; Yeon, S.-H.; Segalini, J.; Presser, V.; Taberna, P.-L.; Simon, P.; Gogotsi, Y. Structure and electrochemical performance of carbide-derived carbon nanopowders. *Adv. Funct. Mater.* **2013**, *23*, 1081–1089. [CrossRef]
6. Gogotsi, Y.; Nikitin, A.; Ye, H.; Zhou, W.; Fischer, J.E.; Yi, B.; Foley, H.C.; Barsoum, M.W. Nanoporous carbide-derived carbon with tunable pore size. *Nat. Mater.* **2003**, *2*, 591–594. [CrossRef]
7. Portet, C.; Lillo-Ródenas, M.Á.; Linares-Solano, A.; Gogotsi, Y. Capacitance of KOH activated carbide-derived carbons. *Phys. Chem. Chem. Phys.* **2009**, *11*, 4943–4945. [CrossRef]
8. Kierzek, K.; Frackowiak, E.; Lota, G.; Gryglewicz, G.; Machnikowski, J. Electrochemical capacitors based on highly porous carbons prepared by KOH activation. *Electrochim. Acta* **2004**, *49*, 515–523. [CrossRef]
9. Sevilla, M.; Mokaya, R. Activation of carbide-derived carbons: A route to materials with enhanced gas and energy storage properties. *J. Mater. Chem.* **2011**, *21*, 4727–4732. [CrossRef]
10. Käärik, M.; Arulepp, M.; Kook, M.; Mäeorg, U.; Kozlova, J.; Sammelselg, V.; Perkson, A.; Leis, J. Characterisation of steam-treated nanoporous carbide-derived carbon of TiC origin: Structure and enhanced electrochemical performance. *J. Porous Mater.* **2018**, *25*, 1057–1070. [CrossRef]
11. Osswald, S.; Portet, C.; Gogotsi, Y.; Laudisio, G.; Singer, J.P.; Fischer, J.E.; Sokolov, V.V.; Kukushkina, J.A.; Kravchik, A.E. Porosity control in nanoporous carbide-derived carbon by oxidation in air and carbon dioxide. *J. Solid State Chem.* **2009**, *182*, 1733–1741. [CrossRef]
12. Kormann, M.; Popovska, N. Processing of carbide-derived carbons with enhanced porosity by activation with carbon dioxide. *Microporous Mesoporous Mater.* **2010**, *130*, 167–173. [CrossRef]
13. Tee, E.; Tallo, I.; Kurig, H.; Thomberg, T.; Jänes, A.; Lust, E. Huge enhancement of energy storage capacity and power density of supercapacitors based on the carbon dioxide activated microporous SiC-CDC. *Electrochim. Acta* **2015**, *161*, 364–370. [CrossRef]
14. Lobato, B.; Suárez, L.; Guardia, L.; Centeno, T.A. Capacitance and surface of carbons in supercapacitors. *Carbon* **2017**, *122*, 434–445. [CrossRef]
15. Barbieri, O.; Hahn, M.; Herzog, A.; Kötz, R. Capacitance limits of high surface area activated carbons for double layer capacitors. *Carbon* **2005**, *43*, 1303–1310. [CrossRef]
16. Chmiola, J.; Yushin, G.; Dash, R.; Gogotsi, Y. Effect of pore size and surface area of carbide derived carbons on specific capacitance. *J. Power Sources* **2006**, *158*, 765–772. [CrossRef]
17. Jäckel, N.; Rodner, M.; Schreiber, A.; Jeongwook, J.; Zeiger, M.; Aslan, M.; Weingarth, D.; Presser, V. Anomalous or regular capacitance? The influence of pore size dispersity on double-layer formation. *J. Power Sources* **2016**, *326*, 660–671. [CrossRef]
18. Heimböckel, R.; Hoffmann, F.; Fröba, M. Insights into the influence of the pore size and surface area of activated carbons on the energy storage of electric double layer capacitors with a new potentially universally applicable capacitor model. *Phys. Chem. Chem. Phys.* **2019**, *21*, 3122–3133. [CrossRef]
19. Bi, S.; Li, Z.; Xiao, D.; Li, Z.; Mo, T.; Feng, G.; Zhang, X. Pore-Size-Dependent Capacitance and Charging Dynamics of Nanoporous Carbons in Aqueous Electrolytes. *J. Phys. Chem. C* **2022**, *126*, 6854–6862. [CrossRef]
20. Mirzaeian, M.; Abbas, Q.; Ogwu, A.; Hall, P.; Goldin, M.; Mirzaeian, M.; Jirandehi, H.F. Electrode and electrolyte materials for electrochemical capacitors. *Int. J. Hydrogen Energy* **2017**, *42*, 25565–25587. [CrossRef]
21. Frackowiak, E.; Abbas, Q.; Béguin, F. Carbon/carbon supercapacitors. *J. Energy Chem.* **2013**, *22*, 226–240. [CrossRef]
22. Pal, B.; Yang, S.; Ramesh, S.; Thangadurai, V.; Jose, R. Electrolyte selection for supercapacitive devices: A critical review. *Nanoscale Adv.* **2019**, *1*, 3807–3835. [CrossRef] [PubMed]
23. Zhong, C.; Deng, Y.; Hu, W.; Qiao, J.; Zhang, L.; Zhang, J. A review of electrolyte materials and compositions for electrochemical supercapacitors. *Chem. Soc. Rev.* **2015**, *44*, 7484–7539. [CrossRef] [PubMed]
24. Dyatkin, B.; Osti, N.C.; Gallegos, A.; Zhang, Y.; Mamontov, E.; Cummings, P.T.; Wu, J.; Gogotsi, Y. Electrolyte cation length influences electrosorption and dynamics in porous carbon supercapacitors. *Electrochim. Acta* **2018**, *283*, 882–893. [CrossRef]
25. Noofeli, A.; Hall, P.J.; Rennie, A.J.R. Ionic liquid based EDLCs: Influence of carbon porosity on electrochemical performance. *Faraday Discuss.* **2014**, *172*, 163–177. [CrossRef] [PubMed]
26. Shi, M.; Kou, S.; Yan, X. Engineering the Electrochemical Capacitive Properties of Graphene Sheets in Ionic-Liquid Electrolytes by Correct Selection of Anions. *ChemSusChem* **2014**, *7*, 3053–3062. [CrossRef]

27. Jäckel, N.; Patrick Emge, S.; Krüner, B.; Roling, B.; Presser, V. Quantitative Information about Electrosorption of Ionic Liquids i Carbon Nanopores from Electrochemical Dilatometry and Quartz Crystal Microbalance Measurements. *J. Phys. Chem. C* **201** *121*, 19120–19128. [CrossRef]
28. Brandt, A.; Pohlmann, S.; Varzi, A.; Balducci, A.; Passerini, S. Ionic liquids in supercapacitors. *MRS Bull.* **2013**, *38*, 554–55 [CrossRef]
29. Kim, E.; Han, J.; Ryu, S.; Choi, Y.; Yoo, J. Ionic Liquid Electrolytes for Electrochemical Energy Storage Devices. *Materials* **2021**, *1* 4000. [CrossRef]
30. Krause, A.; Balducci, A. High voltage electrochemical double layer capacitor containing mixtures of ionic liquids and organi carbonate as electrolytes. *Electrochem. Commun.* **2011**, *13*, 814–817. [CrossRef]
31. Ferrero, G.A.; Sevilla, M.; Fuertes, A.B. Mesoporous carbons synthesized by direct carbonization of citrate salts for use a high-performance capacitors. *Carbon* **2015**, *88*, 239–251. [CrossRef]
32. Lin, R.; Huang, P.; Ségalini, J.; Largeot, C.; Taberna, P.L.; Chmiola, J.; Gogotsi, Y.; Simon, P. Solvent effect on the ion adsorptio from ionic liquid electrolyte into sub-nanometer carbon pores. *Electrochim. Acta* **2009**, *54*, 7025–7032. [CrossRef]
33. Song, Z.; Duan, H.; Miao, L.; Ruhlmann, L.; Lv, Y.; Xiong, W.; Zhu, D.; Li, L.; Gan, L.; Liu, M. Carbon hydrangeas with typica ionic liquid matched pores for advanced supercapacitors. *Carbon* **2020**, *168*, 499–507. [CrossRef]
34. Ortega, P.F.R.; Santos, G.A.D.; Trigueiro, J.P.C.; Silva, G.G.; Quintanal, N.; Blanco, C.; Lavall, R.L.; Santamaría, R. Insights on th Behavior of Imidazolium Ionic Liquids as Electrolytes in Carbon-Based Supercapacitors: An Applied Electrochemical Approacl *J. Phys. Chem. C* **2020**, *124*, 15818–15830. [CrossRef]
35. Härmas, R.; Palm, R.; Härmas, M.; Pohl, M.; Kurig, H.; Tallo, I.; Tee, E.; Vaas, I.; Väli, R.; Romann, T.; et al. Influence of porosit parameters and electrolyte chemical composition on the power densities of non-aqueous and ionic liquid based supercapacitor *Electrochim. Acta* **2018**, *283*, 931–948. [CrossRef]
36. Suárez, L.; Barranco, V.; Centeno, T.A. Impact of carbon pores size on ionic liquid based-supercapacitor performance. *J. Colloi Interface Sci.* **2021**, *588*, 705–712. [CrossRef]
37. Leis, J.; Arulepp, M.; Käärik, M.; Perkson, A. The effect of Mo_2C derived carbon pore size on the electrical double-laye characteristics in propylene carbonate-based electrolyte. *Carbon* **2010**, *48*, 4001–4008. [CrossRef]
38. Presser, V.; Heon, M.; Gogotsi, Y. Carbide-Derived Carbons—From Porous Networks to Nanotubes and Graphene. *Adv. Func Mater.* **2011**, *21*, 810–833. [CrossRef]
39. Urbonaite, S.; Wachtmeister, S.; Mirguet, C.; Coronel, E.; Zou, W.Y.; Csillag, S.; Svensson, G. EELS studies of carbide derivec carbons. *Carbon* **2007**, *45*, 2047–2053. [CrossRef]
40. Härk, E.; Petzold, A.; Goerigk, G.; Risse, S.; Tallo, I.; Härmas, R.; Lust, E.; Ballauff, M. Carbide derived carbons investigated by small angle X-ray scattering: Inner surface and porosity vs. graphitization. *Carbon* **2019**, *146*, 284–292. [CrossRef]
41. Thommes, M.; Kaneko, K.; Neimark, A.V.; Olivier, J.P.; Rodriguez-Reinoso, F.; Rouquerol, J.; Sing, K.S.W. Physisorption of gases with special reference to the evaluation of surface area and pore size distribution (IUPAC Technical Report). *Pure Appl. Chem* **2015**, *87*, 1051–1069. [CrossRef]
42. Elbourne, A.; McDonald, S.; Voïchovsky, K.; Endres, F.; Warr, G.G.; Atkin, R. Nanostructure of the Ionic Liquid–Graphite Sterr Layer. *ACS Nano* **2015**, *9*, 7608–7620. [CrossRef] [PubMed]
43. Afrose, M. Physico-Chemical Properties of Imidazolium-Based Ionic Liquids in Pure and Mixed Polar Solvents. Ph.D. Thesis University of Dhaka, Dhaka, Bangladesh, 2017.
44. Käärik, M.; Arulepp, M.; Leis, J. Partial Oxidation to Extend the Lifetime of Nanoporous Carbon in an Ultracapacitor with Li_2SO_4 Electrolyte. *Molecules* **2023**, *28*, 2944. [CrossRef] [PubMed]

MDPI AG
Grosspeteranlage 5
4052 Basel
Switzerland
Tel.: +41 61 683 77 34

Molecules Editorial Office
E-mail: molecules@mdpi.com
www.mdpi.com/journal/molecules

www.ingramcontent.com/pod-product-compliance
Lightning Source LLC
Chambersburg PA
CBHW041343210926
51850CB00013B/2977
9783725840762